ADVANCE PRAISE FOR

NATURE IN MODERNITY

"Stephen Duguid, a distinguished humanist, has poured a lifetime of teaching, research and reflection into this sweeping and important book. Its premise is that ideas and values are at the root of environmental problems. In other words, the worst form of pollution is that of our *minds!* But, it follows, ideas are also the fix. Duguid uses a wide array of disciplines to explain what needs to be changed in government, science, economics, and religion if humans are to become members, and not masters, of the ecological community. For a very readable one-book overview of the history and future possibilities of human-nature relationships, this is my top recommendation."

Roderick Nash, Professor Emeritus of History and Environmental Studies, University of California, Santa Barbara; Author of Wilderness and the American Mind *and* The Rights of Nature: A History of Environmental Ethics

"Nature in Modernity is a wide-ranging, spirited study of the fate of the nature-culture dichotomy in Western civilization. Stephen Duguid maps the Western tradition against itself, but does not just lament the inchoate present of the environmental catastrophe looming over our creaturely existence. He finds counter-traditions to mastering the natural world by locating new directions and models for legal and ethical thinking and feeling."

Jerry Zaslove, Professor Emeritus and Director Emeritus, Institute for the Humanities, Simon Fraser University, Burnaby-Vancouver, British Columbia

"Stephen Duguid's clear and literate prose style makes this book the best overview yet written on the conundrum that is humanity's relation to nature in its double sense: nature as 'everything else' apart from us, and as our *own* nature. Using both a broad historical sweep and a collection of contemporary authors, Duguid deftly summarizes competing visions of our relation to nature and asks whether it is still possible to imagine altering the trajectory of doom we appear to be on, as we stumble toward a future of catastrophic ecosystem alterations and the collapse of industrial civilization. *Nature in Modernity*'s careful and thorough exposition makes it an excellent introduction to these themes for those who are still unfamiliar with them."

William Leiss, Scientist, McLaughlin Centre, University of Ottawa

"Can humans become 'earth citizens' before it's too late? This wide-ranging and philosophically sophisticated book examines our cultural, literary and even prehistoric evolution for evidence of a shadow or underground tradition of thought and feeling that may save us. Beneath the ruthless march of modernity still lies the tantalizing possibility, as Stephen Duguid puts it, of 'cooperation rather than conflict, mutuality rather than mastery and equality rather than hierarchy'—a vision based on both ancient experience and contemporary ecological realism. In this tenaciously thoughtful work, Duguid argues for a new conception of our true human nature, and shows how values must change if we are to achieve a compelling moral basis for a sustainable world."

Ernest Callenbach, Author of Ecotopia *and* Ecotopia Emerging

"In the stories we tell and the metaphors that guide our understanding, we humans try our hand at the urgent practical and philosophical question of how to live on our one planet. *Nature in Modernity* offers a far-ranging interdisciplinary inquiry into some of the central answers that have been given, testing them for their power to sustain both nature and humanity. In clear, striking prose and through fresh, lively readings of several major figures and episodes, Stephen Duguid peels away nostalgia, tidy common sense and dogma, revealing hopeful elements of tradition and innovation that can, he argues, engage our sympathetic imagination and extend our moral concern to nature."

Ken Smith, Associate Professor of English, Indiana University South Bend;
Editor, Confluence: The Journal of Graduate Liberal Studies,
Association of Graduate Liberal Studies Programs

NATURE IN MODERNITY

PETER LANG
New York • Washington, D.C./Baltimore • Bern
Frankfurt • Berlin • Brussels • Vienna • Oxford

STEPHEN DUGUID

NATURE IN MODERNITY
Servant, Citizen, Queen or Comrade

PETER LANG
New York • Washington, D.C./Baltimore • Bern
Frankfurt • Berlin • Brussels • Vienna • Oxford

Library of Congress Cataloging-in-Publication Data
Duguid, Stephen.
Nature in modernity: servant, citizen, queen or comrade / Stephen Duguid.
p. cm.
Includes bibliographical references and index.
1. Human ecology—Philosophy. 2. Nature—Effect of human beings on.
3. Ethnocentrism. 4. Noble savage. 5. Prehistoric peoples.
6. Enlightenment. 7. Philosophy, French. I. Title.
GF21.D785 304.2—dc22 2009044488
ISBN 978-1-4331-0932-4

Bibliographic information published by **Die Deutsche Nationalbibliothek**.
Die Deutsche Nationalbibliothek lists this publication in the "Deutsche
Nationalbibliografie"; detailed bibliographic data is available
on the Internet at http://dnb.d-nb.de/.

Cover design and typesetting by Jennifer Conroy

The paper in this book meets the guidelines for permanence and durability
of the Committee on Production Guidelines for Book Longevity
of the Council of Library Resources.

© 2010 Peter Lang Publishing, Inc., New York
29 Broadway, 18th floor, New York, NY 10006
www.peterlang.com

All rights reserved.
Reprint or reproduction, even partially, in all forms such as microfilm,
xerography, microfiche, microcard, and offset strictly prohibited.

Printed in Germany

CONTENTS

Preface *vii*
Acknowledgments *xi*
Prologue : A Glance at the Mess We're In *xiii*

Introduction : Reflections on the Primitive and the Modern 1
Anthropocentrism 12
Ecocentrism 14
Knowledge, Values and Action 16
Chapters that Follow 17

ONE   THE FALL FROM GRACE 23

1   **What are We by Nature** 25
A Quest for Human Nature 27
At Home in the Forest — An Anthropological Reverie 36
The Search for Human Origins 40
The Hunter-Gatherers (or Gatherer-Hunters) 44
A Reverie on Savage Children 51
A More Positive Spin 53
Concluding Thoughts 54

2   **What Are We By Choice, Chance and Context?** 63
Divergent Paths 65
The Paleolithic Revolution Reconsidered 67
The Neolithic 70
Rousseau's Neolithic — The Early Transition 78
Athens and Jerusalem 79
The Citizen 83
Socratism 89
Shadowy Greeks and Romans 93
Conclusion 98

TWO   COMING TO SENSE AND ENCOUNTERING SENSIBILITY 109

3   **Thinking the World to Pieces** 111
Oceanic Feelings in Vienna — 1927–1931 113
Mastery 119

4 **Recurrent Sensibility : On Golden Ages, Noble Savages and a Modern Primitivism** 145
Oceanic Feelings Contested Again 146
A Diversion — On Going Back or Going Forward
    by Looking Back 150
Thinking about *A State of Nature* 152
On Golden Ages 161
Sympathetic Noble Savages 167
Conclusion 170

THREE   RE-THINKING CULTURE AND NATURE 181

5 **Finding Ecological Sensibility in a Mechanistic Culture** 183
The Three Epiphanies 186
Oceanic Feelings Again 192
The Biology of Empathy 194
Ecocentrism 201
Some Praxis Implications 205
Conclusion 219

6 **New Insights About Culture and Nature** 229
Four Important Current Trends 235
Re-Assessing Happiness 237
The Contribution of Ecofeminism 241
Nature and the New Physics 244
Asian Philosophies 248
Nature — Mother, Comrade, Queen or Citizen 249

Conclusion 255
Toward Citizen Nature 266
Extending Our Reach 273

Bibliography 293
Index 309

PREFACE

"Two roads diverged in a yellow wood"   ROBERT FROST

Fish stocks collapsing. Glaciers melting. Unpredictable weather patterns. Holes in the Ozone. Disappearing species. Probably it is your concern over these and other much-heralded environmental calamities that has persuaded you to look at this book, a book that offers not a solution but rather a path to understanding and a suggestion for a new way to relate to the beings, phenomena and systems with whom we share the planet Earth.

Like you, I am concerned for the future of human life. I have three grandchildren who deserve to enjoy a vibrant nature, but in a more abstract sense I also care about all the humans existing now and yet to come. And I find myself moved as well by a sense of empathy for all living creatures, especially when I see or imagine their suffering, or see their life project foreshortened. Beyond that, I find I can generate a kind of empathy for the environment *per se*, as our home but also for the *beingness* of all its components — what the environmental writer Aldo Leopold called the Land Community.[1]

In trying to operationalize this empathy Thomas Berry argues that we are beginning to "move beyond democracy to biocracy" and this book explores this assertion.[2] Will trees vote? Will errant pit bulls go to prison? Must rivers pay taxes? What might it mean for humankind if we join with all of nature in a *natural contract*? We are not yet close to seriously considering such questions let alone working out the mechanics of their answers, but we can begin to imagine why another such expansion of the social contract might be both natural and necessary.

I came to write this book after several years teaching university courses that explored the connection between nature and human culture. As an historian and a humanist I was interested in exploring how we have come to understand our relationship with the natural world in order, I hoped, that we might be better able to alter that relationship in directions more amenable to long term sustainability for ecology and humanity. As an amateur anthropologist and psychologist I was also interested in our own human nature; whether we had one or not and, if we did, what it might tell us about our relationship with nonhuman nature.

As you read through these pages you will detect this focus on the humanities rather than science as the preferred means of reaching these understandings. You will also detect early on that this is a sweeping exploration, moving from a review of what we know of our primitive beginnings to speculations about 21st century trends in global culture. As such it is by necessity cursory at times and highly selective. There is no pretension here to cover all the important thinkers or eras or to provide thorough analyses of systems of thought as diverse as Stoicism and Ecofeminism. I have attempted, though, to highlight what I consider the more important of these perspectives and provide suggestions for further reading.

A number of themes have emerged in the writing of this book, themes described in some detail in the Introduction but which bear mentioning here as well. First an admission of sorts. My own professional immersion in the era of the 18th century Enlightenment—the era of Adam Smith, Jean-Jacques Rousseau, Thomas Jefferson and later the Shelleys, Percy and Mary—persuaded me of their conviction that a human nature did exist and that it was in some way fundamentally benevolent. This acceptance of a core "hard-wired" human nature has been reinforced by my reading of contemporary work in neurophysiology and psychology. The conflict between this human nature and our socially derived cultural norms is a basic theme of this book. Related to this is a secondary theme which argues that there is much in our deep past, when our nature was less affected by our cultures, that will be valuable to us now as we attempt to re-think our relationship to the natural environment. Finally, the most hopeful theme I explore is the idea of what I have called a *shadow modernity*, a long tradition within Western culture that runs counter to the more dominant mastery agenda seen by so many today as the cause of our environmental woes. Identifying this alternative modernity and the cultural tools it can give us is one of the primary objectives of this book.

Finally, a word about the title and its implied progression toward building a relationship with nature based on a shared status as citizens instead of master and subject. It originated while I was thinking about the French Revolution and was struck by the leveling power of the French Jacobins' insistence that the ponderous structure of differentiating titles be replaced by the simple prefix of "citizen" to each person, regardless of class, power, gender or position. It seemed to me, though perhaps superficial and largely symbolic, it was, a powerful means of connecting all to the larger whole, the common mission. We are a parsing species, persistently differentiating each of us from a series of others and focusing on parts at the expense of perceiving the whole. Citizen remains a powerful word in human cultures across the globe, extending now to include virtually all humans. Can we use it to build common cause with the rest of creation?

NOTES

1 Aldo Leopold, *A Sand County Almanac* (Oxford University Press, 1949), p 204.
2 Thomas Berry, *The Dream of the Earth* (San Francisco: Sierra Books, 1988), p xiii.

ACKNOWLEDGMENTS

This book has been a long time in gestation and is the product of several years teaching undergraduate and graduate seminars at Simon Fraser University on the theme of culture and nature. I want to thank all the students in my Humanities 350 seminars and my graduate students in Liberal Studies 819 and 812 for the stimulation, questions and ideas they brought to my exploration of this topic. Among those students, special thanks to Carlos Colorado, Justin Klassen and Greg Scutt. The many symposia, conferences and workshops sponsored by the Environmental Citizenship Group in the Institute for the Humanities at SFU provided a rich scholarly environment within which to test ideas and write. In writing the book I profited greatly from on-going criticism and support from Sandra Regan, Nancy Boyd, Ann Cowan, Mary Bissell, Hannah Gay, Clare Heffernan and David Byrnes and from the design and editing skills of Jennifer Conroy.

Finally, I dedicate *Nature in Modernity* to my friend (and cat) George who is my constant reminder of just how slender is that demarcation between humans and other living beings.

PROLOGUE
A GLANCE AT THE MESS WE'RE IN

"We are the environmental crisis"   NEIL EVERNDEN

It is almost a mantra among environmentalists that the earth simply cannot sustain globally the kind of modernity that currently exists in the developed West. What would happen, we are asked, if everyone in China or India wanted an automobile, air conditioning, vacations to Europe or other examples of middle class lifestyles in Los Angeles, Tokyo or London? Environmental observer Lester Brown reminds us every year in his *State of the World* books that the "...fossil-fuel-based, automobile-centered, throwaway economy that developed in the West is not a viable system for the world, or even for the West over the long term, because it is destroying its environmental support systems."[1] And Bill Rees insists that it would take three planet earths to produce the resources and absorb the wastes if everyone lived like today's North Americans.[2] Yet even as these questions are posed we learn in the media that this is precisely what is occurring in China, in India and elsewhere in the once "Third World". And who is going to deny them the right to seek these carnal pleasures of modernity?

And when did it become another commonplace to think of the sun not as a bearer of life but, thanks to ozone depletion, the cause of fatal skin cancers? And how soon will we have to adapt to the already alarming climate shifts that seem to be the result of our modernity? These are only two of the many danger signs we face at the start of the 21st century. Yet despite these portents and the dire warnings from writers like Bill McKibben (*The End of Nature*), alarming data from research bodies like the Suzuki Foundation, Greenpeace, and endless reports from scientists and governments, the political and economic institutions that dominate the industrial nations responsible for most of the environmental problems seem unable to respond in any coherent manner, often lapsing into denial or pursuing only short-term ameliorative policies.[3]

xiv | PROLOGUE

About two hundred years ago, at the start of the ride we are still on, similar kinds of warnings were issued and met with similar mixed responses. In its early stages "Dickensian" became the descriptor of an early modernity that embraced unregulated industry, an instrumentalized science, and a free market culture. Social critics of the time struggled to expose the wrongs of the era, but it was fictional creations the likes of Ebenezer Scrooge, Oliver Twist, Fagin, and Mr. M'Choakumchild that had the greatest impact. The latter's educational technique in the opening scenes of *Hard Times* (1854) is a vivid and depressing account of the mechanization and compartmentalization of language and culture that flowed from the "systems" that had played such a central role in creating this new modernity. But Dickens did more in *Hard Times* than criticize, he also bore witness to the intimate links between the social and environmental catastrophes that accompanied early modernity. Here is Coketown, the home of his *Hard Times* characters:

> It was a town of red brick, or of brick that would have been red if the smoke and ashes had allowed it; but, as matters stood it was a town of unnatural red and black like the painted face of a savage. It was a town of machinery and tall chimneys, out of which interminable serpents of smoke trailed themselves for ever and ever, and never got uncoiled. It had a black canal in it, and a river that ran purple with ill-smelling dye, and vast piles of building full of windows where there was a rattling and trembling all day long, and where the piston of the steam-engine worked monotonously up and down, like the head of an elephant in a state of melancholy madness.[4]

Gone from the public imagination were the daffodils of Wordsworth and the Tuscan hills of Keats or Shelley. Instead there is a grim scene of urban blight, a Bhopal of its time; or a Cleveland with its burning river. Or Chernobyl with its empty streets and lingering radiation. One thinks here of Fritz Lang's still influential 1927 film *Metropolis* as illustrating a second, assembly line and authoritarian stage in the evolution of this modernity, with Rachel Carson's pesticide ridden *Silent Spring* bringing us closer to our own era.

Now we are told that a dramatic "tipping point" is looming. The greenhouse gases we hear about (but cannot see or smell) are reaching levels in the atmosphere that could create changes beyond our ability to manage, especially in terms of ocean water temperature which in turn affects global wind and rainfall patterns.[5] In recent years the dramatic warming of the Indian Ocean has produced droughts in East Africa, the usual monsoon season virtually disappearing. We hear of this via media reports on Darfur where, as Tim Flannery points out, the camel-herding nomads have been forced to drive their herds onto farmland. Hence the "tragedy of Darfur", portrayed as political (or even theological) in origin, is actually an environmental story.[6]

There was an earlier warning as well, not so much descriptive of physical or environmental change but a warning about the dangers of the mastery fixation that had emerged in Western culture in the 17th century and would lead to Coketowns in the 19th. If Percy Shelley the poet kept his focus in the ethereal realm of asking "who am I, whence came I, and why," his novelist wife Mary Shelley chose more pragmatic ground. Barely nineteen when she accompanied Shelley to Switzerland to meet up with Lord Byron in Geneva, she wrote *Frankenstein* or, *The Modern Prometheus* during that rainy summer of 1816. Victor Frankenstein's words in the original opening lines of the novel bring back to any modern reader the scenes from the many filmed versions:

> It was on a dreary night of November that I beheld the accomplishment of my toils. With an anxiety that almost amounted to agony, I collected the instruments of life around me, that I might infuse a spark of being into the lifeless thing that lay at my feet. It was already one in the morning; the rain pattered dismally against the panes, and my candle was nearly burnt out, when by the glimmer of the half-extinguished light, I saw the dull yellow eye of the creature open; it breathed hard, and a convulsive motion agitated its limbs.[7]

Here Mary Shelley seems to anticipate so much of the 20th century's adventures in scientific hubris and roots them firmly in the dualist worldview that was instrumentalized in modernity. Life is composed of "instruments" and that which is without the "spark" of life is simply a "thing". But it is man, not God (and not woman either!) who aspires to provide that spark, to solve the mystery of life.

Science here, at least as practiced by Victor Frankenstein, is disconnected from any moral foundations. Victor practices "bad science", an amalgamation of medieval alchemy and a vitalism popular at the time, but Shelley's critique is wider than just Victor's choice of science. He sought to "enter the citadel of nature" in order to learn the "secrets of heaven and earth" and was willing to take any short cut and violate any personal and social norm to reach his goal. The end, in Victor's mind, justified the means. And, of course, it all goes bad. Victor fails as a parent while succeeding as a scientist. He creates but cannot nurture. He refuses to take responsibility for his creation and instead simply abandons the "Creature" to the wilderness and hence to social persecution.

Here Shelley turns her attention from Victor to the Creature, a mild mannered if ugly orphaned child-human who develops on his own in a kind of "state of nature" as peaceful, vegetarian and optimistic, until driven to violence after being shunned by society because of his appearance. Thus the scientist's creation, full of potential but left to drift, becomes the "monster" we know from the movies and blazes a path of destruction. One can see the links with atomic energy, pesticides, genome research and other potentially useful creatures that have dark sides just waiting to blaze trails of their own.

This is not the place for a full analysis of Shelley's novel which, as Marilyn Butler says, is "...famously reinterpretable."[8] Instead it's a warning of sorts, a warning that throughout this exploration of humanity's relationship with the natural world we will be consulting aesthetic, literary and philosophical sources as much if not more than those in the social sciences and sciences. Products of the imagination, in the spirit of poets like Homer, Dante, Shelley and Snyder, are often more prescient and powerful than the products of reason and in speculations about the future of our species and our home these voices deserve to be heard.[9]

But first we need to review some of the more empirical assessments of what so many now refer to as our environmental crisis. It is not a crisis in the catastrophic sense of the impending nuclear winter that seemed so possible in the Cold War era. Nor is it acute in the sense of a sudden virulent pandemic or tsunami. Instead it is seen as chronic and perhaps systemic. It has the quality, then, of surreptitiousness, an ability to sneak up on us through gradual alterations of atmosphere, species, land and water.

Stephen Boyden refers to this as the "boiling frog" scenario: "If a frog is placed in hot water it will make frantic efforts to escape; it is said, however, that if the animal is put into cold water, which is then slowly heated, it may, after passing through various stages of phylogenetic maladjustment, be boiled to death without so much as a struggle."[10]

While some more phlegmatic types see ecological changes as just examples of natural cycles, extinctions and random events, the dominant voices of our time tend to place the blame on humans — natural in the sense that humans are natural beings, but more importantly cultural because humans have the ability and means to do otherwise. Thus Brian Baxter suggests humans may be "...perpetrating a sixth extinction on this planet, to rival those occasionally produced by natural processes aeons before the human species appeared."[11] Holmes Rolston describes the current scene as "...superkilling by a superkiller.[a] maelstrom of killing and insensitivity to forms of life and the sources producing them."[12] For Rolston the most shocking aspect of the phenomenon is that at the same time humans have more knowledge than ever before about the natural world, a fairly precise ability to choose to predict the ecological consequences of these actions, and the power to do something about it. The fact that we do not, or seem to lack the will and the ethical sense to do so, leads some to the misanthropic conclusion that humans are "evolutionary monsters", "exotics" or "natural aliens".[13] Michel Serres, mathematician and philosopher, concludes that "our culture abhors the world" and Neil Evernden insists, "We are the environmental crisis."[14]

And what does the crisis look like? How does it manifest itself? The lists of looming disasters include:

- Species extinctions at the rate of about 17,000/year, with 10–30% of mammal, bird and amphibian species currently threatened.[15]
- Collapsing fisheries and proliferating "dead zones" in the world's oceans.
- Deforestation at the rate of 17 million hectares/year with special concern for fast disappearing rainforests.[16]
- Pervasive poisoning of air, water and soil, pharmaceutical contamination of watersheds.
- Increasing soil oxidation, desertification at the rate of 6 million hectares a year and erosion of topsoil.[17]
- Gross inefficiency in food production in developed nations, with as much as 10 calories of energy used for each calorie of food produced.

And, of course, the most prominent and far-reaching problems are global warming with its associated increase in CO_2 and ozone holes allowing more ultra-violet radiation to reach the earth. The American Geophysical Union in 2003 declared that "…the world is now warmer than it has been at any point in the last two millennia, and, if the current trends continue, by the end of the century it will likely be hotter than at any point in the last two million years."[18] Further, they concluded that natural influences could not explain the rapid increase in global near-surface temperatures. Edward Wilson in his 2002 book *The Future of Life* concluded likewise, noting that "…the mean surface temperatures of Earth varied by less than 2 degrees F during most of the ten thousand years following the end of the Ice Age. Then, from 1500 to 1900, it rose approximately 0.9 degrees F, and from 1900 to the present time it has increased another 0.9 degrees F."[19] The physicist Marty Hofferty warns us that "We have to face the quantitative nature of the challenge. Right now, we're going to just burn everything up; we're going to heat the atmosphere to the temperature it was in the Cretaceous, when there were crocodiles at the poles. And then everything will collapse."[20]

In 1990, Daniel Botkin predicted that by the year 2000 "…we will begin to see rapid changes over vast areas. In parts of the North, we expect to see stately old trees beginning to die back. The warmer temperature will make many trees vulnerable to insect attacks and different blights. Hikers will increasingly find themselves among dead trees. Loggers will have to choose between harvesting the dead timber and glutting the lumber and paper industries. And the diebacks will affect water supply and erosion rates."[21] Now in 2010, pine beetles proliferate in Western Canada due to warmer winters and the forests are dying with dead trees being harvested exactly as Botkin predicted fifteen years earlier. In addition it seems clear now that rainfall pasterns are changing, sea levels are rising, and glaciers and permafrost are melting.

There are other related issues that are seen as part of the environmental crisis. At the top of the list is the issue of global poverty and persistent equity issues, with 80% of the world's population surviving on less than a quarter of goods produced, resulting in serious ethical dilemmas not only for the over-consuming 20% but also for environmentalists who urge a slowing down of development. Inequality, poverty and high rates of population growth also often contribute to environmental degradation, leading to over-use of marginal lands, deforestation, over-fishing, desertification, pollution and weak social and educational options.

Given these developments, there is no shortage of calls for immediate and fundamental social, political, economic and value change in order to halt a slide into catastrophe. The base assumption of virtually all environmental thinking associated with belief in a looming crisis is "...that no organism or population of organisms can grow and expand its metabolism in a finite environment indefinitely."[22] If it tries—as it is supposed humanity is doing—it will either run out of resources, choke on its own wastes, or destroy the ecosystem of which it is a part. Human activity, it is contended, has grown too large for the biosphere meaning that additional growth will only impoverish. In essence, as James Lovelock, charges, the earth (Gaia in his terminology) is now at war with us. "It is almost as if we had lit a fire to keep warm and failed to notice that the furniture had ignited."[23] While Lovelock's cure for global warming and too many greenhouse gases is a controversial shift to nuclear power, the most outspoken environmentalists counsel dramatic social change as our only hope. Lynton Caldwell sums up the requirements:

> ...the environmental problem is now seen as systemic—multidimensional and inherent in socioeconomic trends, which, if uncorrected, could lead to destructive consequences neither preventable nor remedial by technical or legal means alone. Avoidance would require major social changes, including a reorientation of popular expectations, a redirection of many public policies, and a reformation of institutions adversely affecting the environment.[24]

What this implies, I will argue, is a frontal assault on a deeply embedded value, belief and habit of mind in Western culture; a mastery agenda that relies on a radical separation of humans from nature and a sense of humans as somehow transcendent or, as Neil Evernden puts it in a less flattering way, as "nicheless exotics" who have no need to protect or nurture a home place.

There seems little disagreement that in modern societies grounded in Western cultural traditions, the dominant view is that non-human nature is a "...vast repository of resources, both physical and biological, to be developed, used and consumed by humans for human ends."[25] The mastery or conquest of this nature is, as William Leiss says, one of "our most cherished collective aspirations."[26] It is this "dominion" over nature that we have come to believe will assure social tranquility at last, freeing us from the constant fragility of an existence exposed to disease, disaster and privation. In our wildest dreams we even triumph over death, nature's most potent power over us. Or, as in Mary Shelley's story, we create life independent of the means that nature through evolution has bequeathed to us.

The tendency toward a mastery agenda no doubt has deep roots in Western culture (Felipe Fernandez-Armesto roots it in the millennium prior to Christianity and sees it as only subsequently grounded in Christianity, Stoicism and Renaissance Humanism[27]) and we will be exploring this later. Certainly by the 18th century it was more than a tendency. Adam Smith nodded with appreciation to the often harsh and hostile realm of nature since it had in effect roused humanity to the challenge and in the process "...turned the rude forests of nature into agreeable and fertile plains, and made the trackless and barren ocean a new fund of subsistence, and the great high road of communication to the different nations of the earth. The earth by these labours of mankind has been obliged to redouble her natural fertility, and to maintain a greater multitude of inhabitants."[28]

The ideological riposte to Smith's celebration of mastery was launched in earnest in Max Horkheimer and Theodor Adorno's 1944 book *Dialectic of Enlightenment* in which the triumph of reason and science that was to have liberated humankind instead "...radiates disaster triumphant."[29] Their critique of mechanistic science, dualism, capitalism and modernity's commitment expansion and progress set the stage for the subsequent narratives of environmental decline that dominated the last decades of the 20th century.[30]

The issue of the human relationship to the environment has become a major topic of social, cultural and political concern thanks in part to attacks on conventional wisdom like that of Horkheimer and Adorno, the work of other critics from Rachel Carson to Carolyn Merchant, and the often quite alarming physical signs of rapid ecological change. But given what Leiss and others have said about the centrality of dominion over nature in the Western tradition, these dissenting voices have not resulted in any unanimity of analysis, opinion or policy. From Kyoto to genetically modified foods to forestry practices the debates and disagreements are strident and positions deeply held. Much of this book will be devoted to understanding the diversity of these perspectives and their cultural foundations.

NOTES

1   Lester R. Brown and Christopher Flavin, "A New Economy for a New Century," in *State of the World*, ed. Linda Starke (New York: W.W. Norton & Company, 1999), p 15.

2   Mathis Wackernagel and William Rees, *Our Ecological Footprint* (Philadelphia: New Society Publishers, 1996), p 15. Applying the footprint model, Steve Vanderheiden finds that the "... average Canadian consumes and pollutes at a rate that requires 4.3 hectares of biologically productive land per person while the average U.S. citizen needs 5.1 hectares per person. The average Indian requires only 0.38 hectares, and the worldwide average footprint is 1.8 hectares per person. Setting side the existing 1.5 billion hectares of wilderness to meet the needs of all other terrestrial and aquatic species, Wackernagel & Rees argue that 7.4 billion hectares are available for human use, which comes out o 1.3 hectares of available biologically productive land per person." Steve Vanderheiden, "Two Conceptions of Sustainability", *Political Studies* v. 56, 2008 (pp 435–455) p 449.

3   The feminist philosopher Val Plumwood concludes in her most recent work that "The failure of the dominant national and international political institutions to meet the situation of ecological crisis could not be more clear, a course likely to ensure our demise..." *Environmental Culture* (New York: Routledge, 2002), p 1.

4   Charles Dickens, *Hard Times* (London: Penguin, 1995), p 28.

5   James Speth, *The Bridge at the Edge of the World* (New Haven: Yale University Press, 2008), p 26.

6   Tim Flannery, *The Weather Makers* (New York: Harper Collins, 2005), p 126.

7   Mary Shelley, *Frankenstein or, The Modern Prometheus* (London: Penguin, 1994), p 69.

8   Marilyn Butler, "The First Frankenstein and Radical Science", *Times Literary Supplement* (9 April 1993), p 12. She notes that it can be "...a late version of the Faust myth, or an early modern myth of the mad scientist; the Id on the rampage, the proletariat running amok, more what happens when a man tries to have a baby without a woman."

9   William Leiss calls the contemporary dominance of "scientism" an "attitude of arrogant superiority toward all other ways of interpreting the human experience of the surrounding world...[which] by the 19th century ...had turned its scorn on traditional religious, social and ethical paradigms: The 'scientific method' would suffice, it was claimed, as the sole 'rational' approach to any and every question of values, social justice, and ultimate meaning." *Under Technology's Thumb* (Montreal: McGill-Queen's Press, 1990), p 6.

10   Stephen Boyden, *Western Civilization in Biological Perspective*, (Oxford: Clarendon Press, 1987), p 25.

11   Brian Baxter, *Ecologism* (Edinburgh: Edinburgh University Press, 1999), p 2.

12   Holmes Rolston III: "Environmental Ethics: Values in and Duties to the Natural World", in Lori Gruen and Dale Jamieson, *Reflecting on Nature* (New York: Oxford University Press, 1994), p 76.

13   Peter Zapffe, *Jeg velger sannheten*, (Oslo: Universitetsforlaget, 1983) in Peter Reed, "Man Apart: An Alternative to the Self-Realization Approach", *Environmental Ethics*, vol 11:1, 1989 (53–69); and Neil Evernden, *The Natural Alien* (Toronto: University of Toronto Press, 1993).

14   Michel Serres, *The Natural Contract* (Ann Arbor: The University of Michigan Press, 1995), p 3; Evernden, p 128.

15   Extinctions occur from a variety of causes including habitat destruction, pesticide use, and deliberate human killing. But there are interesting connections to the new global economy as well. For instance, "Fishing policies of rich European countries are directly linked to the extinction of African animals such as elephants, monkeys and warthogs....As

xxii | PROLOGUE

European fishing fleets began depleting the once bountiful fish in Ghanaian waters, local residents—whose population is growing—began hunting more wild meat for market, including animals in nature preserves....That has pushed all kinds of animals into local extinction as Ghanaian hunters scrambled to find enough animal-based protein to go around." Alanna Mitchell, "European Fishing linked to Extinction of African Animals, *Toronto Globe and Mail* 12 Nov 2004, p A9.

16   Edward Wilson calls this "...one of the most profound and rapid environmental changes in the history of the planet." *The Future of Life* (New York: Vintage, 2002), p 58.

17   Evan Eisenberg says 10% of crop land in the US suffers from high saline levels., "Back to Eden", *Atlantic Monthly* (November 1989, pp 57–89), p 59.

18   Elizabeth Kolbert, "The Climate of Man, 2", *The New Yorker*, 25 April 2005, (pp 56–71), p 56.

19   Wilson, (2002), p 67.

20   Elizabeth Kolbert, "The Climate of Man, 3", *The New Yorker*, 9 May 2005, (pp 52–63), p 57.

21   Michael Pollan, "Only Man's Presence Can Save Nature", *Harper's Magazine*, April 1990 (37–48), p 47.

22   Boyden, p 192.

23   James Lovelock, "Gaia must go nuclear", *Toronto Globe and Mail* (10 March 2005). As we will see later, Serres sees this new, more aggressive Gaia as a new 'world object'. On the new planet earth "...the river is threatening the fighter: earth, waters, and climate, the mute world, the voiceless things once place as décor surrounding the usual spectacles, all those things that never interested anyone, from now on thrust themselves brutally and without warning into our schemes and maneuvers. They burst in on our culture, which had never formed anything but a local, vague, and cosmetic idea of them: nature. What was once local—this river, that swamp—is now global: Planet Earth." Serres (1995), p 3.

24   Lynton Caldwell, "Is Humanity Destined to Self-Destruct?" *Politics and the Life Sciences* March 1999, (pp 3–14), p 4.

25   Paul Taylor, *Respect for Nature* (Princeton: Princeton University Press, 1986), p 95.

26   Leiss (1990), p 74.

27   Felipe Fernandez-Armesto, *Civilizations* (London: Pan, 2001), p 548.

28   Adam Smith, *The Theory of Moral Sentiments* (New York: Oxford University Press, 1976), p 183.

29   Max Horkheimer and Theodor Adorno, *Dialectic of Enlightenment* (New York: Continuum, 1995), p 3.

30   Carolyn Merchant, *Reinventing Eden* (New York: Routledge, 2003), p 191.

INTRODUCTION
REFLECTIONS ON THE PRIMITIVE AND THE MODERN

"Our job is to do civilization well, not abandon it."

MORRIS BERMAN

In his last poem, *The Triumph of Life*, the poet Percy Shelley asks a mysterious *Shape all Light* to "Show whence I came, and where I am, and why..."[1] More viscerally conscious than most that he existed in a body destined to die, Shelley drowned off the Italian coast at age thirty in 1822, the poem left unfinished, the questions left unanswered. And despite a death made more tragic by its early arrival, it is perhaps appropriate that Shelley did not have the opportunity to venture an answer.

Such questions — how did we get here and why are we here — are central to a creature destined to acquire an awareness of the end of its time, but likely have no final answer. They remain always lingering, a mystery made necessary by a brain empowered with consciousness and extended foresight. But other even more important and certainly more pragmatic questions — also asked by Shelley — quickly follow in succession and they do warrant answers:

- how ought we to live?
- what is our relation to the rest of that which we see, sense, or intuit as being here with us?

And perhaps most central to our time — how can we reconcile human flourishing and planetary sustainability?

We cannot respond to Shelley's initial questions outside the realms of poetry, myth, religion and science fiction, but we are able to address in a more direct manner these other, more concrete questions and striving to advance

2 | INTRODUCTION

some answers will be the central objective of this book.

Because the question of defining a flourishing life for humans — one that is virtuous, sustainable and respectful of both self, other and whole — is so fundamental to the human condition, it has been remarkably persistent over time, from the primitive and pastoral beginnings of human communities to the anxious modernity of the 21st century. Time has not made these questions any easier to answer, indeed perhaps just the opposite, especially since we now better understand that there is a direct relationship between the nature of our flourishing and the ecological health of the planet. There is no necessary linear progress on issues such as these, no story of a steady improvement from darkness to light, from barbarism to civilization, from formulating questions to discovering answers.

The promise and story of the Christian West was not supposed to turn out this way and it is fitting that the avowed atheist Shelley raises these questions. Nothing is quite what it seems. We humans are at the top of the food chain, master of all the beasts and, seemingly, most everything else. Yet we fear new pandemics, singular terrorists, tsunamis and yet unseen disasters. Despite the envy of the non-moderns we share the earth with (or perhaps because of it) we doubt in some primal sense the wisdom of the materialism that has delivered these masteries. Yet for many no alternative — communism, primitivism, utopianism — seems adequate or possible.

This calls into question the notion of history itself as we have come to define it, as a story of a progressive movement from simple to complex, from primitive to modern, from delusion to truth. I will propose here that we look instead to fluidity rather than mechanics, to circularity instead of linearity and even, if we choose, to "eternal returns" and that by doing so we may discover that there are *shadows* of modernity that have in the past and could in the future lead to a variety of alternative paths.[2] To start thinking in these terms implies that we start by exploring that prolonged era in the human past — perhaps over 200,000 years — that is often called prehistoric or "primitive" not as simply a predecessor or a beginning long since transcended, but rather as an authentic, integral and necessary component of the modern. [3]

This idea, that it would be to our benefit to consider our present selves as integrated with rather than separated from past versions of ourselves — that the adult and child are one — is not a new idea. In the 18th century we find it in Jean-Jacques Rousseau whose work poses the question of whether an enlightened human culture can "...recover some of the advantages of the primitive independent state, without losing the benefits of the social state?"[4] A few years later the query is picked up by William Godwin who asks if one can recapture the autonomy and transparency of primitive society, while continuing to enjoy the benefits of philosophic and material progress.[5] In the 19th century this becomes a central theme of English Romantic poets, German Idealist philosophers and American transcendentalists, with Henry David Thoreau musing about the possibility of combining the "...hardiness of the savages with the intellectualness of the civilized man."[6]

More recently, Herbert Marcuse proposed that the gains of modernity could only be saved if we uncouple that modernity from the repressive civilization that created it.[7] Stanley Diamond, in the same tradition, identified as the central task of our time reconciling the primitive with the civilized, "...making progress without distortion theoretically possible, or, at least, enabling us to experience the qualities that primitive peoples routinely display."[8] Finally, the economists Daly and Cobb, in their influential text *For the Common Good*, place their hope for the planet on recovering "...some of the communal advantages of pre-modern society in a post-modern form, that is, without sacrificing the gains made in individual freedom, human rights, and political equality achieved in the modern period."[9]

These yearnings for synthesis of past and present, primitive and modern, reflect a belief that continuity and change are not choices but rather are intertwined. Perhaps the double helix of DNA provides a model for this interconnection of past and present. In this view the primitive is *embedded within* the modern, always part of the present and therefore subject to the same re-creations, re-imaginings and re-inventions that more contemporary times endure. But it retains as well its authenticity, its reality. We see this complex duality in the study of present-day "primitives" by anthropologists, the objects of their research being at once part of the modern world and yet also antecedent to it.

This line of enquiry posed by Rousseau, Godwin, Thoreau, Marcuse, Diamond and Daly relates directly to the central humane and ecological issues of our era. Any kind of return to pre-modern social forms seems clearly not possible and neither is it enough simply to aspire to locate and nurture aspects of our past, our so-called primitive heritage. Morris Berman reminds us that "...our job is to do civilization well, not abandon it ."[10] If as a species we are to move forward via new adaptations to new conditions we need to do so as beings who are holistic, integrated, complete—not as the halves of an imaginary primitive/modern dualism.

If there is a relationship between our modern selves and our distant origins, it would seem that we need to sort this out. Are the former so constructed and conditioned by the societies and cultures created over the last several millennia that they bear no resemblance to our prehistoric natural origins? Are the two at war, in conflict, in a struggle for dominance? Can the modern self erase the imprint of eons of hominid evolution? Or will the sheer millennial weight of our Paleolithic heritage eventually overwhelm the modern pretence toward mastery and autonomy? Must we revert to a hunter-gatherer or pastoral life for the planet to survive? Or can we reach an accommodation between our selves and the rest of existence, perhaps by constructing or re-discovering an alternative path to human and planetary flourishing, one that has always been with us and may hold the key to reconciling these two aspects of our selves?

These questions are central to what I argue is the key public policy issue of our times; the relationship between human flourishing and planetary sustainability. We seem paralyzed by the possibility that these two may be in mortal conflict, seemingly unable to "fix" the environment yet unwilling to imagine alternative ways to flourish as a species. The Australian feminist philosopher Val Plumwood notes this paralysis as well, fascinated by the fact that we have the means but lack the will.[11]

There are several possible explanations for such an impasse. It is argued by some, for instance, that the economic imperatives of capitalism, which emerged from the 20th century as a potential global system, demand a constant expansion of growth through excessive consumption or destructive conflict. Either of these options seems likely to demand levels of resource use and waste creation that are unsustainable. Yet, there are models of economic and social organization, both existing and theoretical, that seem to offer a means of at least blunting these imperatives. Stephen Boyden has argued that our inability to recognize that our technology-driven cultures as currently run may be ecologically catastrophic stems from some combination of a:

- lack of knowledge of alternatives
- lack of access to the means of change
- lack of an ability to summon the will to change[12]

These are no doubt important roadblocks to building a flourishing culture and a sustainable ecology. But I argue here that they remain problems of a second order, problems that can be addressed only after we consider the more fundamental issue of our collective human nature as it has developed both biologically and historically, and appreciate more completely the impact of that 'nature' on the global environment.

This means for a start a critical engagement with mental and material constructions such as "primitive", "progress", "technology", "nature", "ecology" and "culture". Already in the Prologue the hand is tipped, not only is ecological sustainability a problem, we start from the premise that we are in a "mess" consisting of environmental degradation, individual alienation, species extinction, global warming, ozone holes, resource depletion, decreases in biodiversity, air and water pollution, deforestation, proliferating *dead zones* in the world's oceans, acid rain, erosion of topsoil, fresh water scarcity and the widespread pharmaceutical contamination of watersheds, persistent inequalities, poverty, spiritual apathy and angst, global terrorism, pandemics such as AIDS, Ebola and a variety of flus, consumerism, fragmented family and kinship relations and a general state of precariousness or "insecurity of social standing,"[13] all of which results in what Wendell Berry argues is, even in the most prosperous human communities, "...probably the most unhappy average citizen in the history of the world."[14]

If Berry's conclusion about the general unhappiness of the beneficiaries of modernity has any validity, then there is something very wrong with our modern notion of progress and the world view upon which it is based. Celebrated in the 19th century but problematic throughout the 20th, a culture of growth and progress based on technology and consumption seemed the source of at least as much violence and carnage as it was of advances in more humane directions. As a result, even, or perhaps especially, these privileged beneficiaries of progress are the first to doubt its viability as the basis of culture and the first to reject its utopian promise. Indeed, in Western culture at least, cynicism about consumerism — the most immediate and palpable reward of progress — is rife. Yet at the same time material progress is deservedly celebrated in less developed areas and globalization threatens (promises?) to make it universal even though the utopian beliefs that once sustained it are in remission in its heartland. Thus we move forward into oblivion, morally if not materially. At the start of the 21st century we do not believe in progress, we just do it! We are no longer as confident about the superiority of the modern, the secular and the technological, no longer sure the track we are on is heading upward — yet we have no way to get off, no clearly defined and practical alternative to a modernity as defined in the 20th century.

Our lives are governed, then, by a sense of movement, hopefully forward but always with the fear of sliding backward, a fear grounded in the implicit and explicit dualisms at the base of modernity. Movement must always mean progress and its opposite becomes loss or "stagnation", with the image of a stagnant pond or swamp contrasted with the rushing stream or flowing river. Thus in the popular imagination, dissatisfying as the present might be, it is thought to be better than stagnation. As a result, modern proponents of a "steady-state" economy or of a culture and economy that may sacrifice growth for sustainability are drowned out by the fears that halting or even slowing the rush forward will mean disaster.

It must be acknowledged that this issue of progress is a central stumbling

6 | INTRODUCTION

block for any attempt to re-think human relationships with the rest of the natural world. Common sense tells us that we are better off than our ancestors and that those of us in more developed societies are better off than our less developed neighbours. Edward Wilson warns us convincingly that the "juggernaut of technology-based capitalism will not be stopped," but goes on to insist that it must be "redirected" if the living world is to survive.[15] The philosopher Mary Midgely reminds us that this redirection will not come via purely cognitive choices or scientific innovations, but rather via a "moral choice of a way of life."[16] In the chapters to come we will, therefore, have to think about, interrogate in fact, this idea of progress in an effort to ensure that the so-called mastery agenda that it has become coupled with is indeed more than merely a "legitimizing myth" for a culture gone astray.[17]

The modern critique of progress and mastery is linked, firmly but perhaps undeservedly, with a persistent nostalgia for a more peaceful, steady and less complex time. Some revel in memories of a bucolic past, of small towns and extended families, while others seek to recover an often romanticized pre-modern self-sufficiency and simplicity. As is so typical in times of cultural crisis, fundamentalist religious movements also flourish with calls for a return to the basics of faith. Others even aspire to reach even further back to re-create a pastoral, nomadic life; a new primitivism. J.A. Tainter reminds us that our complex modernity is a very recent phenomenon and that its collapse would not be a fall into "...some primordial chaos, but a return to the normal human condition."[18] But most would argue that going back is not a practical nor even, if looked at closely, a very desirable alternative to "the mess we're in". But if moving forward on our existing path is fraught with danger and voluntarily returning to earlier forms unlikely, what is modern culture to do? The time is clearly right for a reformulation of the modern.

Using a crude chronological formulation of historical eras conforming to millennia, we can date what we often refer to as modernity in the West as beginning around 1500, preceded by the eras we label medieval (500–1500) and ancient (400 BCE–500 CE). Thus in our time we remain firmly within a coherent era which I will refer to as the "modern" to differentiate it from the Neolithic and the earlier Paleolithic. Clearly though, we at the start of a long period of transition to something quite different. Indeed, it is one of the hallmarks of our time to be embroiled in debates about this sense of transition, of a culture having reached certain "limit". Indeed, one may object that we are already in mid-reformulation with the advent of the post-modern. But despite some three decades of experimentation with academic and aesthetic post-modernism, attachment to various "New Age" movements, and a persistent yearning by some to abandon Western modernity and adhere to various

Eastern cultures, these all still seem like false starts or a premature abandonment of the modern.

As serious as our cultural and ecological crises may be, there is a strong consensus that looking outside or beyond our cultural frame will not give us the answers and alternatives we need. Edward Wilson and other scientists warn us that turning against science, technology or even modernity itself will get us no where. Even as committed an environmentalist as Wendell Berry concedes that we must find our way within the realm of our cultural inheritance, arguing that "...inherited forms may be constraining, but they're also enabling; rebellion against such forms, while sometimes necessary, is always dangerous."[19] One of my objectives here is to look to our past to see if we can discover within the Western tradition a pattern of values, practices and understandings — cultural tools — that could form the basis of a reformulated modernity, one built on a new conceptualization of the relationship between humans and the natural world. I argue that such a pattern of values, understandings and practices already exists, that it has surfaced at key points in our past but in each case has failed to carry the day. There is, as it were, a *shadow modernity*, a path taken by few, but by enough to leave a trail.

Neil Evernden in his ground-breaking book *The Natural Alien* set my mind upon this task of cultural reformulation several years ago. He argued that "... the source of the environmental crisis lies not without but within, not in industrial effluent but in assumptions so casually held as to be virtually invisible ...our scarred habitat is not only of our doing, but our imagining, and it will take a profound re-creation of the social world to *un-say* the environmental crisis and constitute a more benign alternative."[20] Lynn White had made much the same point years earlier is his now classic essay *The Historical Roots of our Ecological Crisis*.[21] Scholars such as Carolyn Merchant and Val Plumwood have focused on the importance of linking the harmful effects of anthopocentrism with the long tradition of androcentrism in Western culture.[22] All stress that how we think about nature, our ideas about it, govern our attitudes and our actions with respect to nature, that "...our experience of nature is inseparable from our ideas of nature."[23]

Unlike the modernity we live in, which is characterized by radical fissures and discontinuities, the *shadow modernity* that I want to follow here has deep and contiguous roots not just in Western culture but in the story of the species *per se*. I conceive of it as a kind of doppelganger, a Dr. Jekyll overwhelmed in cultural memory by the more audacious Mr. Hyde. This tradition, an alternative modernity that is whole rather than partial, has as one of its fundamental qualities a relationship between humans and the rest of creation that is deeply subjective, built on ancient and perhaps "natural" ideals of sympathy, mutuality and reciprocity– indeed this is what marks it off from the more pervasive variant of modernity that assumes a human exceptionalism with its accompanying drive to mastery and dominion.

Exploring the features of this authentic, embedded and subjective way of thinking about humans and the natural world requires an interdisciplinary focus and a determination to start at the beginning in order to examine the core assumptions about human nature and about humans and nature that lay at the heart of the incomplete and hence necessarily flawed modernity we have inherited alongside its shadowy alternative. This means, as a start, an examination of the biological origins of humankind, an assessment of the millennia the species lived as hunters and gatherers, and an accounting for the cultural, social and political choices made by humans since that era.[24]

What do we need to know from the start? The key issue must be how we conceive of and describe the relationship between humans and the natural world—animals, plants, landscapes and climate. From at least the Neolithic era of villages, agriculture and pastoralism, we can see that the path of human development has been on a trajectory characterized by ever increasing sociality. This reliance on enlarged networks of human contact—from village to city and tribe to nation—has come at the expense of what we imagine to have been the intimate links with nature that characterized earlier forms of human existence. This ascendancy of the social human has been accompanied by an increasing preoccupation with mastery over nature, a transition from integration and mutuality to fear, antipathy, abstraction and finally domination.

One presumes that such was not always the case, that for early primates and even in the millennia of hunter-gatherer dominance in the Paleolithic era the web of relations between the hominids who would become human and the rest of creation was more balanced, relations with nature assuming perhaps equal rank with hominid social relations. But we are far from being able to offer conclusive evidence that such a balanced relationship with nature was in fact one of the deep structures of human existence. Necessarily we must rely on arguments and suppositions based on tenuous archaeological evidence, on the accounts of early European explorers, and on observations of the contemporary remnants of hunter-gatherer cultures.

Still, it is a reasonable exercise to begin with some "supposings". There are two extreme positions that we can easily identify. The one argues for a kind of "natural" or innate holistic union of humans with nature and the other insists on a fundamental and necessary difference between humans and the rest of nature—a classic example of dualism in practice. The first position often leads to calls for a "return" to earlier practices and ways of thinking while the other can lead either to increasing alienation from nature or to a radical reformulation of the culture/nature relationship.[25] Inevitably, these two perspectives will provide important reference points throughout this book, but I will be arguing throughout, echoing the case made by Val Plumwood, that it is a synthesis we need to arrive at, not a choice. Plumwood insists that we cannot overcome such dualisms as self and other, humans and nature by reversals, by merging or simply conflating difference: "...an adequate resolution of dualism requires recognition of both continuity and difference."[26] But we cannot anticipate synthesis from the start, so let us turn to the first, somewhat less conventional view to open up the discussion.

The Canadian biologist-ecologist Stan Rowe captures the spirit of the holistic position when he insists that "...humans have "...an instinctive affection for the natural world...a love struggling to get out and find expression."[27] And he is in good company with, for instance, Mary Midgely arguing that it is our "natural tendency to love and revere" the natural world because "we are at home in it."[28] Terry Glavin, writing from his island retreat off the coast of British Columbia, sees deep with the human heart "...an ancient and abiding desire to be in the presence of flourishing, abundant, and diverse forms of life that is grounded in human nature."[29] Here the connection has nothing to do with beauty or even utility but is in a strong sense pre-conscious, innate.

But affection and love are difficult words to apply to this relationship. To survive we must consume large portions of the natural world and, to some extent at least, we must exercise some form of mastery in order to assure that consumption. It is no coincidence that among humans there is an almost universal fear of snakes, dangerous heights and noxious odors—we must be cautious to survive in this "nature." On the other hand, we too are consumed and thus are simply one component of a vast natural continuum. The Canadian naturalist John Livingston sees this as an awareness of the "life process" and counsels that our contentment with Being is to be found in "compliance" with the dictates of that process.[30] Taking Rowe's words in a pantheistic sense, to love nature, then, is also to love ourselves.

But it is the quality and character of that love of self that will be at issue throughout this book. In the 18th century the issue was framed most directly by Jean-Jacques Rousseau who carefully differentiated self love (*amour propre*)—a kind of egotistical pride—with love of self (*amour de soi*), a more benign notion akin to the modern idea of self esteem and, more broadly, well-being. Implicitly, *amour de soi* involved loving as well the home the self found itself in. By titling his book *Home Place*, Rowe places some important limits around the claim that we love nature, whether we know it or not. What we love about nature, he implies, is its connection with our home. This may have been the *natural* wisdom of our hunter-gatherer ancestors whose home place

INTRODUCTION

was always so fragile, nature being serenely unresponsive to their crude attempts at mastery. It is with this in mind that in most modern studies of traditional, indigenous, and primitive cultures the salient feature to celebrate is always their capacity for "living with" nature.

This idea, that our affinity with or affection for nature is as much contextual as inherent, tied to our need to nurture a living space, finds support in Edward Wilson's *biophilia hypothesis*, as well as a related *topophilia hypothesis* advanced by Yi-Fu Tuan.[31] Wilson, certainly more scientist than naïve romantic, sees human affiliation with other living organisms as stemming from our ancient home places, with people in Europe, Asia and America preferring "...savanna-like habitats, particularly if there is a bit of peaceful water in the scene....we retain a ghostly preference for safaris past."[32] On the other hand, the great 19th century naturalist and explorer John Muir placed our domestic comfort zone higher up, arguing that "...going to the mountains is going home."[33] Other research claims that we retain an evolutionary preference for certain kinds of trees—broad canopies relative to their height, layered branches, small leaves and a low split in the trunk.

There is a "developmental theme" here with the "child being the father of the man". The childhood of our species—certainly a massively prolonged adolescence—is seen to have hard-wired us to some extent in terms of our home-place comfort zone. And trees and forests may be a special case, one we will explore later, since it does seem that from sacred groves to modern forests, humans have established special relationships with the tree, seeing it not as a mere object, a "...stiff column in a shock of light..." but rather as a "...being astir with life...". And as Martin Buber insists, despite the tree's muteness, the relationship is mutual.[34] It is "the tree itself" that Buber sees, unconcerned about issues of language, soul or consciousness. Within, atop, or even below the tree it is possible—indeed easy for an adult as well as a child—to appreciate this oneness, to gain a sense if not an understanding of that notion of holism that is so central to modern ecological thinking. Sheltered within the tree, it becomes at once an extension of self. No where is this extension more clear than in the musing of the great conservationist and ecologist Aldo Leopold as he beholds his forest of Wisconsin pines:

> It is in midwinter that I sometimes glean from my pines something more important than woodlot politics, and the news of the wind and weather. This is especially likely to happen on some gloomy evening when the snow has buried all irrelevant detail, and the hush of elemental sadness lies heavy upon every living thing. Nevertheless, my pines, each with his burden of snow, are standing ramrod-straight, rank upon rank, and in the dusk beyond I sense

the presence of hundreds more. At such times I feel a curious transfusion of courage.[35]

Leopold is very close here to meeting Martin Buber's criteria for an I-Thou relationship with nature—nature as *Subject* rather than *Object*. In a sense at once Rousseauean, Romantic and mystical, Leopold is "...seized by the relationship..." and the "transfusion of courage" takes place without reflection, without reason.[36]

Such an extension of self or convergence with the natural does not, however, hold in the "real world". In a classic Cartesean dance, we are today caught up in the dualism of culture versus nature and it is being played out in terms of holism versus fragmentation, universals versus particulars, and absolutism versus relativism. The 17th century scientists/philosophers Francis Bacon and Rene Descartes convinced us that understanding and control comes from taking things apart and then dealing with the parts rather than contemplating the whole. The child, the poet and the primitive may claim to sense the holistic nature of nature, but the engine of culture is built on notions of mechanism and fragmentation. Now ecologists, New Age prophets, poets, environmental activists and quantum physics Chaos theorists are urging us to abandon or at least rein in mechanism and reach new understandings of nature based on this sense of holism.

But is this holism merely the naïve holism of a child, the imaginings of a primitive culture or overheated imagination, perceptions necessarily put aside as the cost of maturity, of the ability to "develop", to use the natural world for human ends? And is it as well a dangerous point of view, one which denies the centrality of the human, leading perhaps to a leveling anti-humanist eco-fascism and therefore best set aside or even repressed?

Before answering these important philosophical and political questions, we need to address the empirical/psychological one, namely, is it even possible as adults, indeed as a society, to reclaim this childish and perhaps primordial memory? Our lived experience would seem to answer in the affirmative. The novelist Wallace Stegner makes this point in his autobiographical book *Wolf Willow*, the title referring to a plant unique to his childhood home in southern Saskatchewan. When he returns after an absence of forty years and encounters once again the scent of this gray-leafed bush his senses transport him in a flash and, for the moment "...reality is made exactly equivalent with memory...The sensuous little savage that I once was is still intact inside me."[37]

But this recalling of long forgotten memories is child's play compared with Paul Shepard's radical claim that we are still hunter-gatherers at heart or with the audacity of Freud's determination to root our aggression and our guilt in a species-memory of the killing of the primal father.[38] The anthropologist Richard Lee bases his case for primitive communism on humans retaining a "…deep-rooted egalitarianism, a deep-rooted commitment to the norm of reciprocity."[39] Henry David Thoreau saw the Indians in Maine as "primal human beings" who could lead us back to "realities which only lurk somewhere in the modern subconscious mind.[40]

Dare we imagine with visionaries the likes of Rupert Sheldrake that species memories are embedded at these deep levels and that traces of our hunter-gatherer ancestors remain intact in each of us? And, if so, would this be a threat to the "thin veneer of civilization" we cling to or an opportunity for a renewed relationship with the natural world?

Just as these arguments for innate qualities—biological and psychological—embedded within each human presume them to be repressed, veiled or other-directed by the prevailing culture, so an argument can be made that the historical record of modernity itself contains qualities that could if unveiled, nurtured and promoted lead to a quite different culture—but still within the modern tradition.

It may be perceived in these times of talk of a post-modern culture, that to look into the Western cultural tradition for guidance is a flawed and hopelessly conservative idea, since that tradition itself is rightly seen as the origin of many of the problems we face. It will be, however, to a largely underground or *hidden heritage* that we will turn in this investigation—to, among others, the heritage of Epicurus, Lucretius, Wollstonecraft, Rousseau, Shelley, Thoreau, Muir, Leopold and Carson.[41] In response to a perceived cultural crisis—the inability of our culture to generate and nurture a politics and a society that assures both ecological sustainability and human flourishing—one can either look outside, beyond existing frames, or search for alternative paths within. Writers like the historian Carolyn Merchant[42] and the ecologist Max Oelschlaeger[43] have taken the latter approach, seeking alternative voices within Western culture, voices that combine human well-being and ecological sustainability. It is in that tradition that this book is written.

Perspectives, then, and the relationship between how we think about the natural world and how we act are issues of central concern. While there are many variations to consider, two perspectives or starting points are worth singling out from the start: anthropocentrism and ecocentrism.

ANTHROPOCENTRISM

The social ecologist Murray Bookchin combines a very real concern for the health of the environment with a powerful assertion of the unique and privileged position of human beings. "To be a human animal, in effect, is to be a reasoning animal that can consciously act upon its environment, alter it, and advance beyond the passive realm of unthinking adaptation into the active realm of conscious innovation...the ontological divide between the non-human and the human is very real."[44] By focusing on reason—others might use soul to the same end—Bookchin maximizes the distance between human and animal while acknowledging that humans are still animals. In other hands this strong anthropocentrism becomes openly teleological, with the emergence of reasoning humanity as the central purpose of the cosmos. Humans are seen to be co-creators with the original creative force or impulse, imagining, designing and changing nature being central to our purpose. Thus "...working to change the world around us is not a perversion of human nature, but an expression of it."[45] Max Oelschlaeger, an ardent critic of this perspective, calls this the "..ideology of man infinite or the rise of Lord Man."[46]

Not all human-centered approaches are grounded in teleology or technological utopianism. Advocates of *wise use*, for instance, maintain the view that while natural ecosystems and other species remain "resources" for use by humans, they must be managed carefully to ensure sustainability. Adherents to the popular sustainable development idea advanced by the Brundtland Commission in 1987 accept a moderate version of anthropocentrism, arguing that the pace of using nature must be one "...that meets the needs of the present without compromising the ability of future generations to meet their own needs."[47] Economists like Julian Simon suggest this may be excessively prudent and insist that market forces combined with human skill and ingenuity will solve the resource and waste issues raised by worried environmentalists.

Finally, we will encounter another form of anthropocentrism in the form of what some would call post-modern thinking. Contesting our reliance on sense experience in identifying a nature which is in peril, post-modernists argue that "...there is no single nature, only natures. And these natures are not inherent in the physical world but discursively constructed through economic, political and cultural processes."[48] There is, then, no objective knowledge or nature "out there", but only a humanly created world of words and text. "The world we experience is therefore nothing more than a construction of our senses, and in particular the language we use to describe our experiences."[49] Most post-modernists, however, would concede that a world does in fact exist, but insist that our reliance on language means that we can never really experience it in an objective way.

ECOCENTRISM

While there are no doubt misanthropic voices within the disparate range of voices and positions grouped under the heading environmentalist, most critics of current anthropocentric approaches to nature are focused on creating a new humankind/nature relationship based on mutuality rather than mastery. They start from a very different point, emphasizing existence *per se* as a network of relationships without a hierarchy or teleology. Rachel Carson's notion of the "web of life" epitomizes this approach:

> The earth's vegetation is part of a web of life in which there are intimate and essential relations between plants and the earth, between plants and other plants, between plants and animals. Sometimes we have no choice but to disturb these relationships, but we should do so thoughtfully, with full awareness that what we do may have consequences remote in time and place.[50]

Barry Commoner's equally influential "four laws of ecology" have a similar focus on the interconnections of all matter, insisting that "everything is connected to everything else" and that "nature knows best."[51] He shares Carson's caution that any major human-generated change in the eco-system is likely to be detrimental. James Lovelock's work on the "Gaia Hypothesis has tended to support this idea, insisting that life or biota is a major player in shaping the overall environment of the Earth, its climate and chemistry. Overall, then, ecocentric approaches imply that there is not "life" on the one hand and "environment" on the other, but rather one self-regulating eco-system in which humankind is only one component.

As we will see in Chapters Four and Five, in arguing that a new relationship with nature will require that the moral norms operative in human society be extended to the natural world, many environmentalists have found inspiration in the 18th century belief in an innate moral sense in humans. Popularized by Enlightenment and Romantic writers like Hume, Smith, Rousseau, and Shelley, this idea that we are endowed with a "...moral sense, an intuitive feeling for what is right and wrong", has been integrated with Darwinian thought in more recent times.[52] The supposedly innate moral sense is now seen as having been the result of natural selection, with cooperation, reciprocity and caring being successful evolutionary adaptations.

If the "moral sense" advocates and their desire to extend to the natural world the moral standing that humans enjoy represent an elaboration of the classic values of modernity, the ecocentric approach also has its relativist or post-modern voices. From a blending of Buddhism and modern physics we get a very abstract notion of how the human fits into the natural world:

> Viewed from the point of view of modern ecology, each living thing is a
> dissipative structure, that is, it does not endure in and of itself but only as a
> result of the continual flow of energy in the system...From this point of view,
> the reality of individuals is problematic because they do not exist *per se* but
> only as local perturbations in this universal energy flow....the structures out
> of which the biological entities are made are transient, unstable entities with
> constantly changing molecules dependent on a constant flow of energy to
> maintain form and structure.[53]

Life here is just a flow of energy, reminiscent of the atomist philosophy
of the Greek philosopher Epicurus. The reality of humans, as with all forms
of material being, is thus problematic "...because they do not exist *per se* but
only as local perturbations in this universal energy flow."[54]

While this Buddhist and quantum physics inspired conception of life as
just a variation in constant flows of energy might appear to minimize the significance of humanity and thus be a target for Murray Bookchin's frequent accusation that many environmentalists are misanthropic, that really is not the
case. These perspectives do not single out humans as problematic—all phenomena share that quality. On the other hand, there are ecocentric environmental thinkers who do seem to fit Bookchin's target. The philosopher Paul
Taylor, in a very influential book, argued that while humans are completely
dependent on a healthy biosphere, "...it seems quite clear that in the contemporary world the extinction of the species homo sapiens would be beneficial to the earth's Community of Life as a whole....The Earth would no longer
have to suffer ecological destruction and widespread environmental degradation due to modern technology, uncontrolled population growth, and
wasteful consumption. After the disappearance of the human species, life
communities in natural ecosystems would gradually be restored to their former healthy state...Our presence, in short, is not needed."[55] John Livingston,
in a voice more attuned to eternity, assures us that should this come to pass,
all will be well:

> In nature there are no weird and bizarre notions like egalitarianism and
> the sacred individual to gum up the process. You merely spin out your
> time—eating, voiding, propagating, and eventually rejoining the system at
> a simpler level of organization. But nothing goes away, there is no problem.[56]

This perspective, often condemned as misanthropic, is the subject of a
recent publication that has generated widespread interest, Alan Weisman's
The World Without Us in which humans suddenly disappear and the earth's

potential recovery is explored.[57]

Taylor and Livingston may overstate the case for effect, but the subtle variations surrounding the issue of the centrality of humans versus the centrality of the ecosystem of which they are a part permeates all discussions of environmental concerns from the infamous "jobs vs the environment" debates to population control. And behind, beneath and above the fractious sparring of experts, philosophers, politicians, scientists, activists, media and laypersons the "noise" of climate change, resource depletion, species extinction and industrial pollution keeps up its steady drumbeat.

KNOWLEDGE, VALUES AND ACTION

The nature of the relationship between knowledge, values and action is at the core of this book, seen as the key to understanding the substantive relationship of humankind to nature. The environmental issues discussed earlier, which provoke some to declare a crisis and others to sharpen their debating tools, have led to a number of ameliorative actions (e.g. re-cycling, sustainable development initiatives, research into alternative energy) but little evidence of a fundamental shift in values or an altering of our of faith in a commitment to mastery over nature.[58] Edward Wilson's pressing question remains unanswered: "How best we can shift to a culture of permanence, both for ourselves and for the biosphere that sustains us?"[59]

I argue that such a fundamental shift is necessary, a shift from mastery to some form of reciprocity/mutuality. After all, in a secular version of Pascal's Wager if the predictions of crisis leading to ecological collapse are correct, such a shift might be the only way to save the human species and much else besides (though we know that nature itself will survive without us). And even if the predictions are wrong, the results of such a shift will have beneficial outcomes, certainly better than if we did nothing and the predictions were correct. It has long been a stalwart belief among environmentalists that such a changed consciousness or new set of values will make us "better humans".[60] Prudence, then would seem to dictate that we take these predictions very seriously despite the dissenting views.

But how does one act to encourage a paradigm shift like this? Will a concerted education and public relations effort designed to raise awareness, change thinking, and nurture new eco-friendly values change the behaviour of people? Of nations? Of corporations? Inevitably a focus on ideas, values and ideologies runs up against this issue and readers of this book should consider it carefully. Academics and writers, of course, have a serious stake in seeing the connection between ideas and action as quite robust. But is it?

The evidence is often not very encouraging. Leslie Thiele, a political theorist who studies these issues in some depth, concludes that "...there is little causal relationship between environmental education (knowledge acquisition, increased awareness, and changes in value orientations) and environmentally responsible behavior."[61] Murray Bookchin, arguing from a more materialist or structuralist position, scorns what he sees as the "personalizing" or "psychologizing" of what an ecological crisis that is firmly rooted in the social realm: "If we merely remedy our thinking and living habits, individual by individual, we shall presumably become *plain citizens* of the biosphere with agreeable ecological habits."[62]

Bookchin, of course, has a point. Changes in the area of ideas and values are clearly not enough to bring about the changes in behaviour by individuals, governments and corporations that environmentalists see as essential. But while such changes may not be sufficient, it is reasonable to argue that they remain a necessary condition; that without changes in thought and value we will never be able to even imagine the possible social, economic and political changes that will enable us to view nonhuman nature with wonder and reverence rather than rapaciousness.[63]

CHAPTERS THAT FOLLOW

Following the lead of Neil Evernden and others associated in varying ways with the Deep Ecology approach to environmental issues, I argue here that we will only sort out our relationships with trees and the rest of creation when we stop objectifying nature—when we break through the barrier of anthropocentrism and begin to assign subjectivity to the other-than-human. This idea of the *subject* is structured around an intense individuation in which a self or subject is "...self-subsistent, distinct from everything outside itself, including its own body...a sphere of subjectivity containing its own experiences, opinions feelings and desires, where this sphere of inner life is only contingently related to anything outside itself."[64]

In our dominant Kantian ethics we are thus proscribed, ethically at least, from treating other humans as means to an end. In the past centuries we have made great progress in extending subjectness and the resulting citizenship to virtually the full range of human life, indeed the shift of masses of people from subject to citizen, from object to subject, has been the great heritage of enlightened modernity. However, while some animals and the odd plant, landform or body of water are sometimes treated as if they were subjects, as phenomena they are clearly perceived to be objects that can be owned, used, harvested, mined, consumed, slaughtered, drained, damned or dammed.

INTRODUCTION

Just as the social and cultural revolution in the consideration of fellow humans was driven as much by pressure from below as enlightenment from above, so now the pressure of species extinction, resource depletion, global warming and climate change may be forcing the argument for the continuation of that revolution into the flora and fauna of the planet. An acceptance of some form of a more civil relationship with the natural world may be the only way to avoid a Reign of Terror of cataclysmic dimensions. But do we have the cultural resources to take such a step, to extend the Kantian project of a concern for ends rather than means to include the natural world—a world of beings and phenomena that cannot speak to their interests in the language we demand of subjects? And is that the right path to follow? Can we extend moral concern to nature and at the same time avoid the emphasis on autonomy, abstraction and instrumental reason that pervades the liberal, Kantian tradition of citizenship?

To guide the reader through this interdisciplinary maze of issues, perspectives, ideologies and phenomena the text is structured in three parts. The opening Prologue assessed the problem of ecological sustainability within the context of human cultures increasingly driven by a commitment to material progress. Of particular interest here is an initial assessment of the range and nature of the positions being taken in the public scientific and philosophical debates surrounding ecological sustainability and human development. Subsequently in Part I, *The Fall from Grace*, a first chapter explores the issue of our biological heritage, our human nature as derived from our origins through the long hunter-gatherer era of the Paleolithic. Chapter Two, *Who Are We By Choice, Chance and Context?* explores the interplay between volition and circumstances from the Neolithic period through to early civilization and how this began to shape and alter our responses to the natural world around us.

Part Two, *Coming to Sense and Encountering Sensibility*, turns our attention to the powerful impact of reason, in particular the kind of instrumental reason that came to characterize Western culture from the 17th century. Chapter Three, *Thinking the World to Pieces*, explores the ideas of Bacon, Descartes, Kant and others in relation to the establishment of a distinctly modern (and Western) approach to human/nature relations. Chapter Four, *Recurrent Sensibility: Golden Ages, Noble Savages and a Modern Primitivism*, explores the other side of the modern coin, the persistence of a more holistic, even primitivist, approach to the natural world as the basis for a potential alternative to contemporary modernity.

Part Three, *Re-Thinking Culture and Nature*, returns to the issues raised in the Prologue and seeks to move toward operationalizing a new way of thinking about human/ nature relations. In Chapter Five, *Finding Ecological Sustainabilty in a Mechanistic Culture*, the focus is on the various attempts in the 20th century to begin this re-thinking of humans and nature, including conservationism, preservationism, pantheism, bioregionalism, animal rights and steady-state economics as well as contributions from Buddhist, Taoist and other non-Western traditions. Chapter Six, *New Insights about Culture and Nature*, we return to the weaknesses of a culture built on "reason" and on "rights" and seek to find the cultural tools embedded in our primitive past, in alternative traditions from Epicurus to the Romantics and in modern environmentalism that can be used to create a culture of care, mutuality and reciprocity in which it will be logical to welcome nature in all, its complexity as a fellow citizen. A concluding section addresses more directly the issue of extending moral and legal rights to the nonhuman world.

In his 1990 book *Under Technology's Thumb*, William Leiss outlined the central task facing those of us concerned about the looming conflict between nature and culture.[65] That task was to generate a culture built on "respect for nature", a respect based on a careful reassessment of alternative models—whether primitive, historical, or contemporary—and a parallel reassessment of our use of modern science and technology in attempts to "...manipulate larger and larger sets of ecological interactions for the sake of ephemeral short-run benefits."[66]

In the 1994 re-issue of his *Domination of Nature*, Leiss is more outspoken in his warning, insisting that we face "...a century of global environmental crisis.." which we can only survive by "...rejecting the idea of mastery over nature." This means, he argues, that we must "...find adequate political forms for an appropriate representation of the relation between humanity and nature."[67] This book seeks to build on Leiss's search for a middle ground between faith in harmony and holism on the one hand and progress and technology on the other, finding the path to that middle ground emerging in the coalescence of interest in primitive cultures, romantic sensibility, ecology, sustainable communities, and new approaches to science.

NOTES

1 Zachary Leader and Michael O'Neill, eds. *Percy Bysshe Shelley: The Major Works* (Oxford: Oxford University Press, 2003). p 616.

2 At this point I need only note the importance of contemporary Chaos Theory to this critique of linear ideas of progress. As Hugh Roberts explains, "The state of a nonlinear system at any given time cannot be determined simply by our knowledge of the state of some exterior input; it will always be in part a function of its own preceding state." Hugh Roberts, *Shelley and the Chaos of History* (University Park, PA: Pennsylvania State University Press, 1997).

3 I will use the term primitive to denote this era, aware of its problematic and subjective nature if seen as merely one side of a crude dualism, the other being "civilized". Here, instead of a pejorative term I am using it as descriptive of a time in the history of the human species when a more intimate link was possible between humans and nature.

4 Peter France, *Politeness and Its Discontents: Problems in French Classical Culture* (Cambridge: Cambridge University Press, 1992), p 188.

5 Gregory Dart, *Rousseau, Robespierre and English Romanticism* (Cambridge: Cambridge University Press, 2005), p 86.

6 In Roderick Nash, *Wilderness and the American Mind* (New Haven: Yale University Press, 2001), p 92.

7 John Zernan, "Why Primitivism?", *Telos*, No. 124, Summer 2002 (pp 166–172), p 170.

8 Stanley Diamond, *In Search of the Primitive* (New Brunswick: Transaction, 1974/1993), p 175.

9 Herman Daly & John Cobb, *For the Common Good: Redirecting the Economy toward Community, the Environment, and a Sustainable Future* (Boston: Beacon Press, 1989), p 16.

10 Morris Berman, *Wandering God* (Albany: SUNY Press, 2000), p 16.

11 Plumwood, (2002), p 3.

12 Boyden, (1987), p 24.

13 Zygmunt Bauman, *Community: Seeking Safety in an Insecure World* (London: Polity Press, 2001), p 42.

14 Wendell Berry, *The Unsettling of America* (San Francisco: Sierra Club Books, 1977), p 20.

15 Edward Wilson, (2002), p 156.

16 Mary Midgely, "Criticizing the Cosmos", in Willem Drees, ed. *Is Nature Ever Evil?* (New York: Routledge, 2003) p 24.

17 Zygmunt Bauman, *Modernity and the Holocaust* (Ithaca: Cornell University Press, 1989), p 96.

18 Cited in Bert deVries, et. al. "Understanding: Fragments of a Unifying Perspective" in Goudsblom and deVries, eds *Mappae Mundi: Humans and Their Habitats in a Long-Term Socio-Ecological Perspective* (Amsterdam: Amsterdam University Press, 2002), p 273.

19 "From Standing By Words", cited in Kimberly Smith, *Wendell Berry and the Agrarian Tradition* (Lawrence: University Press of Kansas, 2003), p 77.

20 Evernden, (1993), p xii.

21 Gruen and Jamieson, p 10.

22 Plumwood (2002), p 50.

23 Philip Hefner, "Nature Good and Evil" (pp 189–202) in Drees, p 190.

24 Choice is, especially for social scientists, always a contentious term and more will be said about its use here in Chapter Two. Perhaps the best formulation is that of Karl Marx: "Men make their own history, but they do not make it under circumstances chosen by themselves, but under circumstances directly encountered, given and transmitted from the past. *The Eighteenth Brumaire* of Louis Bonaparte in Karl Marx and Friedrich Engels, *On*

Historical Materialism (Moscow: Progress Publishers, 1972), p 120. Earlier, in his 1845 *Theses on Feuerbach*, individuals are seen as "products of circumstances" but it is stressed that it is individuals who in fact create and alter these circumstances. p 12. Thus while we may, along with Freud, celebrate ourselves as the 'choosing animal', these choices are, by nature and by culture, deeply conditional.

25 See, for instance, Paul Shepard's *Coming Home to the Pleistocene* (Washington, D.C.: Island Press, 1998) for an example of the first position and Gregory Stock's *Redesigning Humans* (Boston: Houghton Mifflin, 2003) for the second perspective.

26 Plumwood, *Feminism and the Mastery of Nature* (London: Routledge, 1993), p 125.

27 Stan Rowe, *Home Place* (Edmonton: NeWest, 1990) p 11.

28 Midgely, (2003), p 21. Prince Charles chimes in on this issue as well, asserting that there is "...a faint memory of a distant harmony...yet sufficient to remind us that the Earth is unique and that we may have a duty to care for it". *Toronto Globe and Mail*, 31 May 2000, p A13.

29 Terry Glavin, *Waiting for the Macaws* (Toronto: Viking Canada, 2006), p 9.

30 *The John A. Livingston Reader* (Toronto: McClelland and Stewart, 2007), p 110.

31 Tuan, Yi-Fu, *Topophilia: A Study of Environmental Perception, Attitudes and Values* (Englewood Cliffs, NJ: Prentice-Hall, 1974), p 93.

32 Mark Ridley, "Do we love nature?", review of Stephen Kellert and Edward Wilson, eds. *The Biophilia Hypothesis* (TLS, 9 Sept. 1994) p 5.

33 in Max Oelschlaeger, *The Idea of Wilderness* (New Haven: Yale University Press, 1991), p 2.

34 Martin Buber, *I and Thou* (New York: Charles Scribner's Sons, 1958), p 8.

35 Leopold, p 87.

36 Peter Reed, "Man Apart: An Alternative to the Self-Realization Approach", *Environmental Ethics*, vol. 11:1, 1989, p 57. This intuitional response to immersion in the natural is explained in more detail by Laurence Buell in his *The Environmental Imagination: Thoreau, Nature Writing and the Formation of American Culture* (Cambridge: Belknap Press of Harvard University Press, 1995).

37 Wallace Stegner, *Wolf Willow* (New York: Viking, 1963), p 19.

38 Sigmund Freud, *Civilization and Its Discontents* (New York: Norton, 1961), p 86–7. "We cannot get away from the fact that man's sense of guilt springs from the Oedipus complex and was acquired at the killing of the father by the brothers banded together."

39 Richard Lee, "Demystifying Primitive Communism" in Christine Gailey, ed., *Civilization in Crisis, Vol. 1* (Gainesville: University of Florida Press, 1992), p 90.

40 Richard Fleck, *Henry Thoreau and John Muir Among the Indians* (Hamden, Conn: Archon Books, 1985), p.2.

41 I am borrowing this notion of a 'hidden heritage' from the book of that title by John Howard Lawson (*The Hidden Heritage*, New York: Citadel Press, 1950), a brilliant but little read upside down history of the Western cultural tradition.

42 Carolyn Merchant, *The Death of Nature* (New York: Harper and Row, 1983).

43 Oelschlaeger, (1991).

44 Murray Bookchin, *Re-enchanting Humanity: A Defense of the Human Spirit Against Antihumanism, Misanthropy, and Primitivism* (London: Cassell, 1995). p 22.

45 Hefner (2003), p 196.

46 Oelschlaeger (1991), p 69.

47 www.are.admin.ch/are/en/nachhaltig/international_uno/unterseite02330/

48 From Phil Macnaghten and John Urry, *Contested Natures* (London: Sage 1998), p 95 in Eileen Crist, "Against the Social Construction of Nature and Wilderness", *Environmental Ethics* v.

22 | INTRODUCTION

26:1, 2004 (pp 4–24), p 6.

49   Robert Foley, *Humans before Humanity* (Oxford: Blackwells, 1995), p 5.

50   Rachel Carson, *Silent Spring* (Boston: Houghton Mifflin,1962), p 64.

51   Barry Commoner, *The Closing Circle* (New York: Bantam, 1972), pp 16–23. The Four Laws are: 1. Everything is connected to everything else; 2 Everything must go somewhere; 3 Nature knows best; 4 There's no such thing as a free lunch.

52   Charles Taylor, *The Malaise of Modernity* (Concord, Ontario: Anansi Press, 1991), p 26.

53   Harold Horowitz in J. Baird Callicott, "The Metaphysical Implications of Ecology" (pp 51–64), in J. Baird Callicott and Roger Ames, eds., *Nature in Asian Traditions of Thought* (Albany: SUNY Press, 1989), p 58.

54   Harold Horowitz, "Biology of a Cosmological Science" (pp 37–49) in Callicott and Ames, p 47.

55   Paul Taylor, p 115.

56   The John Livingston reader, p 63.

57   Alan Weisman, *The World Without Us* (New York: Harper Collins, 2007). Livingston reminds us that in the event of such a demise the true anthropocentric horror will be that "...when man does become extinct, he will never have existed at all. There will be no one, no thing, on Earth to recognize our rubble for what it is, or to make any connection between it and anything that may have passed, so it will never be known that we were here." Livingston Reader, p 99.

58   Lynton Caldwell concludes that since these issues were raised by the Club of Rome starting in 1968, "...no significant change of course has been undertaken by governments nor is evidenced in the global economy." Lynton Caldwell, "Perspectives on the Self-Destructive Tendencies of Humanity: A Symposium Response", from the *Symposium: Is Humanity Destined to Self-Destruct?*, p 269.

59   Edward Wilson, p 22.

60   Christopher Stone is skeptical that a changed environmental consciousness will lead to a reversal of present consumption and exploitation trends, but argues that it would be a step toward that goal and at the same time make us "better". Shelley and others long based their vegetarianism on the notion that it would lead to less violent humans. Christopher Stone, *Should Trees have Standing?* (Los Altos, California: William Kaufmann, 1972), p 48.

61   Leslie Thiele, "Learning the Lesson of Interdependence" from the Symposium: Is Humanity Destined to Self-Destruct?, *Politics and the Life Sciences*, Sept 1999 (257–260), p 258. See also his *Environmentalism for a New Millennium: The Challenge of Coevolution* (New York: Oxford University Press, 1999).

62   Bookchin, (1995) p 109.

63   Jonathan Bate, "Romantic Ecology Revisited", *The Wordsworth Circle* (v. 24:3, 1993, pp 159–162), p 161.

64   Charles Guignon, *On Being Authentic* (New York: Routledge, 2004), p 108–9.

65   Leiss (1990).

66   Leiss (1990) p 87.

67   William Leiss, *The Domination of Nature* (Montreal: McGill-Queen's University Press, 1994) (1972), p. xxvi.

PART ONE
THE FALL
FROM GRACE

CHAPTER ONE
WHAT ARE WE BY NATURE?

"a sentiment of existence"   ROUSSEAU

In September 1765, in the midst of Enlightenment and the vibrant adolescence of modernity, the musician, philosophe and ageing enfant terrible Jean-Jacques Rousseau was spending his days as an exile on a small island in a lake in Switzerland. Were we to seek him out in the early afternoon he could usually be seen offshore, lying on his back drifting in a small boat "...plunged in a host of vague yet delightful reveries".[1]

But Rousseau was no early romantic, content with a life spent viewing memories flashing on an inward eye. In the mornings on the Isle St. Pierre he played the avid empiricist who, having parsed a section of the island into small squares, began a systematic longitudinal study of the island's plant and insect life. Instinctively a democrat, after lunch and following his drift on the lake he helped the local farmer harvest fruit and bring in his crops. In early evening he would sit by the lake and allow the sound of the waves lapping against the shore to ease him back into a reverie which would then "...make me pleasurably aware of my existence, without troubling myself with thought".[2] Each day was then completed via a re-engagement with the social world, sitting with his companion Therese talking over the events of the day, singing old songs and making plans for the next day.

Nor were the reveries a product of will. Rousseau insists in this state "...images are traced in the brain, where they combine as in sleep without the collaboration of the will. All that is allowed to follow its course, and one enjoys without acting."[3] Here, innocently enough, were the early indications of what was to become by the 21st century a crisis for a modernity being built too exclusively on reason and on action.

Rousseau's description of the almost monastic and idyllic completeness of a day spent on the island is punctuated by these interludes of induced reverie in which he strives for insight "without troubling myself with thought" and "without action." Rousseau the modern, the inventor of a musical notation system based on numbers, the proto-behaviourist educator and proto-revolutionary social contract theorist — yearns to combine his modernity with a primitive holism that will allow him to make contact with what that very modernity seems to block, the "feeling of existence." Freud would later describe this as a search for an "oceanic feeling" or what we might today call "oneness with nature" and saw it as stemming from primitive drives that would inevitably clash with the repressive needs of civilization. Modern Darwinians might see these reveries as responses to "memes", ideas, needs or beliefs that are thought to be embedded as deeply as genes in individual humans and to exist independent of particular cultural imperatives.

But what is Rousseau really seeking as he drifts, listens, feels and senses? Is this just another example of cultural primitivism, a hopelessly romanticized longing for a simpler life that probably never existed? It might seem so at first glance, but his objective is not simply to regain the past but rather to seek a unity, an awareness of his being or self in relation to all other being, a unity that is not mediated by the particulars of life, particulars that his reason and his senses constantly remind him of. This "oceanic" awareness becomes, in Rousseau's work and in that of those who follow in this tradition, the starting point for a possible resolution of the contradictions and conflicts implicit in the human condition and made increasingly explicit by modernity. Hence Claude Levi-Strauss calls Rousseau the first anthropologist based on this quest to discover, by using both research and reverie, "...the unshakable basis of human society."[4]

Whatever the source of his need for this "connection" with the whole, Rousseau was not allowed to remain thus disengaged from modernity, being forced by October that same year to leave his island hermitage and continue a wandering, homeless exile. The critique of modernity implicit in his work, however, took root, most notably with the Romantic poets of the following generation and later in the critical work of people like Thoreau, Ruskin, Morris and Muir. In 1965, now well into the apogee of modernity, Timothy Leary, speaking from his own exile on Harvard Square, counseled us to "Tune in, turn on and drop out". Using chemicals rather than the sound of waves lapping on the shore, Leary and his acolytes were on a Rousseauean quest of sorts, still seeking that "feeling of existence" that modernity in its technicist-consumerist mode was making even more elusive than it had been in 1765.

THE QUEST FOR HUMAN NATURE

Floating down the Amazon toward the coast and reflecting on his failure to locate the primitive even in the depths of the Brazilian jungle, Levi-Strauss consoled himself by reading again Rousseau's *Discourse on the Origins of Inequality*. Written in 1755, ten years prior to his sojourn on the Isle St.Pierre, this essay was Rousseau's own anthropological journey to find the primitive. Rousseau sought his own Adam, a first man, and could, of course, only find him philosophically. The actual humans his 18th century sources led him to were, he had to admit, already far removed from their Paleolithic ancestors, just as the Nambikwara in Brazil were when Levi-Strauss met them in person in the 1950's.

Rousseau and Levi-Strauss were each looking to find a baseline for exploring the issue of human nature, or the natural condition of being Homo sapiens. At once critical of the evils they saw in their own societies while rejecting an idealization of the primitive, both were convinced that human cultures and especially civilizations could be improved, that there existed an inherent potential within the human for more cooperation, more individual and collective flourishing, and less reliance on violence and inequality. Without wishing to return to some romanticized past, both saw the primitive, pre-historical era as a site of alternative possibilities, a world of "oceanic oneness" that implied a blurring of the divisions between animate and inanimate, animal and human.[5] But to prove this they needed to find a beginning, an "unshakable basis of human society" that was only minimally affected by the cultures and structures humans create when they live in common over time.[6]

This issue of a return must be confronted before we move to examine the kind of evidence we have for just what the primitive might have been like. Lovejoy and Boas in their classic study *Primitivism and Related Ideas in Antiquity*, describe this romanticizing of the past as "cultural primitivism", the "...discontent of the civilized with civilization, or with some conspicuous and characteristic feature of it. It is the belief of men living in a relatively highly evolved and complex cultural condition that a life far simpler and less sophisticated in some or all respects is a more desirable life."[7] In these times of profound cultural angst it is tempting to locate a better time in the past, perhaps a time of villages, tribes or clans, a time when small was beautiful and humans lived within a covenant with nature. But isn't this delusional?

Nostalgia can be a dangerous course and it is no doubt wiser to see the task as not really one of rejection or of a return to some earlier state of human development but rather one of integration. What was there in the eons of relatively stable human existence that we should strive to retain, to reconcile with our present modern selves, and what is no longer functional, possible or of value? What might we be wise to recapture from our past and what is best left abandoned or transcended?

But the question can be asked a different way if we return to the issue of human nature. Rousseau, the man who introduces this notion to modernity, would argue that the human is inherently isolate, a loner, a self set apart from other selves — by our very nature. Left to himself, he argues, the human self would by choice "...spend all waking hours concentrating on the sweet feeling of his own existence, unconcerned with anyone else."[8] (One thinks here of one's cat on a sunny day, but not the dog) So there is Rousseau drifting on the lake or on the shore listening to the lapping of the waves. It is a sub-human, pre-conscious and pre-social animal that Rousseau admires and insists remains our core, our nature. But is this the real beginning of the "natural human", is this what we were and still are, still should be, or worse yet aspire to be while living compacted in cities? If so, we moderns in the 21st century would seem to be very unnatural indeed.

In the Introduction I raised the issue of the degree to which our organic, inherent, genetic link with our past influences or determines our subsequent cultural creations. Are we, as some evolutionary psychologists claim, still bucolic hunter-gatherers trying to cope with living in urban sprawl? Or were our primitive forebears "...libidinous, rapacious, and generally selfish" and have only become social via severe repression?[9] If we do not experience nature in ways that we were genetically programmed to do, will we cease to value it? Or has our brain and its related consciousness evolved beyond needing these prolonged moments of Rousseauean reverie?

To begin these reflections on the primitive and the modern we will join Rousseau and Levi-Strauss and add to the mix a host of modern archaeologists, scientists, and anthropologists whose work probes the realm of pre-history via artifacts, DNA analysis and informed speculation. The ancient civilizations of the great river systems — Greek, Near Eastern, Egyptian, Chinese and Indian — provide the common bedrock of our various modernities, but they too had a history, the humans in those civilizations being already highly socialized when they began to leave their myths, art, writings and architecture for us to read and interpret.

And so we start at the beginning. We hypothesize that the earth has existed for about 4.6 billion years and think that what we have come to call life (defined as chemical systems capable of replicating themselves) first appeared about one billion years later. Living organisms that we might recognize as such do not appear until much later, some 750 million years ago, followed 300 million years later by vertebrates, then dinosaurs 250 millions later, mammals

150 million years ago and our hominid ancestors a mere six million years ago.[10] There have been, as far as we know, ten distinct species of bi-pedal primates with our own direct ancestors, Homo sapiens, emerging into what we call history about 40,000 years ago — a mere 0.001% of the span of natural history or the history of life — "a bubble submerged in an ocean of deep time."[11]

What are we to make of these virtually incomprehensible periods of time? There have been about 240,000 generations of hominids over the past six million years, and 1,600 generations of Homo sapiens. Most striking at first must be the chronological weakness implicit in the anthropocentric conceit. Still, even with these vast numbers and reaches of time, the hierarchically-minded can still find comfort in the patience and skill required by nature (or God) to finally "get it right" after so many eons of experimentation. My own great-grandfather Ebenezer was alive one hundred years ago and it is possible for me to imagine that the 1,597 generations behind him were all leading to me.

Taking a moment to meditate a bit on these expanses of time will also help us in this attempt to include the primitive era of our own species in our notion of human time or history. This 40,000 years of species history — most of which as we will see is spent as scavengers, gatherers and hunters — seems quite manageable when compared to mammalian time, vertebrate time, or chemistry time.

Still, it may be wise at the start of this exploration to borrow a page from Rousseau or similar "mystics" or romantics and try to grasp these eons via reveries of our own. My own reverie on this issue occurred in 1989, while listening to a talk by James Lovelock at a Bioethics Conference in Brussels. Lovelock, the inspiration behind the idea of the earth as a 'living system' of which we are merely a part (the Gaia Hypothesis), spoke to us about the role of the several eras of planetary glaciation, arguing that these were positive times for the planet. While the northern hemisphere was largely covered by ice, there was abundant life on the continental shelves which had been newly exposed by the drop in ocean levels due to ice. Referring to our current preoccupation with global warming, he argued that:

> ..there is good evidence to suggest that the Earth is healthier during ice ages than during the interglacials as at present. Left alone the Earth would probably recover from its present fever and return to a long period of health in a new ice age...but if it does, perhaps by changes involving the ocean algae, then we may find ourselves on a planet still fit for life, but no longer fit for us.[12]

What a blow perspective can be to human hubris! The complex systems that make up the earth, our home, prefer ice and algae to warmth and humans.

30 | THE FALL FROM GRACE

These thoughts of ice and ages point to the prevalence of change in the life of the planet, change which to human and other mammalian sensibilities is often experienced as catastrophe or disaster. Thus over the past 3.5 million years the earth has gone through twenty-seven complete major climatic cycles, with extended periods of extreme cold sometimes lasting from 20,000 to 60,000 years. One of these cold snaps around 33,000 years ago may have led to the demise of the Neanderthals and while the planet has been steadily warming for the past 20,000 years, we have vivid accounts of the effects of a mini Ice Age within our own era (1400–1850).[13]

But all of this is just the planetary base upon which the history of the hominids is built, and it to this history that we must turn our attention. Looking back from our time, we can create a number of stages or phases of hominid development that lead to us. We have the classic stone, bronze and iron age theory based on archaeological evidence, but more commonly now we tend to create stages based on economic or political indices rather than on the materials of tool construction. Thus we have the now generally accepted four-stage approach:

1. The Primeval or Hunter-Gatherer Stage from the Lower Paleolithic to the Neolithic — this was the longest and most stable phase in hominid history, from roughly 500,000 to 10,000 BCE.
2. The Neolithic or early farming phase from @9,000–5,000 BCE, an era we often associate with the 'primitive' and the pastoral and which is still "... accessible to anthropological scrutiny and reconstruction."[14]
3. Archaic or early urban civilizations based in the Mediterranean Basin, the Middle and Far East and the Americas. Since many people still live in cultures in transition from primitive to peasant and early urban systems, aspects of this phase remain highly visible.
4. Modernity, beginning @450 years ago with large-scale urbanization, politicization of cultures, greatly expanded use of energy, and maximum utilization of natural resources.[15]

The full story of these stages, especially the last two, belongs in Chapter Two, Who Are We by Choice, Chance and Context? But here at the start we need to focus in on the impact of this long period of hominid evolution beginning some 500,000 years ago in order to assess what, if anything, it can tell us about human nature.

In the ancient myths and creation stories humans appear rather suddenly, usually crafted by a creator of some sort with their nature more or less intact from the start. But we moderns have a different story, one based on evolution, and depending on the teller it may or may not include a creator or designer.

Based on the great synthesis of evolutionary thought carried out by Charles Darwin in the 1860's, modern evolutionary theory offers a convincing narrative account of how hominids, for instance, came to be—but the theory remains contentious on the issue of purpose or the why. Just as 18th century thinkers were able to imagine a primitive human existing *after* the Flood but not before, modern adherents of intelligent design are able to reconcile the prolonged primitivism of the archaeological record with belief in a purposeful design.

But did the line of offspring from my great-grandfather Ebenezer Duguid really *lead* to me? Of course we know that this cannot be the case in any truly purposeful sense—there were just too many variables (including grandmothers and mothers, two world wars, etc.) in the way. And is it the case that the seemingly phenomenal complexity of objects like my eye mean there must be design behind creation? No, say the Darwinians, there was more than enough time in vertebrate evolution for an eye to evolve from a primitive light sensitive spot in some early vertebrate species.[16] Darwin himself noted that the human eye was less than perfect in "...correction for the aberration of light", part of his argument that through evolution the elements of nature aspire to be only as perfect as they need to be to in order to compete in an on-going "struggle for existence."[17]

The examples of Ebenezer and my eye bring us to the very crucial issue of progress. Later, when we explore the issue of human culture and the impact of human civilizations on the natural world, the issue of progress will be of central importance, a key justification for human interventions of all kinds. And often this idea of civilizational progress is linked to an over-arching idea of progressiveness in nature *per se*, with humans being seen as a progressive evolutionary "result" and human civilization and its technologies thus being a "natural progression" of an evolutionary imperative. Followers of Darwin beg to differ: "evolution did not imply direction or progress...it did not follow any plan."[18] Darwin argued that natural selection worked upon more or less random individual variations: "The random and haphazard course taken by evolution denied the possibility of any goal for nature or of a plan which culminated in the emergence of the human species;"[19] "...there is no progressive direction...the survival of species is largely the product of randomness...it is all a matter of luck."[20]

But is it just luck? Not according to Darwinian theory. The survival, growth and the physical as well as cultural shape of any species is the result of a complex "struggle for existence" within competitive living spaces. This struggle—carried on by all species, trees in a forest as well as mammals—does not lead merely to the "survival of the fittest" but rather to a process of natural selection via variation and adaptation. For Darwin, natural selection was "...daily and hourly scrutinizing throughout the world, every variation, even the slightest; rejecting that which is bad, preserving and adding up all that is good: silently and insensibly working, whenever and wherever opportunity offers, at the improvement of each organic being in relation to its organic and inorganic conditions of life."[21] There is a lot to unpack in Darwin's words. Most important given our earlier discussion of progress, his use of "bad", "good" and "improvement" should not be taken to be absolute judgments, but instead linked directly to the quality of the beings relation to its environment, its niche or context.

But how does evolutionary theory relate to our search for a possible human nature? The physical shape of hominids, the survival of one branch and the demise of another can be accounted for by different variations, some leading to hereditary traits that are more conducive to survival and reproduction than others.[22] It may be, as we will examine later, that the gradual increase in brain size among hominids—especially in Homo sapiens—was a key variation, one that led to attributes such as language, consciousness, and empathy, and thereby played a major role in the natural selection process. But large brains in hominids also present a physiological problem in that the brain must achieve its size when human infants are very young with the result being that "...the price of human evolutionary success falls most heavily on the mother."[23]

Here, it seems, is a clear case of a decisive inter-relationship of nature and culture. The female hominid is not only the bearer of the next generation, but must also bear the burden of a prolonged period of nurturing before and after birth, a nurturing burden which both increases her value and weakens her ability to survive independently. The importance of this nurturing role in the long-term success of the species can affect males as well, one reasonable Darwinian speculation being that "...men are descended from a long line of men who left on the average more offspring because they were better at calculating the reproductive value of women."[24] Already, then, in the area of relations between the sexes we can imagine how certain aspects of what we have come to see as human nature either are or can be seen to be grounded in the nature of the species *per se*. Darwinians, for instance, would insist that "...our capacity for culture has almost certainly been shaped by natural selection."[25]

Seeing human culture as a mere attribute of natural selection (or nature *per se* in a Darwinian sense) does seem problematic if we return to the notion of evolution being random, directionless. Human cultures seem to have some sense of direction, or at least sets of shared values that have a certain consistency over time and space. Our common cultural habits and values do not seem random. But, of course, neither do they make sense being completely independent of nature. In the long-running nature/nurture debate, the only reasonable ground would seem to lie in the middle. Our genetic inheritance and the cultural and material factors of our lives share in creating whatever we "are", individually and as a species.

Considering for the moment only primates—and especially human primates—there does seem to be a kind of progressive quality to the wandering path of natural selection. The environmental ethicist Holmes Rolston argues that despite a general aimlessness there has been an increase in complexity over time and that this has made possible more creativity in the forms and the sophistication of human cultures. He notes these increases in complexity in the following areas:

- capacity for central control (brain development)
- capacity for sentience or "feeling"
- locomotion
- acquired learning or memory
- communication and language acquisition[26]

This increase in complexity leads, of course, to what we humans call our consciousness, our inner picture of ourselves as the mediating force between that self and time, space and the "other".

Does consciousness, then, create a human nature beyond the building blocks established by evolution? And if so is it a constant or does human nature change as consciousness changes and in the process increase in complexity? If it *is* derivative, then this would seem to be the case. One version of such a perspective on human nature is held by Marxist scholars who would generally claim that "...human nature is a product of society...it is a historically constituted subject and not trans-historical."[27] In this way cultures or states built on slavery, for instance, would produce or create a human nature quite distinct from cultures based on either capitalism or some kind of egalitarian communism.

There is, of course, a good measure of common sense in this essentially historicist (as opposed to essentialist) approach. The behaviours, attitudes and values of people from divergent cultural and political settings do seem to vary considerably. At the height of the 20th century's fascination with the potential of the social sciences to deliberately alter such behaviours, attitudes and values, it became very unpopular to talk about a biologically determined human nature. Today that optimism in the ability of science—social and otherwise—to shape human behaviour and has ebbed considerably, the casualty of too many gulags, wars, Rwandas and Kossovos, the collapse of Soviet

34 | THE FALL FROM GRACE

Communism, and the persistence of poverty, crime and racism in even the most advanced societies. Still, from a post-modern and post-scientistic perspective, contemporary thinkers like the philosopher Richard Rorty still adhere to a kind of relativism, insisting that humans are by nature so malleable there is no sense talking of their having a nature in common.

Into this breach, however, a new paradigm or model of human nature has emerged, one based on a combination of evolutionary theory, psychology and neurobiology. Here the human consciousness that has evolved over the millennia does produce a kind of basic or fundamental human nature, one that is in fact not very malleable and not easily subject to fundamental changes emanating from various cultures. This human nature is instead an evolutionary product, a result of natural selection and has evolved to facilitate complex social interactions among humans.[28]

What kinds of qualities might we associate with a human nature derived via such a "selection for socialization"? Robert Wright, a prolific popularizer of evolutionary psychology, lists reciprocal altruism, compassion, empathy, love, conscience, a sense of justice and a tendency to base status on hierarchy as human qualities with a "firm genetic basis" and as "vestiges of organic history on a particular planet."[29] Of particular importance to Wright is the fusion in Homo sapiens of two quite rare and potentially conflicting qualities: reciprocal altruism and status hierarchy. On the one hand we (particularly males) have a tendency to award status or rank to one another according to specific criteria while on the other, thanks to our reciprocal altruism genes, we moderate this competitiveness via a tit-for-tat process built on empathetic understandings (often associated strongly with females). The result, for Wright and other neo-Darwinians, is a very complicated mammal with species tendencies for radical differentiation and structured inequalities being tempered by an equally innate quality of empathy.

This pairing of opposites or, perhaps more properly a tolerance for dualities, is an even deeper human quality according to Yi-Fu Tuan. Our brain evolved "to pick pairs among segments perceived in nature's continuum.... and to assign opposite meanings to each pair."[30] Thus we commonly juxtapose life and death, male and female, self and other, heaven and earth. This pairing of opposites—what many see as an inherent tendency toward dualism—turns out to be useful for organizing the complexities of the world and perhaps does reflect certain structural patterns in the human brain. But, as Tuan hints, nature itself is much more a continuum than a set of discrete pairings, giving this dualist quality attributes of both a blessing and a curse.

Dualist patterns of thought which focus on sometimes radical differentiations (e.g. self and other, saved and damned) when coupled with an innate drive to arrange phenomena hierarchically can be a recipe for asociality. But most agree that humans are naturally social and if reciprocal altruism is in fact an innate quality, it might be a key factor in encouraging social or cooperative behaviour among humans. But observation also tells us it may be too slender a reed upon which to base human sociality.

But beneath altruism, it has long been argued, lay a much more generic quality: pity, sympathy or fellow-feeling. For a Rousseauean like Levi-Strauss, this fellow-feeling pre-dated human tendencies to stress the differences between self and other or tribesman and stranger. He imagines that in our deep species-past (an era we will be exploring soon) we arrived at a sensual and cognitive point in our evolution at which we became intuitively aware of a connection with any living creature simply "because it was alive."[31]

Sympathy is thus, as Rousseau argued in the 18th century, seen as an original disposition" in humans and is "extended and strengthened as we become more capable of using our senses and more enlightened."[32] This idea of innate sympathy emerges from the 18th century as a powerful antidote to earlier Hobbesian notions of human brutishness and remains a vigorous counterpoint to Freudian ideas of "man being wolf to man" and Jean-Paul Sartre's famous insistence that "Hell is other people". Instead, the proponents of sympathy insist that surface dissimilarities among people are illusions that mask our essential commonalities, our "fundamental unity."[33]

Darwin later extended the argument by insisting that the kind of moral order derived from sympathy permeated nature *per se*, not just human nature. Thus he saw qualities like shame, wonder, curiosity and honor in the "lower species", all part of a process of natural selection that favoured attributes that helped a species survive and prosper.[34] We will encounter this claim many times in the work of environmentalists, animal rights activists and ecocentric philosophers. Indeed, it is the possibility that such a moral quality or foundation can exist within humans, other species and perhaps in nature as a whole that has inspired this book, since it could provide the basis for an assignment of intrinsic value and moral standing to virtually all of the natural world.

But it is to Homo sapiens that we need to look to. For humans, sympathy was the quality that let them acknowledge the presence of the other and enter into their feelings. "All humans....when confronted with the misery of another truly feel a sense of unease, a sense that there is something very wrong, and that they too are suffering" and that this is a "...biologically normal response."[35] More than a heightened awareness of the suffering of others, sympathetic understandings and feelings in humans more often than not include a compulsion to take whatever actions are necessary to alleviate the suffering of the other, and makes the transition from inner feeling to social action.[36]

Before examining in more detail the origin and early history of this sympathetic Homo sapiens, we should hear from Edward Wilson, perhaps the most outspoken contemporary champion of the idea of the moral nature of humans. He takes this position not because he believes in some mysterious implantation of an innate sense of right and wrong, but because we are social, intelligent animals "...fitted by evolution—by God, if you prefer—to pursue personal ends through cooperation....Humanity is the species forced by its basic nature to make moral choices."[37] Because we are by nature social beings, we had to develop moral systems to regulate our behaviours and ensure that cooperation has at least equal standing with an ego-driven competitiveness. We are "...naturally equipped with a universal intuitive rule book...naturally disposed to prefer justice to injustice, to deplore cruelty, to sympathize with distress."[38] The various rules created within human cultures, therefore, are neither delivered nor created out of nothing, they have their source in what for Wilson is a bedrock of human nature. And it is possible, perhaps essential now, that these intuitive preferences that Wilson identifies relate not just to fellow humans, but to nature conceived in the broadest possible terms. But we can start with a more obvious choice: trees.

AT HOME IN THE FOREST — AN ANTHROPOLOGICAL REVERIE

What is about trees and humans? They live at a human scale unlike the mountains and oceans, stronger and more expansive than us and yet seem to inspire intimacy and to evoke feelings of beauty rather than the sublime. Whether in savanna, canopy or mountains, for humans the common denominator for a "home place" seems often to involve trees.[39] Is this affinity for trees based on our shared singularity even in a forest or a crowd, our verticality, our dependence on water, our tendency toward expansiveness? Trees appear to be clothed and adorned like us, have elements of nobility and strength that we like to ascribe to our species, and mirror in comprehensible time frames a human sense of birth, maturity and decay on the one hand and continuity on the other.[40] Like us, they are subject to both visible and invisible disease, the decay and decline of limbs and sudden catastrophes. Trees, even better than our fellow mammals or other sighted creatures, may serve as our "natural" peers, the touchstone for our concerns about the relationship between humans and the natural world.

Paul Shepard, in his persistent attempts to illuminate the biological and behavioural bridge between our ancient and modern selves, saw the compulsive affection for trees among children as the result of a primal memory of our forest origins, with the instinct to climb a "...return to quadrupedal motion, touching a chord in our genetic memory of an aboreal safety."[41] This may be no "elective" affinity, then, but a natural or innate affinity which entails a child-like, romantic, a-rational merging of the human self with the tree. And it is significant that the focus is on the child since, as the Romantic poets reminded us, as children we "...less habitually distinguished all that we saw and felt, from ourselves....", an attribute as adults we allow to decay.[42] Aware at an early age of our extended fragility in a hostile universe and the impossibility of a return to the womb, one of our preferred safety zones as children is in the primate's choice of the tree rather than the early hominid's decision for the depth of the cave or the clear sight-lines of the savanna.

As in so many dimensions of human intimacy, there can be a cost. Forests may have provided a home for our ancient ancestors, a refuge in difficult times for others, and an on-going welcome source of food, fuel and materials, but their passivity and rootedness has placed them in peril as well. As human civilizations developed they tended to consume at will the forests that they found. "When states went to war, entire forests were devastated....as the classical empires spread from east to west along the Mediterranean and north into Europe, the forests were demolished."[43] Jonathan Bate reminds us of the extent of this devastation and its connection with a particular form of human social construction:

- before the Athenian Empire the hills of Attica were canopied
- before the Roman Empire the length of North Africa was wooded and fertile
- before the Venetian Empire, forests stretched from the lagoon to the Alps
- before the American empire the rainforest spread across Brazil[44]

But empires and wars alone are not to blame for the destruction of forests. Across human cultures forests came to be seen as useful, as sources of materials to burn, build with and consume in ever more myriad ways.

But forests were more than just a home place, armory or general store for humans—the relationship was more complex than that. On the one hand in European culture the forests came to be seen by urban and agricultural folk as "an insecure and uncomfortable environment against which civilization had waged an unceasing struggle."[45] For them the former home of humans had become a place of demons and spirits, bandits and outsiders, and therefore needed to be tamed.

> To outsiders, forests are oppressive. The foliage absorbs the light long before it reaches eye level. Trees preside over gloom with knots bared like knuckles....To trespassers from cities and fields, forests are dangerous:

environments which invite you to fight back with fire and ax, because they hide your natural enemies, lose you and starve you. Over a vast stretch of the post-glacial northern hemisphere the forest, while it was still intact, was too big to flee from. You could escape only by chopping it down.[46]

But it was also the place of "sacred groves", hunting preserves for the rich, and for an increasing number as we approach modernity, a romantic link with the past. As the forests were diminished through consumption and clearing they became less fearsome and more sources of beauty and assets in new notions of "landscape". It became a "violation of nature" to witness the destruction of an ancient tree.[47] By the 19th century the issue of preservation of forests had become a powerful force in Western culture, driven by sentiments such as these by the poet Gerald Manley Hopkins when he found a favourite groves of aspens no longer standing:

My aspens dear, whose airy cages quelled,
Quelled or quenched in leaves the leaping sun,
All felled, felled, all are felled;
Of a fresh and following folded rank
Not spared, not one
That dandaled a sandaled
Shadow that swam or sank
On meadow and river and wind-wandering weed-winding bank.
O if we but knew what we do
When we delve or hew-
Hack and rack the growing green![48]

We enter here a world in which trees are truly "friends", individually, in groves and forests, and conceptually. We have left the realm of tree as home that may have helped form our early primate ancestors.[49] We have left as well the dark pagan forest filled with spirits and outcasts and left behind the tree and forest as mere resource, as raw material. We have moved instead toward a kind of deep subjectivity which some might claim is a return to the beginning.

Indeed this moves us close to a central theme of this book, whether or not we can (or should) attempt to regain a range of primal insights and predispositions that fell out of favour over the millennia. Before we return to the text and explore some of these primal times, we should listen to some of the modern voices that claim for trees and, implicitly for nature as a whole, a life or awareness that borders very closely on the Rousseauean sentiment of existence that was explored in the opening reverie.

In 1795, Mary Wollstonecraft traveled with her infant daughter Fanny on a trek through Scandinavia searching out the fate of one of her lover's ships that had gone astray. The journal she kept reflects the almost schizophrenic stance of the Romantics toward the modern and the traditional, but the most moving passages refer to her reaction to the wild nature she observed:

> The continual recurrence of pine and fir groves, in the day, sometimes wearies the sight; but, in the evening, nothing can be more picturesque, or, more properly speaking, better calculated to produce poetical images. Passing through them, I have been struck with a mystic kind of reverence, and I did, as it were, homage to their venerable shadows......I could scarcely conceive that they were without some consciousness of existence.[50]

So here, an admitted fan of Rousseau while at the same time a crusader for the role of reason in human affair has sensed the sentiment of existence in groves of the most common fir tree. In 1837, the young John Ruskin in one of his first books writes of the sense of identification that he feels with the classic ancient oak of the English countryside, wishing to impart to it human-like feelings and sensations:

> A very old forest tree is a thing subject to the same laws of nature as ourselves: it is an energetic being, liable to an approaching death; its age is written on every spray; and, because we see it is susceptible of life and annihilation, like our own, we imagine it must be capable of the same feelings, and possess the same faculties, and, above all others, memory: it is always telling us about the past....The chief feeling induced by woody country is one of reverence for its antiquity....[51]

In 1869, Scottish emigrant John Muir spent his "First Summer in the Sierra". Muir loved trees almost as much as he revered glaciers and mountains — indeed, "of all living things he liked trees the best."[52] What we owe to Muir above all is his keen appreciation, honed through years of wilderness immersion, of the individuality of the natural forms we too often see as undifferentiated or commonplace — to Muir each tree was a unique, vibrant, living self:

> When young it is very straight and regular in form like most conifers; but at the age of fifty to one hundred years it begins to acquire individuality, so that no two are alike in their prime or old age. Every tree calls for special admiration....Were they mere mechanical structures, what noble objects they would still be! How much more throbbing, thrilling, overflowing, full of life in every fiber and cell, grand glowing silver-rods — the very gods of the plant kingdom, living their sublime century lives in sight of Heaven, watched and loved and admired from generation to generation.[53]

40 | THE FALL FROM GRACE

From Ruskin's attribution of human qualities to trees we shift with Muir's more ecological frame of reference to the tree having its own unique "life force". And as cited earlier in the Introduction, in his foundational book for modern environmental thought Aldo Leopold felt "...a curious transfusion of courage" as he stared at his row upon row of snow-laden pine trees standing "ramrod straight."[54]

From home to enemy to resource to landscape to repository of memory to sensate being to fellow-being, trees and humans have a long and varied relationship. The tree, then, begins to attain the quality of a subject, a being that by its nature requires respect. Carrying his demand for an I-Thou relationship between humans to another level, Martin Buber insists that "...it can come about, if I have both will and grace, that in considering the tree I become bound up in relation to it. The tree is now no longer an It. I have been seized by the power of exclusiveness...I encounter no soul or dryad of the tree, but the tree itself."[55]

THE SEARCH FOR HUMAN ORIGINS

"The tree itself". The primal tree. The tree as abstract entity but also as living, visceral being. Buber wants to relate to the tree as a phenomenon, a being living its life as an individual and as species. And we likewise want to think about the first humans, our ancestors, as the beginning of us as species and also as individuals.

In the beginning of us were the primates. Emerging in forested regions during the fading of the dinosaur era about 65 million years ago, these "...small shrew-like creatures...were the ancestors of humankind."[56] Primates that we might more commonly identify as akin to us appear in the fossil record some 30 million years later, eventually settling into six varieties: gibbons, orangutans, gorillas, two species of chimpanzees and the various strains of pre-humans or hominids.[57] The branch of hominids that split off from the chimpanzees (Jolly says that at the molecular level we remain essentially chimpanzees[58]) did so about six million years ago, making us a very young species relative to our primate kin and, of course, mere infants in terms of the span of life on the planet.[59]

While there were many features of these tree-bound primates that were to become important to the development of the hominids that followed, two are of especial interest here: a predisposition to sociality and a highly developed binocular vision. Primates early on specialized in being social, a key aspect of what specialists call "primate core adaptation."[60] Stemming in part from the high reproductive costs borne by females which put a premium on access to energy sources and parental care, this early sociability led to the beginnings of verbal communication, various forms of family life or collective living and a persistent curiosity about and sensitivity to others.

In the realm of the senses the gradual shifting of the eyes to the front yielded the binocular vision so necessary for judging depth and distance and living efficiently within the forest canopy. These early vision-dependent hominids were a "mammalian oddity, a carnivore hunting by sight rather than smell."[61] But among the senses, vision can be the most unreliable which, in a perhaps perverse but certainly beneficial twist, led to a greater reliance on reasoning to de-code the errors that stemmed from an over-reliance on sight.[62] Peter Wilson points out that since primate facial features had also evolved to the point that individuals could more easily mask their feelings and conceal their intentions, reliance on visual cues was particularly dangerous. "Being unsure of the relation between another's appearance and intentions or feelings demands calculation and breeds uncertainty."[63]

Already, then, in early primate evolution the key hominid qualities were present—prolonged gestation and infancy, privileging of vision over other senses, greater sociality and the corresponding need for a more sophisticated cognitive apparatus, i.e. brain. And then came the savanna. It is no small irony that we now live in fear that climate change may destroy the possibility of human existence when most agree that we in fact owe our existence (at least in the form of Homo sapiens) to an earlier dramatic climate shift. Like all other species, we are in the end creatures of climate.

Modern Chaos Theory tells us that weather is one phenomenon that our science will never be able to predict, there being just too many variables for even the most super of super-computers. But if we cannot predict the precise location of tomorrow's blizzard or tornado, we can with some assurance predict the cycle of trends in the planet's climate. The fossil record and geological history tells us the story of a cyclical heating and cooling process at the core of our home planet's unique way of being in the universe. To survive and evolve, the organisms on the crust of that planet must adapt to these shifting temperatures and the climates that follow.

Around 5 million years ago the Earth experienced one of its major cooling and drying cycles which on large areas of its surface favoured the spread of grasslands at the expense of the existing forests. Our primate ancestors, living in the now retreating thick forest canopies and subsisting largely on the fruit and vegetable matter found there, inevitably faced increasing competition for both food and living space. We speculate that around three to four million years ago some of these primates—those less able to adapt to new foodstuffs or less aggressive in protecting a shrinking territory—wandered onto the spreading savanna, a "relatively smooth, grassy ground surface dotted with trees and copses."[64]

Initially the savanna would have been a dangerous environment for a tree-dependent animal, perhaps leading to considerable carnage and dislocation. But in this instance of climate change the process was gradual rather than cataclysmic, with ample time for evolutionary adaptation. The savanna remained a varied eco-system with areas of canopy forests, open bushlands, lakes, and extended grasslands.[65] In one of their key adaptations, the success stories in this forced exodus from the forests came to stand upright and develop into bi-pedal primates, better able to see over the tall grasses, cover longer distances, and reach the food on the trees and shrubs of the savanna.

As well, over time they shed much of their body hair in response to the warmer temperatures of the new more open environment.[66] This was the era of Lucy the Australopithicine, still small-brained like her chimpanzee ancestors, but now able to stride across the plains and survive in the relatively open landscape. But survive only barely. Despite assertions by some that these early hominids like Lucy were aggressive hunters, they turn out to have been more prey than hunter, scavengers and consumers of rough vegetable matter on the new arid landscapes of Africa.[67]

Lucy, the first split-off from the apes that we know of, dates from 3.2 million years ago. From here we can follow a number of other splits or divisions, all under the general rubric of "Homo". The first was Homo habilus (2.5 to 1.6 million years ago) which split into Homo erectus and Homo heidelbergensis (1.6 to 0.5 million years ago). The latter, in turn, divided into the more familiar duo of Homo neanderthalensis and Homo sapiens (co-existing from @400,000 to 35,000 years ago). By about 10,000 years ago, each of the other lines of hominids disappears as a distinct species, not seemingly out-fought by Homo sapiens, but simply out-adapted.[68]

We know from evidence of stone tools that about 2.5 million years ago the chimp-like Lucy was being replaced by the more sophisticated stone tool user Homo habilus. They remained primarily scavengers, but began what we would now call the first human cultural activity; the use of stone tools as scrapers. When Homo erectus appears we find stone tools for digging rather than just scraping and, more important, the stone handaxe makes its appearance. With Homo erectus we find a significant shift, a being who "...had in mind a template which allowed him to envision a tool before he began to remove the flakes from the core."[69] One imagines that the neurophysiology of these early humans developed in highly differential ways and that the evolutionary advantage slowly shifted from the biggest or strongest to those better able to envisage attaching sharp stones to straight sticks. From this cognitive breakthrough the evolution of the hominids proceeds rapidly (in evolutionary terms) with wooden spears appearing 400,000 years ago, compound ballistic tools about 200,000 years ago followed by signs of a spiritual and aesthetic life amid Neanderthals and Homo sapiens.

It's time we looked more closely at the "nature" that was created by this long period of hominid/human evolutionary adaptation. What has proven most striking to those who study these vast epochs of hominid development is the steady increase in the sophistication of the brain. (not a surprising focus, one might imagine, for those who spend their lives studying!) We could perhaps more appropriately be named Homo cogitenensis. Early on hominids became smarter than they needed to be merely to survive, but just smart enough to prosper ever so slowly. To thrive on the savanna our ancestors needed to be able to handle more complex relationships, especially among their own kind and hence their memory capacity and linguistic skills evolved to meet this need — or put in more accurate Darwinian terms, those individuals with greater memory and linguistic skills proved better able to pass on their genes.[70] Even now, we are generally able to monitor successfully relationships with about 150 other individuals, a cognitive repertoire that proved sufficient in the hunter-gather band culture of the savanna.[71]

This acquisition of greater memory capacity and language skills did not come easily or at no physiological cost. As bi-pedal locomotion became more prevalent, hands were freed to experiment with objects which subsequently led to the use of tools. The freeing of hands also lessened dependence on teeth and jaws as means of manipulation and defence, thereby leading to changes in facial structure and shape. Longer legs became an evolutionary asset in the more open savanna and with the increased cognitive demands larger brain size also gave an evolutionary advantage. The larger brain and skull, however, conflicted with the limitations of the pelvis of the now bi-pedal female, resulting in greater birthing difficulties and a lengthening of the infant's post-natal physical development — which in turn, as was mentioned earlier, meant increased dependency and the need for supportive kin groups or other social supports.[72] Morris Berman, citing Bowlby's work, sees this biologically enforced mother-infant attachment as one of the central features in hominid cultural evolution — a clear example of the interplay of nature and culture.[73] But one might also note the role of the male in the provisioning of females and their offspring as a significant contribution to successful reproduction.

It is important that while there are several species of our most ancient ancestors, the primates, existing on the planet, Homo sapiens is the only hominid species that has survived this long period of post-savanna history. Our species branched off about 100,000 years ago, and modern DNA evidence

44 | THE FALL FROM GRACE

confirms that "...we are all descended from a few Africans who wandered out of Africa about 80,000 years ago, presumably in response to the increased aridity resulting from the Ice Age."[74] Thus while there are more or less distinct human groups that we have typically called 'races', thanks to our continual exchange of genetic material we remain a single species.

THE HUNTER-GATHERERS (OR GATHERER-HUNTERS)

As noted, in Darwin's theory of evolution there is no preferred outcome or plan behind the process of natural selection and so hypothetically any of the hominid species could have become the dominant one, or could have co-existed one with another. We know, for instance, that Homo sapiens and Neanderthals co-existed in Europe and the Near East for thousands of years. The latter prospered in Europe before and during the Ice Age and 70,000 years ago were the dominant hominid species in that region. Homo sapiens and their Cro-magnon kin began to migrate into the Mediterranean region about 50,000 years ago, and 20,000 years later the Neanderthals were extinct.

Known as the Middle Paleolithic, this period of Homo sapiens/ Neanderthal co-habitation was an era of gradual expansion of Homo sapiens skills and innovations. While it appears that both groups had fire and primitive language, the soon-to-be modern humans were clearly beginning to behave in more modern ways. Their tailored clothing, better shelters and more efficient hearths "...let them survive the cold glacial Europe, previously the exclusive domain of the Neanderthals."[75]

We don't know what happened during these years of landscape sharing, but there is no compelling evidence of either direct conflict or inter-breeding so we must presume that in the competition for food and other resources, the more cognitively sophisticated Homo sapiens gradually won out over the Neanderthals in the ecological niche they were sharing.[76] In a gradual but sometimes cataclysmic era of climate change those best able to cope with the changes fared best. Hence while the muscular and big-brained Neanderthals had been much better adapted to their original niche than the less robust Homo sapiens, the latter proved to be better "risk managers".[77]

Starting around 30,000 years ago, with this emergence of Homo sapiens as the dominant hominid, we enter the Upper Paleolithic era, the last stone age era before history proper begins. This is the classic hunter-gatherer era, one that we know quite a bit more about and the one that perhaps most clearly marks out the deep structures that will be decisive in shaping modern humans. Some call this the first "great leap forward": a "quantum leap in the range and complexity of technological behavior", and a "...major watershed in cultural development."[78] Now at last, three million years after Lucy and her fellow Australopithicines emerged from the forest canopy, we can begin to see ourselves, beings in community, with a politics, a language, a complex economic system and "...leading a recognizably human way of life."[79]

This high point of stone age hominid culture was built on at least 100,000 years of an increasingly complex hominid social life structured in small bands of perhaps 30–50 individuals that lived on scavenging, gathering and subsistence hunting. The successful ones in these bands were those whose predispositions, emotional capacities and social skills let them survive and reproduce, thereby passing on these predispositions and temperament via their genes and the skills via language and memory. Given the slow pace of evolution, it is reasonable to presume that these predispositions, temperaments and cognitive abilities have not changed very much from their time to ours. Since, as Stephen Pinker insists, all significant evolution of the human brain was completed during the hunter-gatherer era of the Pleistocene, it is not at all unreasonable to assert that they had an "...intelligence equal to our own."[80]

These mental and physical resources, predispositions and inherited skills manifested themselves in a sophisticated and stable culture that we can see echoed in the remnants of hunter-gatherer groups that survive to this day. The basic attributes of this hunter-gatherer culture include the following:

1. kinship of paramount importance in social and economic organization
2. widespread, diffuse social and economic functions for each individual
3. access to rudimentary technologies
4. egalitarian communities — leaders persuade but do not command
5. nomadic or mobile
6. sharing as a central attribute of social interaction
7. nature is spiritual home and source of all good things
8. property held in common
9. able to meet material needs without great expenditure of energy[81]

Morris Berman concludes: "Our kind started with band society, in which the typical size was about 25 to 30 people, leadership was context-specific, war (as opposed to aggression or impulsive homicide) was nonexistent; and ritual was largely ad hoc.[82] The archaeological evidence suggests that most of the world's population was organized in this way prior to 10,000 BCE."[83]

This image of small bands of self-sufficient, peaceful groups related by kinship with minimalist leadership based on reciprocity rather than power begins to look a lot like "primitive communism" and is so labeled by Richard Lee, one of the foremost experts on hunter-gatherer culture.[84] Is this another great irony of our times? The communism we struggle with in the modern era

THE FALL FROM GRACE

is seen as an ambitious (and improbable?) feat of social engineering contrary to, many argue, the core elements of our human nature. But here we seem to have an argument that Homo sapiens was in fact hard-wired via natural selection over thousands of generations to have a "nature" adapted to an egalitarian, propertyless and near anarchic culture. As Richard Lee insists: "By any dictionary definition of communism, our ancestors were communists."

Assuming for the moment that there is merit to the hard-wired argument, that we are essentially "Pleistocene hominids keyed with infinite exactitude to small-group, omnivorous life in forest/plain edges of the wilderness,"[85] what can we learn about ourselves, our own human nature, by examining political, economic, social and spiritual aspects of Paleolithic culture? While no Rousseauean "State of Nature", (that, presumably, was further back in Lucy's time) the Upper Paleolithic does probably correspond to the pre-state, pre-civilized era that Thomas Hobbes and countless moderns since have envisaged as the time when human life was "nasty, brutish and short". But according to Lee and most other anthropologists who study this era, Thomas Hobbes got it wrong.

The first factor that comes to mind given the extraordinary growth of human population and social overcrowding that has come to characterize modern culture since at least the 1500's is the issue of the size of human communities in the Paleolithic era. The foraging and hunting bands that existed throughout these millennia averaged between 25–50 individuals, linked with a linguistic or tribal unit of perhaps 500 people.[86] For us perhaps the most striking feature of the social system built around this band structure and the one that contrasts most directly with modern life is that for any individual human being almost all contacts with other humans would be with persons well known to him or her.

This intimate familiarity of one with the other may have had the positive effect of lowering stress by minimizing the presence of the "stranger", but it must also have lowered the amount of stimulus, which perhaps was a factor in the very slow pace of innovation that characterizes the Paleolithic.[87] The stability, then, that characterizes the Paleolithic is in part a function of size. There were nations, tribes and perhaps even societies existing in what Pierre Clastres calls an "extraordinary patchwork", but the local bands that comprised them took great care to preserve their autonomy and thus prevented the establishment of any kind of large state apparatus.[88]

Certainly humans—a word we can now begin to use with some assurance—by around 70,000 years ago had moved to some middle ground of social organization. No longer dispersed in isolated groups of foragers, most were linked in loose aggregates bound by intermarriages essential to maintaining band strength and genetic diversity. Others in more ecologically generous areas such as southern France or Spain may have already begun the shift toward ranked societies with hereditary chiefs or shamans.[89]

Based in part on the study of contemporary hunter-gatherer groups, modern scholars presume that even given these early forms of social organization, individual and band autonomy remained the most prized value. The bands remained egalitarian in terms of wealth and consensus-driven in terms of power and decision-making. Eleanor Leacock, noting that males and females played equal roles in this consensus, says "...the basic principle of egalitarian band society was that people made decisions about the activities for which they were responsible."[90] High value, then, was placed on trust and empathy, on reading the interests and concerns of others. "Leaders depend upon followers to uphold their reputations. But followers join the band of a leading individual...because they trust him."[91] And what they trust is that the leader will respect the autonomy of the follower.

Since hierarchy and kinship are two aspects of social organization that early on become central aspects of modernity, it is important to explore their roots in Paleolithic culture. While kinship was important in these small bands of hunter-gatherers, it was not determinative of loyalty. Marshal Sahlins argues that a generalized reciprocity held sway over kinship ties, a general recognition of the necessity of sympathy toward those in need, independent of both kinship and the record of that person's past contributions.[92] Leacock insists that the band as a whole, not the individual nuclear families of which it was composed, was the key economic and social unit.[93] Kirkpatrick Sale calls this a system of *Hetarchy*, a system based on "...complimentarity, or mutuality, under which the members of a single species within an eco-niche act reciprocally and non-hierarchically to promote and defend their community."[94]

This quality of reciprocity and the mutual respect it implies was, of course, a survival tool and hence to some extent merely utilitarian in origin and thus not a sign of some inner goodness or even innate sympathy. But it does allow us to soften the "nasty, brutish and short" version of human life in these so-called primitive times. A social ethic, a specific way of living in harmony with each other evolved over thousands of generations so that it, like binocular vision, large brains and bi-pedalism became part of human nature.

As to hierarchy and the issue of leadership, it is no doubt true that even in egalitarian band cultures chiefs existed "...because in every human group there are men who differ from their fellow-beings in that they like prestige for its own sake, are attracted by responsibility and for whom the burden of public affairs brings its own reward."[95] But in most cases this leadership is dispersed, with individuals called upon to serve not holistically, but in specific fields of endeavour. Thus it remains a transient phenomenon. Clastres makes a persuasive case that this dispersal of power was the result of "...a very early premonition that power's transcendence conceals a mortal risk for the group...."[96] Thus power becomes a negative, the leader being forced to give over his wealth (via, for instance, the Potlatch) and thus become poor in order to affirm his role as leader.

But what about conflict? We are getting a clear picture now of small bands of between 25–50 individuals linked by kinship, egalitarian and cooperative by nature, valuing self-sufficiency, largely non-gendered in terms of social roles and valuing at the same time personal and group autonomy.[97] While there is considerable evidence that these qualities resulted in a relatively stable and peaceful life, it would be naïve to think that no conflicts occurred both within and between bands.

Within the band environment life was lived "...in full view of the group", a kind of "involuntary intimacy" which made privacy impossible.[98] This is certainly the case today in most hunter-gatherer cultures and must have been so in the Paleolithic era. In such a hothouse environment conflicts between individuals and between families are bound to occur but the most common response seems to be movement—not violence or attempts to work things out, but rather by "throwing in the towel."[99] These were nomadic bands and they existed in what for us would seem a luxuriously under-populated world. And so, when trouble arose within the band, one faction could simply move on "...without difficulty and at a moment's notice."[100]

Turning to conflicts that might arise between bands the solution appears parallel to that for internal conflicts. With no tradition in hunter-gatherer bands of strong leaders, no defined territoriality nor a warrior caste, "moving on" rather than violent struggle proved the preferred option. Living as we are in a historical tradition so fraught with incessant conflicts driven by possession, vengeance or pride, to understand this earlier era we need to imagine a much more spacious earth populated by small groups of humans only minimally if at all motivated by these modern drives. As Rousseau warns us in his explorations of pre-history, too often when we speak of "savage man", we depict in fact "civilized man."[101]

This pattern of conflict resolution by withdrawal instead of either clash or resolution leads to some important insights about potential human predispositions. The band culture that persisted for these thousands of generations provides support for the innate propensity to cooperate thesis advanced by Darwinians like Matt Ridley and Frank Sulloway. But as Sulloway points out, this cooperative spirit relies heavily on trust and intuition, both of which require intimacy which in turn means living in small groups of known companions.[102] And, conversely, this intimacy of the band generally results in a kind of exclusivity, a deepening mistrust of those outside the circle of intimacy where trust and intuition are less effective.

Morris Berman explores this tension between autonomy, cooperation and exclusivity in his Rousseauean argument that human beings may not be all that social, and especially not naturally communitarian. He sees an historical—perhaps genetic—predisposition for humans to thrive only in small family or kin groups rather than in what we have come to call communities.[103] Thus Berman/Rousseau would argue that small groups of intimately connected but stubbornly autonomous individuals who prize self-sufficiency of both group and self and the freedom that stems from easy mobility may be the "happiest" and most natural state of being human.

Stability. Autonomy. Equality. Here, then, are the important qualities that the vagaries of nature persuaded our primitive ancestors to value and to create systems and structures around. The central structure, of course, had to be the economy—how they organized the acquisition and distribution of food and other essential resources—and here they devised what we have labeled an immediate-return system. Small was truly beautiful for these semi-nomadic hunters and gatherers who satisfied their needs by relying on constant movement in a world of abundance—living on the day-to-day flows from gathering and hunting, eating their way through the environment, moving on and returning again. Despite the traditional label of "hunter-gatherer", between 60–80% of the food consumed was in fact vegetable matter stemming from gathering, a role presumable dominated by females and children.

Assured of open space and secure in the ability of a properly balanced human/nature density, the bands could live forever in the present. No need for storage, security of property, speed or a rigid division of labour by gender or by rank. The "going was the goal", avoiding treading on others, avoiding the need even to nurture nature, or to develop more efficient means of production.[104] The most studied contemporary hunter-gatherer people, the !Kung, have lived for thousands of years in "equilibrium with their environment."[105]

Should an imbalance threaten—such as excessive numbers—there were means of demographic restraint via infanticide, senilicide or extended nursing.[106] Any hint of ambition or greed was more than blunted by the consensus that leisure was the most desired way to pass time. Based on the study of contemporary hunter-gatherer bands, Lee concludes that work occupied about two and a half days of the week—hence the frequent attribution of "affluence" to hunter-gatherer culture.[107] In this leisure time they sleep, gamble, play music, dance and engage in games and other amusements. Life is easy if, by our standards, dull.[108] Physical effort was minimal and, according to

Boyden, " The innate tendency in humankind toward indolence thus plays an essential role as a part of the set of mechanisms, behavioural and physiological, which have the effect of ensuring that human beings in their natural habitat do not over-exploit their resources, internal or environmental."[109] And so our touchstone Rousseau, a modern "primitive", confirms this predisposition by finding his greatest pleasure in *far niente*, doing nothing, becoming a man dedicated to idleness.[110] And this minimalist human brings us, finally, to what must be our central concern, the relationship between the Paleolithic, primitive human and the natural world.

If our primitive ancestors simply ate their way through their environment and if a natural abundance forestalled the development of any nurturing drives vis-à-vis the environment, what might this tell us about our modern predispositions toward nature? The jury seems divided, or at least hopelessly caught up in debates that are conditioned by some very contemporary prejudices. On the one hand there is evidence that humans in the Paleolithic were as rapacious as we toward the natural world, and on the other there are claims for a now largely lost oneness with nature.

Edward Wilson, always pragmatic and to the point, argues that the eating habits of humans have always been a threat to biodiversity and especially to those animal species that were "the big, the slow and the tasty."[111] Thus some estimate that it took only 300 years for the first human hunters in North America (the Clovis people) to wipe out all the megafauna — mammoths, mastodons, ground sloths, horses, camels — every animal on the continent weighing over 200 lbs or so.[112] Likewise in Australia and later in New Zealand over 80% of large mammals disappeared after the arrival of humans.[113] Jonathan Kingdon, a major proponent of the humans-as-aggressive-predators position, attributes these kill-offs to the introduction of dogs and rats into new eco-systems, to the use of fire to clear land and herd animals, to rampant over-hunting, and to a "...gigantic meat-eating bonanza."[114] On the other hand, recent evidence that bison in China, Siberia and North America began a precipitous decline about 37,000 years ago due to climate change, long before we think humans were in the Americas.[115]

Related to the evidence-based case for the excesses of primitive human predation are the more philosophical/anthropological arguments that place humans necessarily in opposition to nature in general and other animals in particular. In this view there is a solid line separating humans from other animals. Social ecologist and prominent opponent of the oneness with nature argument Murray Bookchin dates this line form at least the era of Homo erectus, citing a persistent hominid desire to alter the wild environment by fire, by settlement, by mass slaughter of animals, "...a decisive break with the largely passive and adaptive nature of animal behavior."[116]

In support of Bookchin's position we can cite Levi-Strauss's argument that the prevalence of face-painting and adornment by humans symbolizes their sense of independence from the realm of animals, a "...fundamental opposition between nature and culture."[117] Freud too identifies a primal drive to mastery over "man, beast and things" as indicative of an "unshakable confidence in the possibility of controlling the world."[118]

These are arguments for a primal drive to mastery, before the religions and cultures others see as the true source of the separation of nature from culture. There is little room here for Rousseauean states of nature that treasure idleness, no noble savages nor holistic world views. Instead the mastery drive is "present at the creation", a genetic drive calmed or negated only by the poverty of means and the remarkable fecundity of nature. Speaking of the hunter-gatherers of the Paleolithic, Kingdon says: "In general they lived within ecosystems where regulation of numbers was more a property evolved by the prey than a product of prescient self-control exercised by the predator."[119]

A REVERIE ON SAVAGE CHILDREN

"Kill the beast! Cut his throat! Spill his blood!" Who can forget the children's dystopia William Golding created in *Lord of the Flies*? Jack the hunter with his burgeoning band of wild boys versus Ralph, Simon and Piggy. The slaughter of the pig and the malevolent movement from pig to Piggy to Simon as the object of the hunt. Anticipating with some chagrin the concept of oneness, of the harmony of one with all, Golding describes the boys in the climactic scene as embodying "...the throb and stamp of a single organism."[120]

The boys on the island had become primitive, retreated in a Hobbesian nightmare to an existence that could only be nasty, brutish and short. Ralph and his conch were the only signs or symbols of the fragile social contract remembered from the adult world and its power faded easily once the jungle, hunger and aloneness came to dominate. We're told that the human species is unique in its extension of the years of childhood and that "...analysis of early stages in the development of an individual should be a real guide to our intrinsic nature."[121] Golding's vision can obviously lead to some grim conclusions about that nature.

And if the children are our guide, it's easy to side with the notion that in the human psyche nature really is the "other". My own childhood memories of de-lanterning fireflies, smashing nests of garter snakes, shooting at birds with pellet guns and torturing no end of insects finds ready witness in the literature of abusive and predatory children. The 18th century poet John Clare tells us of the sad fate of a badger pitted against youth:

He falls as dead and kicked by boys and men
Then starts and grins and drives the crowd agen
Till kicked and tirn and beaten out he lies
And leaves his hold and cackles groans and dies[122]

Keith Thomas recounts similar tales of seemingly innate and purposeless cruelty by boys:

Small boys were notorious for amusing themselves in the pursuit and torture of living creatures. In the grammar schools cock-throwing was a widely observed calendar ritual. On Shrove Tuesday the bird was tethered to a stake or buried in the ground up to its neck, while the pupils let fly at it until it was dead....Outside school, children robbed birds' nests, hunted squirrels ....caught birds and put their eyes out, tied bottles or tin cans to the tails of dogs, killed toads by putting them at one end of a lever and hurling them into the air by striking the other end, dropped cats from great heights to see whether they would land on their feet, cut off pigs' tails as trophies inflated the bodies of live frogs by blowing into them with a straw.......[123]

And Wallace Stegner confesses his own inhumanity to the 'other' in his memories of a Saskatchewan boyhood:

Nothing...could have prevented us from hunting, fishing, trapping, and generally fulfilling ourselves as predators....we had our own savage feasts out in the willows, dining upon sage hen or rabbit broiled on sticks over the fire. When larger game failed we netted bullfrogs, or caught them on a fish hook baited with a scrap of red flannel, and hacked off their legs and roasted them.......We lived an idyll of miniature savagery, small humans against rodents. Experts in dispensing death, we knew to the slightest kick and reflex the gophers' ways of dying....We were as untroubled by all our slaughter as early plainsmen were by their slaughter of the buffalo.[124]

None of this implies any universals, but only a potential that seems to be disturbingly prevalent at this early period of the human male's life. Young girls seem more immune from predatory pleasures and perhaps cross cultural studies would be illuminative. Most of the boys either grow out of the need for such slaughter, have it repressed via wise counsel, or make it their profession. For the latter adults these predispositions become respectable, organized and therefore no longer merely cruel.

A MORE POSITIVE SPIN

What a difference another perspective can make! Many anthropologists and ecologists insist there is no firm evidence for Paleolithic plundering of the environment. They argue instead that hunter-gatherer culture was based on a harmony with the natural world that developed logically and inevitably from their life situation. Felipe Fernandez-Armesto in his exploration of "civilization" points out that the human claim to superiority over the rest of nature can only be traced back to the immediate pre-Christian era.[125] Max Oelschlaeger, echoing the exhaustive research of Marshall Sahlins and Richard Lee on contemporary hunter-gatherer groups, insists that "Paleolithic people had few wants yet lived an affluent and satisfactory life in harmony with nature."[126]

The carnivorous nature of humans, at least from Homo erectus on, does create a problem for the advocates of harmony. Hunting is, after all, the deliberate, direct and violent killing of wild animals, "...an armed confrontation between humanness and wildness, between culture and nature."[127] It is as well just part of the predation that virtually all animal species engage in. Accordingly the defenders of natural harmony point out that this hunting is for food, not for governance, control or mastery.[128]

Still, the brutal circumstances of human predation seems to have persuaded Homo sapiens at least that hunting had to be processed or ritualized in some way in order to soften the stark contrast it seemed to impose between humans and the other. In effect, the advocates for harmony insist, the reciprocity, sharing and respect that characterized human-to-human relations in the band culture had to be transferred to relations with the natural world. The novelist J.M. Coetzee, commenting on advocates of hunting as a means of getting in touch with our true nature, sees bullfighting as a contemporary vestige of this primitive oneness:

> Bullfighting, it seems to me, gives us a clue. Kill the beast by all means, they say, but make it a contest, a ritual, and honor your antagonist for his strength and bravery. Eat him too, after you have vanquished him, in order for his strength and courage to enter you. Look him in the eyes before you kill him, and thank him afterwards. Sing songs about him.[129]

Morris Berman concludes that this ritualization of the relationship between human (culture) and animal (nature) eases the anxiety, fear and resulting aggression that would otherwise accompany a merely predatory subject/object relationship and thereby eliminates or at least softens any sense of alienation from nature. "Rocks, trees, rivers, and clouds were all seen as wondrous; alive, and human beings felt at home in the environment."[130] In a

54 | THE FALL FROM GRACE

challenging paper on three contemporary hunter-gatherer groups (in India, Africa and Malaysia), Nurit Bird-David describes their attribution of life and consciousness to natural phenomena such as forests, river sources and specific trees. She calls them agents or natural agencies who socialize with the tribe, provide food and gifts and stand in a parent-like relationship with the self-described "children of the forest."[131] And so we return via this version of the human story to Stan Rowe's "home place" and to Martin Buber's tree "in itself", with our primitive hunter-gatherer ancestors in touch with each element of their world as wondrous things in themselves.

CONCLUDING THOUGHTS

This is no simple nature versus nurture debate. We know much more now, indeed learn more every day, about the impact of our genetic heritage and the workings of natural selection. Indeed, with the global retreat of the two grand narratives constructed in large part on the power of nurture over nature—Marxism and Freudianism—we are left with Darwin as the source of modernity's last dominant narrative. The combination of the theory of evolution with the empiricism of science has demonstrated the powerful role of inheritance in defining at least some components of a human nature. And the accompanying persistence of endemic social problems has caused us to question our faith in the potential of nurture.

Hence I began this exploration into the sources of our thinking about and acting upon the natural world with what I have called our predispositions. By this I mean our *deep* predispositions, those we have inherited not by cultural transmission but by the process of natural selection over eons of time. The predisposition argument is a powerful one, and no one makes the case more persuasively than Edward Wilson. Here he draws the link between our evolutionary past and our seeming inability to plan for the long term:

> The relative indifference to the environment springs, I believe, from deep within human nature. The human brain evidently evolved to commit itself emotionally only to a small piece of geography, a limited band of kinsmen, and two or three generations into the future. To look neither further ahead or farther afield is elemental in a Darwinian sense. We are innately inclined to ignore any distant possibility not yet requiring examination…it is a hardwired part of our Paleolithic heritage. For hundreds of millennia those who worked for short-term gain within a small circle of relatives and friends lived longer and left more offspring…[132]

This has the feel of logic to it. If this, then that….

But there is some doubt that the founding premises of the argument are all that strong. The fossil record is weak and "...there are no fossils of social organization."[133] The utility of using contemporary hunter-gatherer cultures as sources for speculating on the Paleolithic is not clear. These anthropological observations provide clues, but as Freud warned us, "...primitive races are not young races.[134] We will not find pristine hunter-gatherers of 10,000 years ago among the Hazda or the !Kung, who have been living side-by-side with agriculturalists for millennia and are now adjacent to urban modernity.

Even more interesting than the suspect nature of the data is the fact that we are aware now of these predispositions as predispositions. What is the impact of this awareness which our culture (i.e. nurture) has given us on these predispositions given us by nature? We are, as Rousseau argued, a perfectible species in that our cognitive abilities have developed to the point that we are liberated from being determined by our environment and, possibly, by our predispositions. One might argue that all beings, perhaps all phenomena, are perfectible in the sense of having the potential to achieve fully what is in their "nature". I have advanced the case in this opening chapter that one aspect of that perfection for humans had been achieved or at least come close to achievement, namely a sense of being "at one" with one's home environment and having accomplished this by sustaining a balance between means and desires. The exceedingly slow development of means over the long Paleolithic era, the gradual and halting development of tools, inhibited any rapid rise in desires and led to stability and, Rousseau argued, a kind of happiness. We will see how a more rapid alteration in means gives rise to an even more rapid increase in desires with the result, Rousseau argued again, being unhappiness. Few counsel going back to that earlier happiness since it resulted from fulfilling only one dimension of human perfectibility. The rapid development of means signals another, equally important aspect of that perfectibility.

So this initial foray into exploring our relationship with the natural world is empirically inconclusive, as it must be given its prehistoric nature and the paucity of evidence. But it does open an important door; the strong possibility that harmony with nature and the other, if not clearly the norm was at least a strong part of the human repertoire from the "beginning". This at least checks though it cannot checkmate the darker "man is wolf to man" (and nature) theory that has held sway for so long. The work of Lee, Sahlins, Jolly, Berman, Shepard, James Wilson, Woodburn, E.O. Wilson, Leacock and others must open our minds to the possibility that cooperation rather than conflict, mutuality rather than mastery and equality rather than hierarchy might come to characterize our relations with each other, with other species and with the Earth.

56 | THE FALL FROM GRACE

NOTES

1   Jean-Jacques Rousseau, *Reveries of the Solitary Walker* (London: Penguin, 1979), p 85

2   Later, in the 20th century, Martin Buber will refer to this as the "I-Thou" relationship, the essence of which implies "…that there is no reason, no time, for reflection. We are seized by the relationship" Peter Reed, p 57.

3   Jean-Jacques Rousseau, *Rousseau, Judge of Jean-Jacques: Dialogues*, ed. by Roger Masters and Christopher Kelly (Dartmouth: University Press of New England, 1990) p 143.

4   Claude Levi-Strauss, *Tristes Tropique* (New York: Atheneum, 1978) p 390.

5   Marianna Torgovnick, *Gone Primitive: Savage Intellects, Modern Lives* (Chicago: University of Chicago Press, 1990), p 217. In the 7th Walk in Rousseau's *Reveries of A Solitary Walker* (London: Penguin, 1974) p 111), he claims to "…feel transports of joy and inexpressible raptures in becoming fused as it were with the great system of beings and identifying myself with the whole of nature."

6   Levi-Strauss, p 390.

7   Arthur Lovejoy and George Boas, *Primitivism and Related Ideas in Antiquity* (Baltimore: Johns Hopkins, 1935/1997), p 7.

8   Harry Neumann, "Philosophy and Freedom: An Interpretation of Rousseau's State of Nature", *The Journal of General Education* (v. 27:4, 1976, pp, 301–307), p 305.

9   Robert Wright, *The Moral Animal* (New York: Vintage, 1994), p 321.

10   Foley, p 59.

11   Tim Megarry, *Society in Prehistory: The Origins of Human Culture* (London: Macmillan, 1995), p 1.

12   James Lovelock, "Gaia and the Balance of Nature" (pp 241–252) in P. Bourdeau et al, *Environmental Ethics: Man's Relationship with Nature* (Luxembourg: Commission of the European Communities, 1990, p 249.

13   Jonathan Kingdon, *Self-Made Man: Human Evolution from Eden to Extinction?* (New York: John Wiley, 1993), p 69.

14   Diamond, p 127.

15   Boyden, p 5.

16   The geneticist R.J. Berry insists that "…the vertebrate eye could have evolved from a primitive light-sensitive spot in 364,000 generations; it is estimated that about 15 times as long was available". Cited by Alan Battan, "Is there an absolute plan—or just a desire for one?", *Toronto Globe and Mail*, 7 June 2003, p D6.

17   Charles Darwin, *The Origin of Species* (London: Penguin Books, 1985), p 229.

18   Adam Kuper, *The Invention of Primitive Society* (London: Routledge, 1988), p 2.

19   Megarry, p 37. Thus giraffes did not gain long necks in order to graze at the tops of trees, but rather because their ancestors with extended vertebral columns had tended to survive and concentrate this feature in future populations.

20   Foley, p 9.

21   Darwin, p 133.

22   Wright, p 23.

23   Foley, p 190. "Certainly there may be benefits in large brains, both socially and ecologically, and there are selective pressures in favour of an increase in brain size, but these are usually outweighed by the costs. Most animals are better off putting their energy into muscles or large stomachs…What is unique among humans as an evolutionary event is….the occurrence somewhere in their evolutionary history of the ecological conditions

that....allowed the benefits of greater intelligence to greatly outweigh the costs. It is these ecological circumstances that are rare." p 171.

24   A. Zee, "On Fat Deposits around the Mammary Glands in the Females of Homo Sapiens", *New Literary History*, v. 32, 2001 (pp 201-216)

25   Randolph Nesse, "Natural Selection and the Capacity for Subjective Commitment", in Randolph Nesse, ed., *Evolution and the Capacity for Commitment* (New York: Sage, 2001), p 10. James Q. Wilson, one of the most prominent contemporary advocates for there being biological roots to a common human nature, cites a norm of reciprocity which has been found in every culture "...for which we have the necessary information." As with the comparable innate sense of justice, Wilson agrees with Piaget's conclusion that these biological roots are a necessary but not sufficient condition for the full development of the quality. *The Moral Sense* (New York: Free Press, 1993), p 58.

26   Holmes Rolston III, "Naturalizing and Systemizing Evil", in Willem Drees, ed., *Is Nature Evil?* (New York: Routledge, 2003), p 69.

27   Lee (1992), p 82.

28   Megarry, p 117.

29   Wright, p 13 and p 328.

30   Tuan, p 16.

31   Claude Levi-Strauss. "Rousseau: Father of Anthropology", *UNESCO Courier* (March 1963), p 12. "Prior to hunting, the relations of our ancestors to other animals must have been very much like those of the other noncarnivores. They could have moved close among the other species, fed beside them, and shared the same waterholes. But with the origin of human hunting, the peaceful relationship was destroyed....". from S.L. Washburn and C.S. Lancaster,"The Evolution of Hunting" in R.B. Lee and I. DeVore (eds) *Man the Hunter* (Chicago: Aldine, 1968 pp 293–303), p 12. Tim Megarry concludes that "...it is not legitimate to propose that modern humans possess a biological basis for killing which stands as an unreformed mental residue from a savage past." Aggressive behaviour did exist, but it was contextual, not innate." p 211.

32   Jean-Jacques Rousseau, *Emile* (Chicago: University of Chicago Press, 1979), p 39.

33   Lauren Wispe citing Schopenhauer, *The Psychology of Sympathy* (New York: Plenum Press, 1991), p 72.

34   Donald Worster, *Nature's Economy* (Cambridge: Cambridge University Press, 1977), p 182.

35   Erich Lowey, *Suffering and the Beneficient Community* (Albany: SUNY Press, 1991), p 30.

36   Wispe, p 68. The novelist J.M. Coetzee, reflecting on the seemingly "inhuman" actions of some humans toward the other notes that while there "...are no bounds to the sympathetic imagination...there are people who have no such capacity...and there are people who have the capacity but choose not to exercise it." *The Lives of Animals* (Princeton: Princeton University Press, 1999), p 35.

37   Edward Wilson, (2002) p xxii.

38   Alan Ryan, "Reasons of the Heart", *New York Review of Books*, 23 September 1993, p 52.

39   The Biophilia hypothesis developed by Edward Wilson starts from the premise that there is an innate affiliation between humans and other living organisms. Research to support this hypothesis is based on gauging certain Pavlovian responses to various natural stimuli (e.g. fear of snakes). This research indicates that respondents in Europe, Asia and America prefer "savanna-like habitats, particularly if there is a bit of peaceful water in the scene.... The biophiliac interpretation is that we spent most of our history in African savanna and probably evolved to prefer that kind of habitat, as safest and most supportive for living.

58 | THE FALL FROM GRACE

Now we retain a ghostly preference for safaris past." Other research claims that we retain an evolutionary preference for certain kinds of trees—broad canopies relative to their height, layered branches, small leaves and a low split in the trunk. Mark Ridley, "Do We Love Nature", review of Stephen Kellert and Edward Wilson, ed. *The Biophilia Hypothesis, Times Literary Supplement*, 9 September 1994, p 5.

40 Keith Thomas, *Man and the Natural World* (London: Penguin, 1984), p 222.

41 Shepard, (1998), p 42.

42 Percy Shelley, "Essay On Life" in David Clark, ed., *Shelley's Prose* (New York: New Amsterdam, 1988, pp 171-175), p 174. Even though as adults we rely on harvesting and clearcutting the forest, we still strive to 'get to the top', climb ladders, seek to 'branch out' and comprehend our complex bureaucracies via flowchart trees.

43 Robert Harrison, *Forests, the Shadow of Civilization* (Chicago: University of Chicago Press, 1992), p 56.

44 Jonathan Bate, *The Song of the Earth* (London: Picador, 2000), p 88.

45 Roderick Nash, (2001), p 8.

46 Fernandez-Armesto, (2001), p 144.

47 Thomas, (1984) citing William Marsden in 1783, p 213.

48 *Binsey Poplars* (Felled 1879) in Kim Taplin, *Tongues in Trees: Studies in Literature and Ecology* (Bideford, Devon: Green Books, 1989), p83.

49 Paul Shepard argues that our eyes, our most efficient and valuable sense, we owe to our forest origins: "Animals have become binocular for two reasons: predation and jumping. Either or both. That much of the human eye is the product of life in the crown of a tropical forest is indicated by its similarity to the monkey eye. Our binocularity developed in jungle treetops where primate ancestors jumped from limb to limb." Paul Shepard, *Man in the Landscape* (College Station: Texas A&M University Press, 1991, 1967), p 6.

50 Mary Wollstonecraft, *Travels in Sweden, Norway and Denmark* (London: Penguin, 1987), p 119.

51 John Ruskin, *The Poetry of Architecture* cited in Kim Taplin, p 72.

52 Stephen Fox, *The American Conservation Movement: John Muir and His Legacy* (Madison: University of Wisconsin Press, 1981), p 109.

53 John Muir, *My First Summer in the Sierra* (London: Penguin, 1978), p 52.

54 Leopold, p 87.

55 Buber, p 8.

56 Boyden, p 6.

57 Alison Jolly, *Lucy's Legacy: Sex and Intelligence in Human Evolution* (Cambridge: Harvard University Press, 1999), p 159.

58 Tim Megarry notes that Chimp studies show that conceptual thought, conscious awareness, reasoning ability, symbolic communications and culture all exist in chimpanzees though in rudimentary form. They have used sign language, showing an ability to associate, categorize and use concepts and abstractions. Chimps have good memories and "...cannot be said to live permanently in the present......Tool-making and using by this ape, as well as what appeared to observers as pre-planned group foraging, constitute evidence of an ability to anticipate the future." Megarry, (1995), p 120.

59 Robert Foley, p 68. These dates are open to dispute. Tobias argues that "we now know that Hominids are much older than was previously thought, emerging so 10 million years ago instead of the 5-7 million years ago previously thought". Philip Tobias. "Twenty Questions about Human Evolution", *Human Evolution* (v. 18:1-2, 2003, pp 9-50), p 45.

60 Foley, p 175. Orangutans are, of course, the exception.

WHAT ARE WE BY NATURE? | 59

61 Frances Barnes, "The Biology of Pre-Neolithic Man", in Boyden, p 2.
62 Peter Wilson, *The Domestication of the Human Species* (New Haven: Yale University Press, 1988), p 15.
63 Peter Wilson, p 21.
64 Edward Wilson, (2002), p 134.
65 Merchant, (2003), p 179.
66 Stephen Clark, "The Moral Animals", *Times Literary Supplement* (6 September 1996), p 25, and Jolly, p 323.
67 Matt Cartmill, *A View to a Death in the Morning: Hunting and Nature through History* (Cambridge: Harvard University Press, 1993), p 17.
68 Berman, (2000), pp 39-40; Andrew Smith, "Archaeology and evolution of hunters and gatherers", in Richard Lee and Richard Daly, eds., *Cambridge Encyclopedia of Hunters and Gatherers* (Cambridge: Cambridge University Press, 1999), p 384-386.
69 Andrew Smith, p 385. Jolly calls Homo erectus a "hugely successful species, almost human, perhaps with conjugal sharing and love, but somehow still lacking imagination." Jolly, p 368.
70 In primates the neo-cortex area of the brain controls the information-processing required to maintain these relationships. Environments that require larger groupings of humans "...will only be successfully colonized after the evolution of larger neo-cortices". Clive Finlayson, *Neanderthals and Modern Humans* (Cambridge: Cambridge University Press, 2004), p 207.
71 Stephen Clark, p 25. We can obviously 'recognize' more than 150 faces, but we are more limited in our ability to 'know' more than around 150 individuals.
72 Megarry, pp 123-138; Alison Jolly, p 323.
73 Berman, (2000), p 345.
74 David Thomas, "Our Father: Adam the Bushman", *Vancouver Sun*, 3 February 2003, p A6. Clive Finlayson says there were probably three major out-of-Africa expansions, starting about 1.9 million years ago, followed by another 840-420 thousand years ago and the last 150-80,000 years ago. p 207. Recent research at Cambridge University is said to show "definitively" that modern humans originated from a single area in Sub-Saharan Africa. www.reuters.com, 18 July 2007.
75 Rick Gore, "Neanderthals: The Dawn of Humans", *National Geographic Magazine*, January 1996 (pp 3-35), p 30. The recent discovery of 75,000 year old necklace beads in a cave in South Africa is evidence that Homo sapiens were "acting like modern humans" some 30,000 years prior to their move into Europe, thus enhancing the edge they had over the less innovative Neanderthals. Anne McIlroy, "World's Oldest Necklace Discovered", *Globe and Mail*, 17 April 2004, p A9.
76 We do know that no Neanderthal genes have survived. Clive Finlayson, p 74.
77 Finlayson, p 73.
78 Paul Mellars, "The Upper Paleolithic Revolution", in Barry Cunliffe, *The Oxford Illustrated Prehistory of Europe*, (New York: Oxford University Press, 1994).
79 Megarry, p 276.
80 Oelschlaeger (1991), p 6.
81 Berman, (2000) p 57; Stanley Diamond, p 108; Tim Ingold, "On the Social Relations of the Hunter-Gatherer Band" in Richard Lee and Richard Daly, p 404; Richard Lee and Richard Daly, "Foragers and Others", *Cambridge Encyclopedia of Hunters and Gatherers* (Cambridge: Cambridge University Press, 1999), p 1; Marianna Torgovnick, p 21; Richard

Lee, "Hunter-Gatherer Studies and the Millennium: A Look forward (and Back)", Keynote Address, 8th Conference on Hunting and Gathering Societies, National Museum of Ethnology, Osaka, Japan 1998, p 826.

82 James Wilson and others argue that cooperation or prosocial behaviour among early Homo sapiens (and other hominids) was the key to their survival. Collective hunting bands were more successful; and women who could attract men were better able to nurture children. "And so by natural selection and sexual selection, individuals with prosocial impulses had greater reproductive success." Wilson, (1993), p 70.

83 Berman, (2000) p 350. Alison Jolly argues that social behaviour is the key factor in explaining the evolution of ape/human brain size and corresponding intelligence. "Living in a troop of 12 or 20 others means dealing with 12 or 20 separate relationships on a daily basis....troop life preceded the development of more complex relations, and ....these in turn shaped the evolution of intelligence....The capacity to keep track of social relations may even set an upper limit on troop size." p 214.

84 Richard Lee in Ingold (1988), p 255.

85 Shepard, (1998), p 137.

86 Some claim they were even smaller, John Calhoun arguing that the optimum size of adult members of a band was between 7-19. p 131.

87 Frances Barnes, "The Biology of Pre-Neolithic Man", in Boyden, p 6.

88 Pierre Clastres, *Society Against the State* (New York: Zone Books, 1987), p 213.

89 Megarry, p 278.

90 Eleanor Leacock, "Women's Status in Egalitarian Society: Implications for Social Evolution" (pp 139–164) in John Gowdy, ed. *Limited Wants, Unlimited Means* (Washington, D.C.: Island Press, 1998), p 143.

91 Ingold in Lee and Daly, p 407.

92 Ingold, p 400.

93 Leacock, p 145.

94 Kirkpatrick Sale, *Dwellers In the Land* (Philadelphia: New Society Publishers, 1991), p 98.

95 Levi-Strauss, (1978), p 316. This assessment is supported by J.A. Tainter in his work on complex societies. See Bert deVries, et. al. "Understanding: Fragments of a Unifying Perspective" in Goudsblom and deVries, p 273.

96 Clastres, p 44.

97 The general proscription of women from hunting may be an indication of the greater importance of gathering as opposed to hunting rather than discrimination based on gender. Megarry, p 308.

98 Peter Wilson, (1988), p 26.

99 Berman, (2000) p 69.

100 James Woodburn, "Egalitarian Societies" in Gowdy, p 92. J.A. Tainter stresses the ability of hunter-gatherers to "...simply withdraw from an untenable social situations." J.A. Tainter, *The Collapse of Complex Societies* (Cambridge: Cambridge University Press, 1988), p 24-5.

101 Jean-Jacques Rousseau, *Discourse on the Origins of Inequality* (London: Penguin, 1984), p 78.

102 Sulloway, p 39.

103 Berman, (2000) p 66.

104 Berman, (2000) p 81.

105 John Gowdy, "Hunter-Gatherers and the Mythology of the Market", in Lee and Daly, p 391.

106 Marshall Sahlins, "The Original Affluent Society" in Gowdy, p 32.

107 Richard Lee, "What Hunters Do for a Living, or, How to Make Out on Scarce Resources", in

Gowdy, p 52.

108 Jack Harlan, Crops and Man, (Madison, WI: American Society of Agronomy, 1992), p 8.

109 Boyden, p 70

110 Rousseau, *Reveries of A Solitary Walker*, p 83.

111 Edward Wilson, (2002), p 92.

112 John Terborgh, "The Age of Giants", review of Tim Flannery, *The Eternal Frontier: An Ecological History of North America and Its Peoples*, in *New York Review of Books*, 20 Sept 2001 (44–46), p 45.

113 Donald Grayson, "The Archaeological Record of Human Impacts on Animal Populations", *Journal of World Prehistory*, v. 15:1, 2001, p 9.

114 Kingdon, p 92. Tim Megarry finds no evidence for these claims, arguing that "Studies of hunting peoples have frequently found that a calm uncompetitive personality usually prevails in such societies and foraging normally implies a non-abusive accommodation with nature." p 265.

115 *Vancouver Sun*, 26 November 2004, p A9.

116 Bookchin (1995), p 124

117 Levi-Strauss, (1978), p 234.

118 Sigmund Freud, *Totem and Taboo* (London: Routledge & Kegan Paul, 1960), p 89.

119 Kingdon, p 311.

120 William Golding, *Lord of the Flies* (New York: Perigree Books, n.d.), p 152.

121 Kingdon, p 42.

122 From Clare's "Badger", cited in James McKusick, "A Language that is ever green: The Ecological Vision of John Clare", *University of Toronto Quarterly*, v. 61:2, 1991 (pp226–249), p 239.

123 Thomas, p 147.

124 Stegner, p 259 and p 275.

125 Fernandez-Armesto, p 548.

126 Oelschlaeger (1991), p 25.

127 Cartmill, p 30.

128 Kirkpatrick Sale, p 93.

129 J.M. Coetzee, *The Lives of Animals* (Princeton: Princeton University Press, 1999), p 52.

130 Morris Berman, *The Re-enchantment of the World* (New York: Bantam, 1981), p 23.

131 Nurit Bird-David, "Beyond The Original Affluent Society: A Culturalist Reformulation", (pp 115–137), in Gowdy, p 123.

132 Edward Wilson (2002), p 40.

133 Kuper, p 8.

134 Sigmund Freud (960), p. 3.

CHAPTER TWO
WHO ARE WE BY CHOICE, CHANCE AND CONTEXT?

"Two roads diverged in a yellow wood"   ROBERT FROST

Animals, we say, act on drives and instincts. Beasts reason not is our mantra. Landforms and trees are shorn even of drives and instincts and make no choices. But we humans are the choosing animal. And in the wake of choice comes the burden of responsibility. So far in our story there seems to have been little evidence of hominid choosing, but that will change as Homo sapiens becomes the modern human and creates the cultures of hunters, pastoralists, farmers and villagers.

But choice is such a contentious word! Of course we choose all the time thanks to our highly developed quality of consciousness, but these choices are always conditioned by context (what Marx called circumstances) and subject to chance. One thinks of Machiavelli, that great advisor to princes, and his acknowledgment that in the end all the Prince's choices were still subject to *fortuna*.[1] We are also reminded by writers with a sociological bent that we are social beings born into conditions of interdependency and social configurations that make it impossible to reduce our actions to individual choice or motivation.[2] With these twin caveats of fortune and circumstance always kept in mind, we can nonetheless pursue the matter of choice. However it came about, by the late Paleolithic era Homo sapiens was far less determined by its environment, more able to manipulate it by choosing responses to both circumstances and fortune.

It is in these early more deliberate responses to being-in-the-world that the paths to the modern are first charted and it is here that we can begin to speculate on alternatives, on paths of development that could have resulted

THE FALL FROM GRACE

in more ecologically sustainable human cultures. This shift toward choosing implies the risk of upsetting the balance between humans and their environment seemingly so characteristic of that long era of hunting and gathering, bringing to mind the fate of Coleridge's ancient mariner. Standing in perhaps for the whole of humanity, he makes a fateful and decisive choice, a most intemperate and ill-advised act of irresponsible violence, and hence begins his fall into environmental and social hell:

> With my cross-bow
> I shot the Albatross

We know, and he soon discovers, that in doing so he acts to sever the bond between humanity and nature, a nature which in being part of him in fact loves him but must respond to the mariner's assault with a grim and dangerously apocalyptic lesson in death, decay and destruction. The bird had been their guide through a fog-bound and ice-clogged sea, and now they were lost:

> Down dropt the breeze, the sails dropt down,
> 'Twas sad as sad could be;
> And we did speak only to break
> The silence of the sea!
> All in a hot and copper sky,
> The bloody Sun, at noon,
> Right up and above the mast did stand,
> No bigger than the Moon.
> Day after day, day after day,
> We stuck, nor breath nor motion;
> As idle as a painted ship
> Upon a painted ocean.
> Water, water, everywhere,
> And all the boards did shrink;
> Water, water, everywhere,
> Nor any drop to drink.
> The very deep did rot: O Christ!
> That ever this should be!
> Yea, slimy things did crawl with legs
> Upon the slimy sea.

The mariner, his mates all dead, his ship becalmed and dreadful, must make amends, must rekindle the bond with nature that had been manifested in the Albatross. To do that he must become an eternal messenger, the one who warns us of the perils we face, perhaps an early version of the modern environmental activist. And it is Coleridge's insightful understanding of the natural world that has him base this renewed bond on an expression of love, not for the beauties of nature, but for the slimy water snakes that befoul the turgid water upon which the mariner's ship floats:

> *Beyond the shadow of the ship, I watched the water-snakes:*
> *They moved in tracks of shining white,*
> *And when they reared, the elfish light*
> *Fell off in hoary flakes.*
> *Within the shadow of the ship*
> *I watched their rich attire:*
> *Blue, glossy green, and velvet black,*
> *They coiled and swam; and every track*
> *Was a flash of golden fire.*
> *O happy living things! No tongue*
> *Their beauty might declare:*
> *A spring of love gushed from my heart,*
> *And I blessed them unaware*

It is a pantheistic vision that the mariner must communicate to all those he encounters, a Franciscan view that celebrates the holistic realm of sustainability:

> *He prayeth well, who loveth well*
> *Both man and bird and beast*
> *He prayeth best, who loveth best All things great and small;*
> *For the dear God who loveth us,*
> *He made and loveth all.*

DIVERGENT PATHS

It is at the fading of the Paleolithic era of hunter-gatherers that we, in retrospect only, encounter a decisive turning point in human history. To cite another poet, Robert Frost, "Two roads diverged in a yellow wood" around 10,000 years ago. And, like the poet, most of humankind chose the path "less traveled by", in this case a path that led to agriculture, settlement, property, hierarchy and civilization instead of the well trod path of hunting, gathering and self-sufficiency. And as Frost concluded, such choices make "all the difference."[3]

We will examine here the process of this choosing and in later chapters review just what these differences were and how they shaped the way humans interacted with each other and with their natural surroundings. But since choices are involved here it is important to speculate a bit on the path *not* taken—what we call the path of foraging, hunting and pastoral grazing—a path that may well have led humans to a more benign and sustainable if less developed future.[4] And to provide some illumination, we turn again to Rousseau.

Rousseau floating on Lake Bienne relaxing in his reveries of simple existence—mere being—served to alert us to one of the more powerful modern critiques of civilization. To be aware of, to appreciate fully the mere fact of existence, the "sweet experience of being alive", was asserted to be a central quality of the Paleolithic hunter-gatherer—the "savage" in Rousseau's language.[5] Rousseau and many others might argue that to achieve such an awareness of existence may actually require a kind of cultural and structural simplicity and communalism similar to that thought to have prevailed in earlier eras.

If, as some claim, disenchantment, alienation and loss of wonder are implicit in modern civilization, and if we are hence no longer able to sense the wonder of our own being and its connection with the natural world, we lose as well our ability to appreciate and value the simple existence of other parts of that world. If other humans, and often even ourselves, become in modernity abstract objects, then must not all of nature be objectified as well? After all, if we cannot see and feel the wonder of us, how can we be expected to see and feel the wonder of a tree, a spider or a river?

We will return to these thoughts later in looking at the work of contemporary deep ecologists, nature poets and others, but for now we need to turn our attention to yet another Rousseauean insight that will inform our venture. As his *Reveries of the Solitary Walker* attests, Rousseau by the end of his life seems to have become reconciled to being alone, isolated and alienated from other humans—though not from nature. Indeed he seems to wear his social isolation as a perverse badge of honour, starting the first of his ten reveries by asserting aggressively rather than pathetically, "So now I am alone in the world."[6]

In fact, though, in his rejection of the modernity he lived amidst he was far from alone. Nor did he idealize the misanthropic life of the solitary. For Rousseau, as for the following generations of Romantics, communitarians, anarchists, advocates for small being beautiful, and other members of these shadow perspectives, the road not taken led not to stagnation or isolation but instead took humanity from the hunt to the field and the village—and stopped there.[7]

In one of his earlier works, an essay probing the origin of the gross inequalities he saw as plaguing 18th century modernity, Rousseau issued a powerful challenge to the bourgeois culture of the day. After waxing eloquent on the unreasoning happiness of his "savage" in the state of nature he posits that this happiness should have been eternal—or as eternal as anything can be on a mutable earth. It was, he insisted, the natural condition of humans. But an accident, an external force that "need not have occurred" intervened and this primitive happiness came to an end.[8]

And here we might reasonably ask, if there was in fact no teleology—Darwinian or Holy—that determined that the final version of the hominid line—Homo sapiens—must leave the affluence of a hunting and gathering life for the city and civilization, then why would they make this shift? Rousseau's question finds many a sympathetic ear in our time. We still wonder what would compel our distant ancestors "...to stop wandering freely from one food supply to another in order to settle in one place and eat the same food every day?"[9] Why would our ancestors "...abandon equality and justice for all and choose instead vertical relations in which a small number of elites run the show?"[10] What accident or external event would lead to a politics in which "...a few individuals...persuade the vast majority of the community to moderate or even abandon rigid egalitarian ethics, to give up communal access to any important resources, and let a few individuals exert disproportionate influence on the decisions and behavior of the majority of the community?"[11] Or, as Rousseau put it over 200 years ago, how did it happen that "...a handful of people should gorge themselves with superfluities while the hungry multitude goes in want of necessities?"[12] It cannot have been an easy transition.

In this second chapter we will end where we began, with the prospect of an eternal return to the "sentiment of existence" in the here and now, that perhaps unique feature of human consciousness that continues to assert its claim in modern times, albeit in a subdued voice.[13] But first we must assess the actual process of the shift from a preoccupation with this immediate quality of being and integration with natural surroundings toward a range of human cultures all focused more on both past and future and on heavens and hells, cultures built on the primacy of language, reason, sedentary living, hierarchy and a division of labour. And in close companionship with all these developments, was the emergence of a radical disjuncture between humans and the rest of the natural world.

THE PALEOLITHIC REVOLUTION RECONSIDERED

In our anti-teleological era many are sympathetic with Rousseau's insistence that the equilibrium of the long Paleolithic era required an external, extraordinary intervention to bring about its demise. Sensitized now to a kind

THE FALL FROM GRACE

of relativism that seeks to respect diversity in human achievement (or even lack thereof), the remains of the Paleolithic that we find among the existing so-called primitive cultures demand to be seen as possible alternatives rather than mere remnants of an earlier phase. Often exhausted by an endless chain of changes, it is easy for moderns to be seduced by the image of equilibrium that seems to characterize the Paleolithic era. Michel Serres offers perhaps the most eloquent set of imagined images of this Rousseauean natural state of mankind:

> The vagabond life of the first men, the state of nature, as we will say, is first a state, in other words an equilibrium. Tough race, with big bones and strong muscles, on a hard earth, with big fruit...what the earth produces of its own pleased these men, to satiety. Equilibrium is evaluated on the scales of exchange: the sun and the rains gave, they were happy with the gift. They wrapped themselves in mother earth just as, at night, they wrapped themselves in branches or foliage. Acorns, water and caverns, to eat, drink, sleep. Their wandering knew no discrepancy between yield and need.[14]

This natural condition—a proverbial Garden of Eden—only becomes "primitive" of course, when viewed by advocates of the cultures that followed.[15]

In the case of Adam and Eve there certainly was a dramatic event that caused the expulsion from their Garden of Eden. Many have assumed that an Ice Age about 10,000 years ago played a similar role in smashing the long equilibrium of the Late Paleolithic. What geologists call the Holocene followed, an unusually warm era that provided a "...window of opportunity for the making of history."[16]

The shift from timeless equilibrium to the development and change involved in the making of "history" is embodied or personified in the shift in human activities from hunting and gathering to agriculture. This was not as dramatic a change as has sometimes been imagined, but rather proceeded at its own glacial speed. Indeed scholars are increasingly more interested in signs of continuity in the evolution of Homo sapiens than in indications of radical shifts.[17] Looking back over 50,000 years of climate changes and successive waves of human migrations from Africa across the Eurasian land mass, the stress is now on seeing modern humans emerging "...through a process of accretion of new behavioral systems...a reflection of the increasingly sophisticated social systems of the people who entered environments such as those of the Eurasian Plains."[18]

These arguments call into question the idea of a revolution in the Late Paleolithic, suggesting instead that many of the features associated with this shift toward a more recognizably human culture—eg. language, symbolic representations, more complex tools and technologies — can be found as early as 400,000 years ago in the African archaeological record.[19] And if the shift from hunting, gathering and foraging to cultivation began 10,000 years ago, it takes at least 5,000 years for any significant archaeological record of settlement to appear, first in Mesopotamia, Egypt and the Indus Valley around 6000 BCE and a thousand years later in China and Crete.[20]

Are these issues of periodization, continuity versus change or "revolutions" really important? What purpose do they serve? A Rousseau might argue that they give us a discrete view of another human way of living, in this case a different model of human relations with the natural world, a Paleolithic sustainable eco-system. They also, however, often persuade us to dismiss different ways of living as "in the past", of a different era. And one of the prime examples of this latter habit is the idea of a fundamental divide between the Paleolithic and the Neolithic.[21]

The dualist primitive/civilized approach required a revolution to overcome what appeared to be an eternally sustainable and lethargic hunter-gatherer culture, an idea advanced most powerfully by Gordon Childe in his widely read *What Happened in History?* In Childe's formulation the Neolithic Revolution came to be seen as the source of civilization — the beginning of history. Morris Berman and other "cultural gradualists" insist, however, that such an approach masks millennia of cultural developments which only very gradually lead to settled human communities that embody early forms of the key building blocks of modernity: a strong anthropocentric world view, a politics that affirms social inequality as the human norm, and spiritual belief systems that institutionalize the perhaps natural acceptance of a vertical split between heaven and earth.[22] There was, then, no single Neolithic Revolution, "…but rather a series of independent, albeit relatively rare, events that developed separately in a few parts of the world and then diffused to other regions."[23]

The gradualist version of human cultural development starts from our earlier question: given the absence of an evolutionary or teleological imperative, why should such changes occur? Population pressure may have played a crucial part, making it essential for communities to shift to food production.[24] Some anthropologists have linked the beginnings of agriculture to a seemingly inevitable aspect of human economic life: the emergence of a surplus. They suggest that over the past 10,000 years human culture experienced a gradual shift in the nature of social and economic organization, with two ways of being hunters and gatherers existing side by side: one functioning in an immediate return (IR) economy and others living in delayed return (DR) economies.[25] The IR hunter-gatherer groups were the classic form that had existed for millennia, accumulating only what they needed to consume, hence with no need for the technologies of storage or the politics of surplus. The DR groups, even if dependent on just minimal accumulation, began to develop more complex hierarchies and significant social inequalities. Surplus, then as now, became the engine for change, leading eventually to conflicts, obligations, hierarchies and ideologies.[26]

So, with an engine in place—the disposition of surplus—we can step back from the hunter-gatherer continuum and watch our Homo sapiens ancestors approach us. The approach is slow and steady and has been marked off in retrospect into several stages.

THE NEOLITHIC

A false start of the Neolithic appears about 12,500 years ago, called the Natufian period by archaeologists. Indicative of the precarious nature of the shift from hunting and gathering to growing, the Natufian settlements in what is now Israel and northern Jordan saw some degree of sedentary living (evidence of substantial architecture in the remains of small groups of six to seven structures housing about 60 individuals) but with domestication limited to rye and to dogs. Contrary to the expected pattern of steadily increasing complexity, over the 1500 years of the Natufian settlements the movement goes in the opposite direction toward simplification and a gradual return to a hunting and gathering existence.[27] This same precarious nature of the civilization process is illustrated later in the Baluchistan/Indus Valley region where and agrarian and urbanized culture appeared about 7000 years ago only to eventually collapse. Cities disappeared and the population reverted to scattered rural sites and subsistence farming.[28]

The period starting around 10,000 years before the present (BP) and lasting for about 5,000 years, is known by archaeologists as the Mesolithic Era—roughly from the end of the last Ice Age to the start of what can be seen as a primarily agricultural economy in the several gathering places of humans. The Neolithic proper overlaps with the Mesolithic, variously dated at @9,000 to 6,000 years BP. It is here that we can discern the appearance of hierarchical cultures, the increasingly widespread presence of agriculture, and the beginnings of significant human alterations of the natural world to serve specific ends. Still, however, there has been no dramatic shift, no discernible moment of change. The humans of the Neolithic were still "...expert in the arts of foraging and hunting...", but were adding to those skills the ability to "...manipulate and perhaps control the native resources of animals and plants at their disposal, and of living more intensely social lives."[29] This was a time, then, when humans in the Near East were "making important decisions regarding their futures."[30]

It is toward the end of the Neolithic era, about 4500 BP, that we approach the time in human history that Rousseau might have designated his Golden Age, that period "...between the indolence of the primitive state and the petulant activity of our own [which]...must have been the happiest epoch and the most lasting."[31] Claude Levi Strauss agrees, offering this rosy assessment of the Neolithic as Golden Age:

> In the Neolithic period, man knew how to protect himself against cold and hunger; he had achieved leisure in which to think; no doubt there was little he could do against disease, but it is not certain that advances in hygiene have had any other effect than to transfer responsibility for maintaining demographic equilibrium from epidemics, which were no more dreadful a means than any other, to different phenomena such as widespread famine and wars of extermination. In that mythic age, man was no freer than he is today; but only his humanness made him a slave.....Rousseau was no doubt right to believe that it would have been better for our well-being had mankind kept to a happy mean between the indolence of the primitive state and the irrepressible busyness of our self-esteem.[32]

Also known as the Copper Age, it was an era of small houses, small things and modest village-like settlements. Even with the existence of nascent hierarchies and the resulting inequalities, the domestic sphere of the family and household far outweighed any kind of public sphere in terms of human preoccupations. Even Paul Shepard, ardent advocate for a hunting and gathering culture, calls this hamlet society era the "best life humans ever lived", one in which "...regenerative, subsistence economies blended the cultivated and the gathered, the kept and the hunted."[33]

Of course, generalizations are difficult. While human populations were relatively sparse on the planet, they were unevenly distributed. In difficult, forested terrain settlements remained small and the impact on the natural environment was limited to clearing patches of land. In other areas such as the larger river systems and plains, the settlements were larger and more sophisticated and the impact on the natural environment correspondingly greater.

Clearly, this was one of those transitional eras. In some areas such as the regions drained by the Yangtze, Indus and Nile rivers social structures that we can reasonably call "systems" arose to govern human settlements. New ideas of "...leadership, hospitality and negotiation with the implicit threat of force" began to emerge.[34] Remnants of walls are found at many settlement sites, implying a social hierarchy, potential threats from neighbours, and an economic surplus. At the same time thoughts about nature and the relationship of humans to it began to change as well

> In unprecedented and innovative ways, groups of Neolithic people manipulated cosmology and the states of consciousness that provided them with the fundamental building blocks for religious experience, belief and practice. Seers thus gained political power and economic influence as well as religious domination. Therein lies the real, innovative essence of the Neolithic: expression of religious cosmological concepts in material structures as well as in myths, rather than the passive acceptance of natural phenomena (such as caves), opened up new ways of constructing an intrinsically dynamic society.[35]

If the folkways, totems and taboos of the so-called "covenant with nature" that characterized the Paleolithic had in large part been designed to ward off a powerful and capricious nature, now in the Neolithic ideas of control and transcendence were beginning to emerge.

Our sources are thin here, but Robert Harrison points out that the Babylonian hero Gilgamesh (@2700 BCE) in the epic of his name looks out on the earth with despair: "dumb, inert, insurmountable, revolving her relentless cycles, turning kings into cadavers, waiting impassively to draw all things into her oblivion. Gilgamesh, standing in here for all humans to seek their own transcendence, resolves to challenge oblivion."[36] In the Indus Valley around the 8th century BCE, the texts known as the Upanishads advanced a creation story that clearly set humans apart from the rest of nature. Likewise the first books of the Old Testament (@1200 BCE) tell the story of humans being created in the image of the Creator and given souls and precedence over all nature. Elements of what would become civilization were arriving and Rousseau's Golden Age would soon drift away on the winds of size, complexity, metallurgy and power.

And so now we can return to the theme of choosing. If anything can be natural in a realm of beings subject to evolution, the hunter-gatherer existence of Homo sapiens seems a reasonable image of how this species was "by nature"—egalitarian, subsistence living, vulnerable, migratory, adaptable, responsive more than pro-active but with "...an intelligence equal to that of today's humans."[37] But over millennia of responses to ever shifting environmental contexts the forces of natural selection changed the way this human nature manifested itself in predilections and behaviours, how it mobilized that intelligence into new categories of cognitions.

While acknowledging the arguments for evolutionary continuity and that the specific cognitive changes in humans took place over hundreds of millennia, it seems clear nonetheless that about 10,000 years ago these changes began to manifest themselves in dramatic enough fashion to provide a real benchmark—hence the case for a divide between the Paleolithic and the Neolithic. In contemporary speculation on this process Rousseau is often singled out as the cartographer of this shift in focus or priority from nature to culture. But his influential insights into these stages of human evolution have deep roots in earlier reflections on this change. The Roman writer Varro in 37 BC noted that:

> ...it must be true that step by step from the most remote period human life has come down to this age, as Dicaearchus writes, and that the earliest stage was a state of nature, when men lived on those things which the virgin earth bore; from this life they passed to a second, a pastoral life, and as they plucked from the wild and untrammeled trees and bushes acorns, arbute-berries, blackberries, apples and garnered them for their use, in like manner they caught such wild animals as they could and shut them up and tamed them.... Finally, in the third stage, from the pastoral life they attained the agricultural, in which they retained many of the features of the two earlier periods, and from which they continued for a long time in the condition which they had reached until that in which we live was attained.[42]

What caused those shifts that even the Ancients were so keenly aware of? Of course the shifts are really long periods of over-lapping transitions, with each mode of living neither disappearing nor remaining pristine. Early farmers clustered in clearings and villages would obviously continue to hunt as long as game was available. However, the gradual increase in food

At the risk of being repetitive, it is worth re-stating some of the key issues we confront in assessing our human nature. The argument from continuity says that the mental consciousness and spiritual dimension we now consider so central to the nature of humans were only latent—not original or transcendent qualities—cognitive capabilities or possibilities rather than eternal verities or inevitable developments. The specific forms that this consciousness takes are developed, not determined. This was Rousseau's basis for arguing that "what we see not necessarily being what we are."[38] Hence nature and nurture are inextricably linked. The large-brained species being one with many possibilities.

In its most optimistic formulation, anthropologists, archaeologists and philosophers assert that the limits we so often observe to our capacities for cooperation, sustainable living, peace and social equality do not derive from a rigid human nature, but are in fact socially constructed by the combination of market relationships and an excessive focus on individualism.[39] New cultural forms, then, could negate or alter these seeming limitations, freeing up other latent possibilities.

On the other hand, we have a "realist" view that the limits we observe are in fact grounded in a human deep structure or nature and that it is falling prey to sentimental or wishful thinking to deny this. The supposed precedent of relative peacefulness and lack of ecological destruction during the Paleolithic was simply the result of poor technologies and low levels of population—not some essentially benign or holistic nature.[40] This realist view implies a kind of Promethean trajectory to human cultural development, beginning with the domestication of fire and proceeding to the extinction of competing large mammals on all continents and then on to our current determination to transform and manipulate all non-human nature.[41]

production that came with improved technologies and more deliberate cultivation meant larger populations could be sustained which in turn decreased the amount of wild game in the vicinity. Clive Finlayson, in an impressive analysis of relations between Neanderthals and Homo sapiens, argues that the competitive advantages conferred by farming would have made the process of transition from hunting and gathering to agriculture "autocatalytic" (i.e. self-sustaining).[43]

It is important to remind ourselves, however, that this decline in reliance on hunting and foraging would not have been seen at the time as a good thing: "The sustained work of digging, weeding, guarding and processing was unlikely to have been adopted by nomadic foragers except under duress."[44] Robert Harrison offers an ecological answer to our question which resonates with our own 21st century confrontations with forced change. As noted in Chapter One, Homo sapiens had evolved in ways adaptive to the great Ice Ages, a plains creature thriving on hunting large herds of mammals in good times and relying on gathering in lean times. As the glaciers retreated 10,000 years ago the terrain quickly reforested and the great herds of the plains fled from the "...inhospitable density of the forests." Some human groups died out as a result, others followed the herds to the North (and remain with them to this day), while still others managed to accommodate themselves to the new ecological realities via agriculture.[45]

It is difficult, pondering these responses of our distant ancestors to their own experience of global warming, to imagine how a future observer might comment on the choices 21st century Homo sapiens made in response to climate change. Taking the long view, Philip Tobias sees this history of living organisms on earth as swinging between sudden periods of growth followed by periods of extinction. Finlayson in his assessment of the Paleolithic sees the survival of modern Homo sapiens as more a matter of luck than destiny.[46] We return, then, to the issue of choice vs chance and context. Norbert Elias in his work on the "process of civilization" found it to be characterized by neither order nor disorder, but instead "merely blind and indifferent to the fate and feelings of men". Likewise, the development of civilization via settlements was not a purposive movement, but set in motion "...blindly, by specific changes in the way people are bound and live together."[47] While the natural world may operate according to certain patterns, cycles or even laws, these are not related in any way to human interests, needs, or plans.[48] For Elias, human mastery of nature can be correlated to the sense of "detachment" following a decreasing fear of nature thanks to the random accumulation of knowledge and technologies. This does cause one to wonder whether the gradual increasing fear of nature in our time might, ironically, lead in turn to a decrease in that sense of detachment and a return to "involvement".

While no doubt a useful means of survival in a changing world, Harrison calls our ancestors' shift to agriculture a "deep humiliation of the human species—a helpless surrender to the laws of vegetation that had choked the land with forests, depriving the nomads of their prior freedom of mobility."[49] A battle lost, but a war on-going. As a "savanna species", we can see now that humans responded to this humiliation by attacking the forests, creating anew the ancestral open spaces.[50]

But hunters and gatherers did not become farmers overnight. In the early period of the Neolithic, three adaptations co-existed: farmers, pastoralists, and foragers. And to a limited degree the three groups still co-exist today, with the addition of a fourth, the urban dwellers. Then, as now, there are tensions between the groups and the most marginal are still the foragers. Farmers and pastoralists, on the other hand, often have quite close and symbiotic relations; intermarrying, trading goods and even trading places.[51] As mentioned earlier, we see evidence of this in the current conflict in Darfur.

And both farmers and grazers share in the dramatic increase of the human impact on the environment that began with the Neolithic. Paul Shepard, an advocate of a radical return to hunter-gatherer culture even in our own time, insists that pastoralism is the worst offender, calling it "...a kind of domesticated hunting" which acts like "...sandpaper on the skin of the earth. It first opens and then kills forests, at first converts grasslands to wealth and then gradually reduces them to an indigence which is reflected in the animals and people they contain."[52] This sanding process and its agricultural counterparts began at different times in different regions with sheep and goats the first to be domesticated around 8,000 years ago, about the same time grain crops appear in the Fertile Crescent.[53]

Morris Berman sees these parallel "tamings" of plants and animals as reflecting a fundamental change in human thinking and valuing.[54] Wild and tame became the core dualism that guided human culture, eventually converted into a host of subsidiary dualisms: civilization/barbarism, culture/nature, reason/emotion, self/other, male/female. For writers like Berman and Hugh Brody, who celebrate the wild, the development of agriculture is the equivalent of the Biblical Fall.[55] And a quick fall it was.

Those of us with a predilection for Darwinian or evolutionary lenses through which to view and assess human history are necessarily concerned with the large expanses of time required for species adaptation to new circumstances. An exceedingly complex animal, humans evolved to their present form over many millennia, with body and brain becoming hardwired—to use a modern analogy—by evolutionary adaptation to fit into a specific ecological niche. Then, in an evolutionary "blink of an eye", the combination of climate change and a highly evolved set of mental tools led humans to make a number of quick choices in order simply to survive. But there was no time for any natural or biological adaptations.

Reflecting on this phenomenon, evolutionary psychologists warn us that "...it is improbable that our species evolved complex adaptations even to agriculture, let alone to postindustrial society."[56] For outspoken critics of the modern like Paul Shepard, the result of this mis-match between biology and culture is "..a kind of madness."[57] He speculates that we remain essentially hunter-gatherers trying to cope with modern urban living.[58]

We will return to this theme in later chapters but for now it is enough to assess just what this transition to a settled, village, farming culture might have meant for our distant ancestors. As is the case for many animals, our hunter-gatherer ancestors banded together as a means of survival. We are highly vulnerable animals, "...dependent on the free care and concern of others." We require networks of support based on generosity, justice and gratitude, qualities that "...enable dependent beings such as we are to live decent human lives together.[59] For millennia the band fulfilled this need but with settlement the band gave way, gradually, to the family, the village, the town. What can we say about the implications of this transition to the settled life? What kinds of things change when people used to life being a moveable feast or famine settle and begin to root themselves to place?

What must first confront us is the dwelling—the hut or house that creates for the first time a private sphere. Hunter-gatherer bands were distinctly open communities with no real distinction between public and private.[60] The village, on the other hand, is more like a collective made up of interdependent individuals serving very specialized ends. Complex role differentiation begins to appear and the rigid rhythms of the harvest takes over from the flexible—indeed often serendipitous—rhythms of the hunt.[61] Above all, work appears, leisure time begins to fade and the "affluence" described by Richard Lee and others begins to wane.

As noted in Chapter One, hunter-gatherers (and their contemporary kin) devoted little time to what we call work.[62] This presence of leisure time implies a lack of incentive or drive to employ what we might call more productive ways of living and working. Jack Harlan tells a story from his travels in contemporary rural Anatolia that captures the essence of the resistance that may have once slowed the transition to settled life, much as similar resistances have impeded the transitions to industry and now post-industry:

My interpreter and I had seen a family harvesting a field and stopped. He talked to the people while I collected some samples. My interpreter later told me that he had commented to the farmer that he could harvest the field in half the time if he would use a scythe and cradle. The farmer looked at him in astonishment and said: *Then what would I do?*[63]

Max Weber raised this issue in his work on the Protestant Ethic, noting that "...a man does not by nature wish to earn more and more money, but simply to live as he is accustomed to live and to earn as much as is necessary for that purpose."[64] Thus even today this perhaps innate impulse or wisdom that resists the culturally derived pressure to work survives on the peripheries of modernity, indicating that we may not be by nature a homo economicus.

A range of physiological changes also accompanied the move to a more settled culture. Humans shrank in size, life expectancy decreased, teeth decayed, nutrition declined and diseases increased.[65] And there were physical changes to the ecosystem as well. Edward Wilson dates the current "extinction spasm" from the Upper Paleolithic with its "...improved tools, dense populations and deadly efficiency in the pursuit of wildlife."[66] Cultivation itself wrought havoc with what had been a natural vegetative condition of the earth. Freeman Dyson cites Vladimir Vernadsky's 1926 book on the biosphere in which he celebrates the "...waist-high growth of feather grass, a continuous clothing of the earth, protecting it against the heat of the sun.....Man alone, violates the established order and, by cultivation, upsets the equilibrium."[67] Aldo Leopold adds the fascinating insight that by plowing the land farmers blunted the effect of fires that at one point swept the grasslands and as a result made it possible for trees finally to triumph over grass.[68]

There were other equilibriums upset as well. We do not know much about family life or, put more broadly, gender relations in the Paleolithic. But based on the observations of contemporary hunter-gatherer groups it would appear likely that males and females existed in a complementary and roughly equalitarian relationship. With the arrival of village culture and an increasing dependence on growing food as opposed to hunting or simply gathering, relations between the genders shift. It is presumed that women took the lead in the early experiments in cultivation. More tied to a home base than their hunter mates and more practiced at gathering, they were more likely to discover the benefits of tending plants, weeding, and harvesting seeds.[69] For some, this central role in cultivation resulted in a shift in power and influence. Shepard sees cultivation as "inescapably feminine" and as a "blow to the male ego."[70] Wilson sees women as controlling the private domestic sphere and thereby exercising "...considerable, if not total, power over the male."[71]

THE FALL FROM GRACE

But here we enter a realm of speculation on specific cultural phenomena for which the data available cannot provide firm evidence. It seems just as reasonable that the early speculations of Friedrich Engels and Lewis Henry Morgan were correct and that the specialization of labour implied by agriculture and the solidification of the family unit in fact reduced the status of women.[72] To explore with more clarity the issue of what humans "chose" to become, we need to shift our gaze from this prolonged period of transition from hunting, gathering and pastoral grazing to farming to an era for which we have more evidence.

Perhaps the one salient feature that can be highlighted to prepare us for the cultural complexities of the societies of the Ancient World is one that our guide Rousseau has singled out as the primary legacy of the Neolithic: the increasing prevalence of "comparison" and the focus on "difference". Worth citing in some detail if for no other reason than to enjoy a speculative tour de force, here is Rousseau's summary of all we have covered so far:

ROUSSEAU'S NEOLITHIC — THE EARLY TRANSITION

Men who had previously been wandering around the woods, having once adopted a fixed settlement, come gradually together, unite in different groups, and form in each country as particular nation, united by customs and character—not by rules and laws, but through having a common way of living and eating and through the common influence of the same climate. A permanent proximity cannot fail to engender in the end some relationships between different families. Young people of opposite sexes live in neighbouring huts; and the transient intercourse demanded by nature soon leads, through mutual frequentation, to another kind of relationship, no less sweet and more permanent. People become accustomed to judging different objects and to making comparisons; gradually they acquire ideas of merit and of beauty, which in turn produce feelings of preference. As a result of seeing each other, people cannot do without seeing more of each other. A tender and sweet sentiment insinuates itself into the soul, and at the least obstacle becomes an inflamed fury; jealousy awakens with love; discord triumphs, and the gentlest of passions receives the sacrifice of human blood.

To the extent that ideas and feelings succeeded one another, and the heart and mind were exercised, the human race became more sociable, relationships more extensive and bonds tightened. People grew used to gathering together in front of their huts or around a large tree; singing and dancing, true progeny of love and leisure, became the amusement, or rather the occupation of idle men and women thus assembled. Each began to look at the others and to want to be looked at himself; and public esteem came to be prized. He who sang or danced the best; he who was most handsome, the strongest, the most adroit or the most eloquent became the most highly regarded, and this was the first step towards inequality and at the same time towards vice. From these first preferences there arose, on the one side, vanity and scorn, on the other, shame and envy, and the fermentation produced by these new leavens finally produced compounds fatal to happiness and innocence.

But still, for Rousseau, this remained a Golden Age, an era "...between the indolence of the primitive state and the petulant activity of our own pride...", a road not taken as it turned out. Instead, the engines of comparison and difference that accompanied settlement continued apace and the choice made by humans to attach significance to these comparisons produced what we have come to call society:[73]

> As long as men were content with their rustic huts, as long as they confined themselves to sewing their garments of skin with thorns or fish-bones, and adorning themselves with feathers or shells, to painting their bodies with various colours, to improving or decorating their bows and arrows; and to using sharp stones to make a few fishing canoes or crude musical instruments; in a word, so long as they applied themselves only to work that one person could accomplish alone and to arts that did not require the collaboration of several hands, they lived as free, healthy, good and happy men so far as they could be according to their nature and they continued to enjoy among themselves the sweetness of independent intercourse; but from the instant one man needed the help of another, and it was found to be useful for one man to have provisions enough for two, equality disappeared, property was introduced, work became necessary, and vast forests were transformed into pleasant fields which had to be watered with the sweat of men, and where slavery and misery were soon seen to germinate and flourish with the crops.[74]

ATHENS AND JERUSALEM

And so we arrive at Civilization; at History. In these days of global cultural sensitivities it must be asked why Athens and why Jerusalem? Why not an Egyptian, Chinese, or South Asian centre of culture and civilization? There were, after all, contemporary cultures just as advanced and whose descendants avoided obliteration. The Shang dynasty in the Yellow River region of China, replete with writing, bronze, chariots and other symbols of civilization, flourished @1600 BCE. An Aryan civilization arose in the Indus Valley @1500 BCE, following the collapse of an even earlier culture. But it is the

Greeks and the Judeo-Christian cultures that continue to attract in this attempt to sort out the attributes and pathologies of the modern.

I have had to draw boundaries throughout and here it seems clear that our focus must remain on the culture based in the West, grounded in Judeo-Christian spiritual and moral beliefs and a rationalism that remains stubbornly rooted in Greek classical thought. It is this culture that has contributed the most (so far!) to the current global environmental crisis and it is also the culture that has led the way in refining the concept of the citizen and in extending the legal, ethical and political rights and responsibilities of citizenship beyond the rich, the powerful and those whose standing derives solely from kinship or tradition.

Almost as if by design, Athens and Jerusalem face each other across a sea, symbols of the dualism that lies at the heart of Western culture. The home of first the Jewish and then the Christian faith, Jerusalem remains the spiritual home for many in the West and beyond (even if that notion of home is contested by Islamic culture), while Athens remains the symbolic home of Western reason and science. Both of these cultural "home places" played central roles in the anthropocentric world view that became so dominate in Western culture and as well contributed to the more holistic counter tradition that I have identified as a persistent *shadow modernity*.

Athens. The Acropolis. The pantheon of heroic individuals—indeed the first individual humans we know: Plato, Socrates, Aristotle, Euripides, Alcibiades, Sophocles, Epicurus, Alexander, Sappho. All men except for one. All historic beings who lived in a known time and space. And before and amongst them, the more diaphanous and mythical selves: Achilles, Hector, Athena, Zeus, Dionysus, Medea, Antigone, Oedipus, Creon, Helen. A star-studded cast, a full house representing all the qualities—from Apollonian to Dionysian, from wild to tame, reason to passion, civilized to barbarian.

These airbrushed images of the Classical World centre on Athens above all as the iconic polis, the model for the democratic, enlightened community we continue to promote and cherish. Michael Ignatieff conveyed this sense in his *Needs of Strangers* when he described the power of this dream of the polis:

> Its human dimensions beckon us still: small enough so that each person would know his neighbour and could play his part in the governance of the city, large enough so that the city could feed itself and defend itself; a place of intimate bonding in which the private sphere of the home and family and the public sphere of civic democracy would be but one easy step apart; a community of equals in which each would have enough and no one would have more than enough; a co-operative venture in which work would be a

form of collaboration among equals. Small, cooperative, egalitarian, self-governing and autarkic: these are the conditions of belonging that the dream of the polis has bequeathed to us.[75]

And so we leap from Rousseau's huts with humans dancing and declaiming in front of a communal fire to an idealized symbol of the modern: the city, its walls and its citizens—Athens, Sparta, Thebes, Corinth. Ignatieff's polis—arguably drawn more from Rousseau's *Social Contract* Geneva than Plato's Athens—remains only an ideal, albeit a very important one. The citied culture that characterized the Ancient World was in fact highly stratified, often xenophobic and provided the perfect context for alienation instead of integration. And it was this citied, literate, hierarchical culture at the base of what we now see as the Western Tradition that provoked the choices—both individual and social—that in turn were decisive in creating a modernity constructed on a fundamental duality of subjects and objects.

But—and this is central to the argument I am making—the ideal of the city or polis as a community of equals, perhaps one in which even the other-than-human could assume a certain "subjectness", remained embedded in these early chapters of the story of the modern, even if only a shadow, underground or under siege.

This hierarchical, anthropocentric world view with its celebration of the singularity of human consciousness and cognitive talents was contested ground from the start and has remained so ever since. This is no where more clear than in the Judeo-Christian tradition as codified in the Bible. The decisive book of Genesis, a collection of "...poetic or cultic narratives which previously had circulated orally and without context among the people", contains within it the notorious competing versions of human relations to the natural world.[76] In Genesis 2 and 3, thought to be from about the 8th century BCE, God places Adam and Eve in the Garden of Eden and directs them to "dress and keep it and live peacefully with the animals." This is the Biblical basis for the potential of a "stewardship" role for humankind. This is the human as a responsible governor, a vision of "...benevolent, non-violent management". The human is in charge within a context of mutuality with nature.[77]

Even for St. Augustine (354–430), who placed humans next to God in the hierarchy of life, thereby affirming the fundamental anthropocentrism of Christian theology, all of nature—including what we call the inorganic—had value in the eyes of God and hence had a "claim on our love".[78] But like the Stoics, Augustine insisted that since animals lacked reason and soul, they had no claim in the area of legal or moral rights—we must love them but need not respect or protect them.

82 | THE FALL FROM GRACE

For Christians in the tradition of St.Francis (1182–1228), the Bible endows humans with the responsibility to tend the Earth in the absence of its true owner, God. It was St. Francis who insisted on a radical extension of Augustine's definition of valuing nature and in doing so created a powerful counter-tradition within Christianity. For St. Francis the relation between humans and animals was one of mutuality, there being an "...indissoluble kinship and community between humans and their fellow creatures under God."[79] His reading of the New and Old Testaments convinced him that animals share with humans a kind of soul and as such are equal in God's eyes.

Genesis 1 on the other hand, originating several centuries later than Genesis 2 and 3 (roughly contemporary with Classical Greece), takes a very different tone, with God urging Adam and Eve to be fruitful and multiply, subdue the Earth and take dominion over nature. This dominion becomes overtly violent by Genesis 9 as the animal world "...lives in fear and terror of man."[80] But even this potentially definitive justification for mastery can have a softer interpretation. Robin Attfield points out that the notion of mastering and subduing the Earth in Genesis 1 can be interpreted as being consistent with the specific Hebrew notion of kings being responsible to God for exercising only a "...circumscribed mandate."[81]

Clearly, Genesis 1 has carried the day in the Christian approach to the natural world, though thanks to the struggles of many Christians the possibility of an ecological Christianity has remained yet another "path not yet taken". Likewise in Athens, the anthropocentric views of Plato and Aristotle are the Greeks most listened to in the creation of modernity, but the more eco-friendly Epicureans along with the Stoics and the Cynics have retained a presence, albeit largely underground, in Western culture.

What the Greeks bequeathed to us was a political story, a determination by Greek philosophers and dramatists that humans were the *zoon politikon*, the political animal; the animal who must live in communities not just by preference, but as the means of survival. They understood the fundamental dilemmas raised by such a centrality of the political. Or, if they did not "understand" them in the absolute way we use that word, they at least intuited the problem. They were close enough, far closer than we can ever be, to the hunter-gatherer era to sense intimately the crucial role of the sensual and emotional in the human condition and the challenges it posed to the *zoon politikon*. But for them only beasts, barbarians or gods would choose to live outside the city.[82]

The Greeks understood that humans had been sensual beings long before they became the sophisticated cognitive beings that they understood themselves to be. We know now that the sensual drives are embedded, hard-wired by millennia of evolutionary development. And we know that the conscious, reasoning, cognitive apparatus that is such a central part of the modern human has its own circuitry. Acknowledging these two aspects of the human condition—sensual and cognitive—the Greeks in their artistic and mythic productions explored this fundamental Dionysian and Apollonian dualism.

This Ancient Greek story as conveyed in its literature (Medea versus Jason, Antigone versus Creon...) and its history (Athens versus Sparta) and its philosophies (Plato versus Epicurus) plays off this dualist pattern and becomes entrenched in the intellectual and cultural traditions that flow from it. And, sadly for the rest of us still governed by that tradition, the story is almost always one of conflict rather than resolution, victory instead of compromise, alienation rather than integration. And the result, since the Dionysian and the Apollonian remain culturally embedded in each of us, must be the repression of one or the other and, as so many theorists of mind and culture have reminded us, such repression eventually results in violence, illness or other pathologies. And nature, untamed and inhuman, remains always at the centre of the conflict, fueling our Dionysian sensuality.

But for the Greeks, at least most of them, the power of human reason—seen by them as a gift from the gods—would carry the day if nurtured carefully. Plato, the Lancelot of reason in his time, believed that prior to the gift of fire and the arts humans had lived a rough and primitive life for thousands of years, an "...altogether miserable" life. The playwright Aeschylus agreed, describing people in these unreasoning eons as living "...buried in the ground ...like little ants in the recesses of sunless caves."[83]

But the primacy of reason is getting ahead of the story itself. What really appeals to us about the so-called ancient or classical Greek culture is that we are able to observe at close quarters for the first time the process of human choice. In the previous 10,000 years when humans were also perhaps consciously making choices, they were doing so anonymously, without leaving clear written records. But from about 500 BCE, we suddenly have voices, debates, rationales and a history of the tensions and struggles that characterize the terrain between human intention/choice and the conditions that surround it.

THE CITIZEN

If now, in the 21st century, I am to make a case for giving the natural world in all its varied components some kind of "standing" as citizens of a global eco-community, how am I to derive a meaning or understanding of citizen and create a conception of nature that allows it in some way to assume the

THE FALL FROM GRACE

rights and obligations of citizenship? It seems inadequate simply to make up a definition or assign meanings arbitrarily, as if we lived in a world of blank slates. The Greeks knew that "nothing comes from nothing" and that every idea has its sources.[84] Any critical meaning of the term citizen has to involve more than just voting, or paying taxes, or obeying the law, and nature must be much more all-encompassing than simply views of landscape or parts of the world that are useful for humans to appropriate. The Greeks were among the first to wrestle with these two words and their debates, conclusions and disagreements remain very much our own.

After millennia of living in small, consensus-driven communities of pastoralists, farmers and hunters, the humans that settled and prospered in the Eastern Mediterranean from Egypt to the Fertile Crescent to Greece found themselves in communities that were too large to be intimate and too complex to be easily governed by consensus or by the loose hand of tribal leaders.[85] Lacking the inborn restraints that enabled some social organisms to live harmoniously together in groups, the humans in these more compressed and populous communities had to develop necessary structures and patterns of self-restraint that would enhance stability.[86]

Quite naturally perhaps, it seems that linking traditional ties of kinship and power with spiritual claims was the initial response in the ancient world to organizing these new urban centres. When we encounter the Greeks, however, we find a more sophisticated approach to the issue of living in community, an approach based on law rather than kinship or religious faith. This increasing reliance on secular laws, with disputes adjudicated through reason and precedent, came to symbolize the triumph of the polis or state over kinship.[87] While the importance of kinship in politics obviously lingers on well after the Greeks, the family itself becomes increasingly relegated to a private or domestic sphere and is seldom now the basis of government or even the core institution of society.[88]

This shift to a law-based, bureaucratically run state was no doubt in many ways a progressive development but in an important way it furthered as well the increasing separation of humans from nature. For all of their weaknesses, kinship based human communities displayed for all to see an intimate reliance of natural functions, in this case sexual reproduction. It was that most natural and universal of all phenomena, sex, that created the sinews of these early human communities and asserted their legitimacy and continuity. The replacement of this legitimizing force by abstract reason and abstract law must have helped persuade the citizens of these communities that their culture was outside of, indeed above, nature.

Just as the transition from hunting to farming was slow and filled with

WHAT ARE WE BY CHOICE, CHANCE AND CONTEXT? | 85

resistance, so did the linking of power with kinship resist being sundered. The citizen, that abstract, self-sacrificing stalwart of the polis that Plato wrote about in the *Republic* and Aristotle celebrated in his *Politics* and his *Ethics* was also a kinsman, and not far removed from the tribesman of more primitive times. The tension implicit in this era of transition was a central feature of many of the great dramatic works of classical Greece, but no where more clear and pristine than in Sophocles' *Antigone*. An early Greek text (@442 BCE) that is now often taught as a core text to undergraduates, it sets the baseline for the debates about the competing obligations of citizenship (culture?) and kinship (nature?).

The play opens with a scene of tribal carnage. A civil war has just ended, a war driven by competing family and tribal ambitions to rule the city of Thebes. For the proto-modern urban Greeks in their several and often competitive city-states, civil war was the greatest calamity, weakening the body politic in its on-going conflicts with neighbouring cities and empires to the East. Family and tribal claims, asserted overtly instead of in a more subtle politics, were seen as a throwback to barbaric, primitive times and part of the great cycles of historical crises and collapse that the Greeks were determined to bring to an end.

So here, with Antigone's confrontation with her uncle Creon, we can locate a primal insight about culture and nature that will become central to Western culture. Antigone, the teenage daughter of Oedipus, sister to the warriors Polyneices and Eteocles, and niece to Creon represents kinship writ large — even larger when we know that the entire family of Oedipus is scarred by his own commission of incest, the primal prohibition in kinship relations. And perhaps stemming from this deep flaw, her ambitious brother Polyneices has provoked the civil war, invading Thebes to unseat his brother Eteocles. The result is the death of each by the other and the transfer of the city to their uncle Creon.

Creon, wiser than the brothers, declares the primacy of the civil over the familial. As punishment for his transgression, Polyneices will not be buried. Disgraced, dismembered, allowed to decay by the forces of nature and thereby denied entry to the afterworld, he will be the reminder to all of the passing of an era and the beginning of civil society. Antigone, the next oldest in the line of Oedipus, assumes the counter position, defending the fundamental primacy of kinship obligations. She will choose to die in the effort to honour her brother, despite or independently of his actions. And, of course, the intensity of her kinship ties is reinforced by a set of spiritual beliefs that require her to take action. Contemptuous of her younger sister Ismene's reluctance to defy Creon's order, Antigone takes her stand:

I will bury him. I will have a noble death
And lie with him, a dear sister with a dear brother.
Call it a crime of reverence, but I must be good to those
Who are below. I will be there longer than with you.
That's where I will lie. You, keep to your choice:
Go on insulting what the gods hold dear.

Ismene, anticipating the Aristotelean idea that life in community with others is the only life possible, responds defensively to her sister's challenge:

I am not insulting anyone. By my very nature
I cannot possibly take up arms against the city.[89]

Ismene will later change her mind and seek to join her sister in the final *gotterdamerung* of the family of Oedipus.

Here Sophocles, unwittingly perhaps, is giving us an anthropology, the story of the shift of human affiliation from the intimacy of family to the more abstract realm of the state. A few years earlier Aeschylus in the *Oresteia* had attempted a similar story, shifting the basis of justice from vengeance by kin to legal judgments by a court. In *Antigone*, Creon speaks to this more abstract idea of state and community, and explains why he must place order over the rights of his niece:

And I will make my word good in Thebes —
By killing her. Who cares is she sings "Zeus!"
And calls him her protector? I must keep my kin in line.
Otherwise, folks outside the family will run wild.
The public knows that a man is just
Only if he is straight with his relatives.

So, if someone goes too far and breaks the law,
Or tries to tell his masters what to do,
He will have nothing but contempt from me.
But when the city takes a leader, you must obey,
Whether his commands are trivial, or right, or wrong.

Traditionally, kinship ties were the source of safety from the perils and mysteries of the natural world. Norbert Elias argues that this position of constant danger forced humans unto a deep involvement with nature. Kinship ties being emblematic of that involvement as were various religious and cultural beliefs. For Sophocles, this era of danger is largely past and humans now could enjoy a feeling of detachment from nature (and the accompanying reliance on kinship) that Elias sees as the beginning of the modern.

Creon's state will therefore base its rule on reason and laws rather than passions or traditions. The problem of order, of citizenship, was central for the Greeks who saw themselves as surrounded by barbarians who were without laws, without civilization. They were, in that sense, wild and dangerous nature—the nature of earthquakes, floods and shooting stars, rather than the rational nature of diurnal movements or placement of the stars. And Sophocles has Creon add a crucial few lines later in his speech which will have echoes well down to the present:

> Anarchy tears up a city, divides a home,
> Defeats an alliance of spears.
> But when people stay in line and obey,
> Their lives and everything else are safe.
> For this reason, order must be maintained,
> And there must be no surrender to a woman.[90]

And so we have the crucial linkage, pervasive in the Greek classics, of passion, unruly nature and the female.

But we listen to the ancient Greeks today because they were in fact too sophisticated to see things in such simplistic ways. They recognized that the world they observed was not so easily understood as Creon claimed. Hence his reason was flawed and corrupted by his own passions. And Antigone's case was well argued, supported by many and in many ways quite reasonable. And so, if the obligations of kinship, feelings of oneness whether with another person or even another manifestation of nature are reasonable as are the more abstract notions of citizen and law, then we remain in the same conundrum that confounded the Greeks.

There was no winner in *Antigone*. Creon's son, betrothed to Antigone, dies in her arms. Creon's wife kills herself in grief. A lesson is learned perhaps, but subsequent Greek history shows it did little good. Yet the rule of law, reason, community and the state did eventually establish its hegemony in philosophy if not always in fact and the tradition of the citizen began its long birthing.

And what of nature?

Greek drama and philosophy is littered with clues about the flowering of a new, more abstract conception of nature in the ancient world. The Greek *oikumene*, or realm of civilization, was already deeply hierarchical, following a trend established earlier by neighbouring cultural traditions. As noted earlier, Felipe Fernandez-Armesto in his sweeping essay on the history of humankind cites the *Upanishads* (written between the 8th and 4th centuries BCE) in India as the first evidence of humans clearly separating themselves from nature. In the Vedic creation myth, humans were fashioned from the Creator's limbs

THE FALL FROM GRACE

while other animals were made from a "...kind of chaos of milk and ghee. It suggests a hierarchical model of creation, with humans clearly ranked higher than other creatures."[91]

For Aristotle, "..all animals must have been made by nature for the sake of men."[92] Building on this acceptance of a hierarchical view of creation, Greek culture in particular began to focus on reason, the perceived superior rational powers of the humans, as the proof that they belonged at the apex of that creation. And there were the already mentioned barbarians, the non-Greek-speaking outsiders. Seen as Antigone-like in their willingness to let their passions hold sway, or without reason and hence not really human, these beings were in effect wild nature in human form. For the Greeks it was controlling this "beast within" that marked the human, the citizen, off from the essentially child-like, undeveloped and unreasoning barbarian.[93]

In the literature of the period one of the classic prototypes for the barbarian was the notorious Medea. From Asia, dark, clever, unscrupulous, possessed of magic powers and passionate, Medea was in many ways the Greeks' worst nightmare. As sure as the sirens had tried to lure Odysseus from his path, Medea led Jason astray and made him love her in return for her crucial help in the theft of the golden fleece. The sexual appeal of the "dark other" thus has a long tradition in Western thought.[94]

Like the myth of Oedipus and his family, the story of Jason and Medea has many versions. The most influential, though, is the one created by the great playwright Euripides, the one in which Medea, the alien wife betrayed, wreaks her cold and passionate revenge on the inconstant Jason by murdering their two children. Infanticide by mothers is often seen as the most a-human of acts, unnatural, defiant of all civilized norms and expectations, in our time often only explained by reference to chemical imbalances in the brain or serious psychic aberrations.

One might imagine that for the Greeks the play was received in a different way. Medea—beautiful, maternal, and intelligent enough to reason cleverly—is barbarian enough to ignore that reason and take the knife to her children in order to satisfy her passionate anger toward Jason. The lesson for the contemporary audience, I wager, was less how cruel was Medea but more how foolish Jason had been to link his life and fortunes with a barbarian. The woman Medea is volatile, violent, capricious, beautiful, dangerous, unpredictable and even sublime—a tsunami about to break onto a civilized beach, a volcano about to bury a city. And as such she is nature as well, a force which one no longer dares try to appease or live with, but rather to avoid or control at all costs—Jason should have known better! Medea, like nature, was inaccessible to reason and following the lead of Plato, the Greeks put more faith in abstract

WHAT ARE WE BY CHOICE, CHANCE AND CONTEXT? | 89

values and in fashioning rules for the citizen's proper moral and political life than in trying to understand the more subtle and veiled qualities of nature.[95]

But of course the Greeks lacked the technical and material means to control either nature or the barbarians that surrounded them. Medea, after all, escaped the wrath of the city in a chariot drawn by dragons! Those tasks would be bequeathed to the Romans and the cultures that followed them down to our own. Despite their failure, the response of the Greeks to the crises and dilemmas posed by these raw forces left a powerful philosophic tradition, and in concluding this discussion I turn now to two dominant approaches that emerged from ancient Greek culture and continue to shape our own; the Socratic/Platonic/Aristotelean focus on mastery through reason, and the Epicurean and Stoic counsel of passivity and acceptance.

SOCRATISM

We know of Socrates (469–399 BCE) through the words of his student Plato (427–347 BCE). And until the Renaissance of the 14th century we knew of Plato largely through commentaries about him rather than through his own words. But parts of one of Plato's famous dialogues, the *Timaeus*, did survive the collapse of first Greek and then Roman civilization and it was the dialogue that dealt most directly with the nature of the natural world and the position of humans within (and above) it. And so Socrates via Plato has had a direct and powerful impact on the way Western culture has come to think about the natural world, an impact that stretches back some 2,500 years.

While the *Timaeus* is the text that has the most direct relevance to our story, I want nonetheless to start out with an excerpt from another dialogue, the *Phaedrus* because it contains what for me is the most telling and illustrative line from Socrates concerning the Greek understanding of the place of the natural world in human affairs. The dialogue opens with Socrates and the young Phaedrus walking outside Athens along a river shaded by plane trees. Phaedrus is going to recount for Socrates a speech he had recently heard about love which, of course, will give Socrates an opportunity to give an even greater speech in turn. As they walk and talk they decide to head toward a tall plane tree so they can sit in the shade:

> Socrates: It is indeed a lovely spot for a rest. This plane is very tall and spreading, and the agnus-castus splendidly high and shady, in full bloom too, filling the neighbourhood with the finest possible fragrance. And the spring which runs under the plane; how beautifully cool its water is to the feet. The figures and other offerings show that the place is sacred to Achelous

and some of the nymphs. See too how wonderfully delicate and sweet the air is, throbbing in response to the shrill chorus of the cicadas — the very voice of summer. But the most exquisite thing of all is the way the grass slopes gently upward to provide perfect comfort for the head as one lies at length. Really, my dear Phaedrus, a visitor could not possibly have found a better guide than you.

Phaedrus: What a very strange person you are, Socrates. So far from being like a native, you resemble, in your own phrase, a visitor being shown the sights by a guide. This comes from your never going abroad beyond the frontiers of Attica or even, as far as I can see, outside the actual walls of the city.

Socrates: Forgive me, my dear friend. I am, you see, a lover of learning. Now the people of the city have something to teach me, *but the fields and trees won't teach me anything....*[96]

Plato here has drawn a Socrates who can appreciate the beauties of the natural world and describe them with an unmatched eloquence. Who cannot hear the cicadas and feel the warm breeze in reading the first passage? But Socrates has already told Phaedrus in an earlier passage that his sole interest in life is to respond to the oracle at Delphi and strive to "know himself" — not for narcissisistic enjoyment, but in order thereby to know humankind. And so, while he can appreciate the other-than-human natural world, because it does not reason, he cannot engage in dialogue, and is missing what Socrates calls "soul", it can teach him nothing. And like the barbarians who also reason not, if one cannot learn from something, its value is problematic.

As Shakespeare will later attest, man is the measure of all things and that measuring will be done, since the Greeks, via reasoning and language. To know oneself and to know others is the central quest of humankind. One could have said much the same thing for Paleolithic hunters and gatherers perhaps, but this knowing is no longer accomplished by intuitive connections formed in small bands linked by kinship and lives lived in communal intimacy. As well, to know oneself then was to know a being who felt in close communication with the natural world, without a reliance on language or reason. Life in cities is more complex and human relations more abstract, employing an ethics formed by formal reasoning or formal theologies mediated by language. Besides this very anthropocentric knowing, the rest of nature and the cosmos is of interest of course, but for Socrates and Plato it remains subsidiary. Aristotle's broader interests will contest this fixation on the human, but will never stretch as far as Aldo Leopold's identity with his pines or Rousseau's merging with the sounds of waves lapping on the shore of the Isle St.Pierre.

And what was it that Socrates felt he (and we) needed to learn? Here we enter the *Timaeus*. In Plato's version of the creation story the Creator/God is a "good" force who makes the universe out of pre-existing material that was chaotic, in "...a state of inharmonious and disorderly motion."[97] In bringing this chaos to (a still imperfect) order the Creator imbues it with "...reason in soul and soul in body." But it remains imperfect, only partially conforming to the good intention of the Creator.[98] Perfection, then, lies beyond the perceived realities of temporal life. Flawed by chaos from the start, even the powerful God creator cannot create perfection from chaos and disorder. But, "out there" the forms of all bodies exist in perfection and it was the task of humans, bodies with reason and souls, to attempt in their lived lives to attain an understanding of those perfections. Here, then, we encounter Plato's famous theory of forms. Peter Nicholson explains it this way:

> Plato had been trying to answer Socrates' questions, such as *What is beauty?* And *What is justice?* which Socrates had left unresolved. Plato did not think that answers could be found in the beautiful objects we see and the just actions we witness, because they are so numerous, diverse, and variable according to changing circumstances, that there is no common feature which can be beauty or justice. He claimed that beautiful objects and just actions should be understood as particular and differing instances of true beauty and true justice, which appear partially and imperfectly in each object and action. Beauty and Justice themselves –and all similar ideas — are forms, and exist whole, pure and perfect in another world, a heaven. In our world — the *real* world we mistakenly call it — forms can appear only imperfectly, as particular instances.[99]

So, Plato can appreciate the cicadas and the plane tree, but he learns nothing from either about beauty or even about the truth of cicadas or plane trees. The good life is to be pursued through contemplation of the forms, through efforts to return as it were to understandings of perfection that are embedded in the soul. We are immortal souls housed in mortal bodies and ideally we spend our temporal lives searching for that immortality. And the tool we use is reason, which we need to combat the desires of the temporal body, which as appetites are "...like a wild beast which must be fed."[100]

Charles Taylor notes that Plato stands "...at the head of a large family of views which see the good life as a mastery of self which consists in the dominance of reason over desire."[101]

By turning away from the vegetative and animal origins of life, the stuff of so many earlier creation stories, Plato portrayed life as being passed downward from an "...ideal realm of disembodied form...", a new idealism that

THE FALL FROM GRACE

served to abstract humans and the human enterprise from nature, displacing everything upward.[102] Perfection would require not integration or, to use a modern term, sustainability, but rather a radical separation from the world.

It may be worth a side trip from the Greeks to contrast Plato's creation story with the widespread Turtle Island creation story of indigenous peoples in North America. Here there is a Creator, a "Great Originator", who tries three times to make a world. The first two are failures, the last being destroyed by the Creator by a great flood. Floating on the endless water, the Creator asks several animals to dive into the water to find some mud to build a 3rd Earth. All fail except the turtle who, after a prolonged submersion, returns to the surface with the mud, which the Creator uses to create land. The land is placed on the back of the turtle's shell so that "...it would be forever a reminder for the people of the animal that carries it's home on it's back throughout it's life, and must protect and preserve it's home because it's home protects and preserves it."[103] The points of contact—the concept of the Creator, the mud, the flood—are fascinating, but more important is the difference in connection between humans and nature—the turtle is something the humans can learn from.

Plato, of course, was not the only Greek philosopher to place such high value on reason, but his most well known colleague Aristotle (384–322 BCE) used reason in a more earthbound manner. Raphael's famous fresco "The School of Athens" shows the two philosophers walking side by side arguing their positions; Plato pointing upward with raised finger and Aristotle extending his arm forward with the palm facing the ground. Aristotle was more "grounded", more interested in investigating the phenomena of "this world" instead of seeking insight into truth-bearing forms residing in another world.

But Aristotle was no environmentalist. His deep interest in the workings of nature had behind it the intention to control it, not live in harmony with it. All bodies had a "nature", a form if you will, and the natural life process was to strive to achieve that perfection and it was the natural destiny of humans to control all other species and phenomena. And so like Plato and Socrates, Aristotle gave pride of place to human reason as *prima facie* evidence that humans were at the peak of the hierarchy of creation, as near to god the creator as a being could be. This idea was by Roman times to be taken as Natural Law, the way things were by nature. Here is Cicero (106-43 BCE) on what came to be seen as "humanism":

> That animal which we can call man, endowed with foresight and quick intelligence, complex, keen, possessing memory, full of reason and prudence, has been given a certain distinguished status by the supreme God who created

him; for he is the only one among so many different kinds and varieties of living beings who has a share in reason and thought while all the rest are deprived of it.[104]

And so one could say the essence of what was to become modernity was born, just the details needing to be worked out. But, of course, even the history of the West is not that simple. There is the collapse of the Roman institutionalization of Greek philosophy to consider and the creation of a millennium of medieval culture — what we so egocentrically call the Middle Ages, the era between us and the ancients.

But even more important for the argument being made here is the joker in the Greek and Roman deck; the existence of a radically different and often underground philosophical tradition. I refer here in particular of Epicurus, Lucretius, and Seneca and it is with them that this exploration of human choosing will conclude.

SHADOWY GREEKS AND ROMANS

The Epicurean and Stoic traditions, based on the works of the Greek philosopher Epicurus (341–271 BCE), his Roman disciple Lucretius (99–55 BCE), and the Stoic Seneca (4–65) is our second "road not taken". The two schools of thought share a great many qualities, especially in terms of their determination to control needs and desires and not be distracted by the whims of culture and politics. But for our purposes, the Epicureans will take centre stage. Just as the hunter-gatherer tradition, though marginalized in the Neolithic preference for settlement, retains its presence in our global cultures, so does Epicureanism remind us of adjacent paths of social policies and personal lifestyles that still await exploration. Though not as "pregnant with the future" as the dominant Greek schools of Plato or Aristotle, the school of Epicurus (known as the Garden) has for the past two millennia proved an especially powerful attraction for those who question the basing of human happiness on either material acquisitions or spiritual transcendence.

If Plato was the model idealist, then Epicurus was his materialist opposite number. For the Epicureans there were no Platonic forms "out there" and no Aristotelean perfections waiting to be found here on earth. There was, as Democritus (460–370 BCE), the founder of materialist philosophy said, nothing but atoms and space. "The world is in fact not what it appears to be, but behind the world of appearances there lies a reality utterly unlike the world of the senses."[105] Suddenly, as Michel Serres argues, the gap between 400 BCE and 2100 virtually disappears: "There is in Lucretius a global theory of

94 | THE FALL FROM GRACE

turbulence.... His physics seems to me truly very advanced. Along with the contemporary sciences, it holds out the hope of a chaotic theory of time."[106]

Turbulence? A chaotic theory of time? Serres sees great promise in the atomism of the Epicureans, pointing toward a path that deserves our attention. It is important to stress that the Greece of Epicurus and the Rome of Lucretius and Seneca were both cultures in crisis. This sense of crisis or imminent collapse is always manifested first in the public sphere; the realm of community, politics, and citizenship. But it becomes pervasive as well in the private or domestic sphere with concerns by individuals for the well-being of the body. If the polis is in crisis, if one can no longer presume citizenship as a primary identity, if the health of the body-self is in constant jeopardy, then reason and philosophy often lead us to privilege the private sphere, the realm of family, the household, the self and concerns for the well-being of the soul.

This path to the private as traced by Epicurus, Lucretius and Seneca has often proved an attractive alternative to the more public traditions established by Socrates, Plato and Aristotle. In her writings on Epicureanism and Stoicism, the political philosopher Martha Nussbaum has placed this tradition in the context of therapy, a kind of ancient therapeutic counseling which targets not the ills of society or state or citizen, but rather the ills of the self when faced with chaos, collapse or corruption in the public sphere.[107] Or, equally important, Epicurean moods surface in cultures when the reason-based philosophies and technologies of the public sphere seem unable to respond to intractable problems, whether insistent barbarians at the gates or ecological crises generated from within.

The Epicurean intervention (much of which is shared by the Stoics), strongly therapeutic in Nussbaum's construction, is not, however, based on mysticism but rather consists of a set of rational philosophic premises which in turn rest on some very complex science. The key to happiness or contentment is seen as freedom from disturbance and anxiety in the soul and freedom from pain in the body. To achieve these freedoms involves acceptance of the following basic premises:

1. there are no divine beings who threaten us
2. there is no next life
3. what we actually need is easy to get
4. what makes us suffer is easy to put up with[108]

What we actually need then is limited to three things: the necessities of life, bodily health and peace of mind. The latter, which is really the central concern of Epicurean philosophy, requires:

WHAT ARE WE BY CHOICE, CHANCE AND CONTEXT? | 95

a. avoiding unpleasantness from fellow humans
b. escaping the pangs of conscience
c. avoiding worry about the future (including death)

It is important to note, given the earlier discussions of human nature, that for Epicurus and Lucretius happiness and/or contentment are the natural or default conditions of humankind. Perhaps we could imagine a truly epicurean life as a refined version of the hunter-gatherer life of the Paleolithic era, at least as described by the likes of Marshall Sahlins and Richard Lee.[109]

In the early practice of this philosophy the focus on the inner peace of the self meant avoidance of the public sphere in favour of membership in a small and individualistic community of friends. Since in this kind of setting the choice of needs is primarily the responsibility of the individual, self-fulfillment or contentment can be guaranteed by setting oneself as few needs or desires as possible, indeed only those that one can realize independently, without being helped or influenced by others. The ideal state restricts the truly valuable to that which is readily attainable and to condemn what is unattainable as valueless and indifferent.[110]

As Nussbaum says, in Epicurus's Garden they "...do not so much show ways of removing injustice as teach the pupil to be indifferent to the injustice she suffers."[111] Likewise in the Stoic tradition the objective was always self-sufficiency, being content with what was essential. These philosophic approaches to the "good life" have obvious links to the kind of minimalist lives of the hunter-gatherers described in Chapter One and, were they more prominent in our time, would provide a powerful philosophic grounding for environmentalists advocating lowering levels of consumption and abandoning aspirations for endless development.[112]

Arriving at this kind of indifference to accumulation is, however, a complicated process. Humans tend to be preoccupied with issues such as of the meaning of life, immortality, and a conviction that there must be more to being than just material existence. Even more fundamental is the keen awareness of pain and death and the fear of both. How might one become indifferent to such obvious calamities? How do we get from the Epicurean conception of the here and now — meaninglessness, pain, chaos — to there — contentment? Lucretius, following the lead of Epicurus, insists that we postpone such difficult questions and start at the foundation, noting that:

Garments of piecework came before garments of woven cloth.[113]

The piecework of the Epicurean system, in essence its physics, starts from the materialist base of the atom, the smallest particle of being first envisaged by Democritus. In this ancient atomic theory:

96 | THE FALL FROM GRACE

- Atoms fall downward with the same velocity in empty space (void) owing to their weight
- Interactions between atoms which result in the formation of bodies take place as a result of *swerves* which occur by chance and lead to collisions.
- While there are a limited number of shapes of atoms there are an infinite number of atoms within each shape
- On collision, atoms become interlocked by little branches or antlers, with only the atoms of the soul being spherical.
- The infinite number of atoms produce an infinite number of universes in infinite space.[114]

As Carolyn Merchant summarizes an Epicurean universe, it "...postulated the existence of an infinite number of unchanging atoms of different shapes and sizes moving ceaselessly through infinite void space, falling, swerving, combining, and separating to form the objects of the changing sensible world."[115] The earth, like all objects, is therefore a result of accident. The universe is a material system governed by the laws of matter and filled with structures, some stable and others unstable. "The stable ones will persist and give the appearance of being designed to be stable, like our world, and living structures will sometimes develop out of the elements of these worlds."[116]

One of the most salient aspects of post-Newtonian understandings of the natural world is the idea that the occurrence of random fluctuations within natural systems have the potential to make them unpredictable. Thus the future need not, indeed does not, derive deterministically from the past because within any natural system apparently random choices and catastrophic bifurcations can spread quickly and thereby radically alter the existing system.[117] These unforeseeable directions, of course, result in a new kind of order — hence the conclusion that from chaos comes order — from the swerving atoms of Epicurus comes Creation, the Universe, ourselves. As Michel Serres imagines it in self-immolatory eloquence: "I accept my dissolution in the burning plasma of matter. And the rest is turbulence. The eternal silence of these infinite spaces soothes me."[118]

Epicurean science was based on creating a middle position between the helplessness or resignation implied in a world run by gods and the determinism of a world governed by immutable Laws or Necessity by allowing for spontaneity and chance in the natural world and thereby a form of free will in the human world. The spontaneity that Epicurus envisaged in the atom was inside the atom, not given to it by an outside force (God).[119] Thus Chance is the form under which the spontaneity inherent in the atom reveals itself to us. But that creative spontaneity was not in the atom, but rather in its swerve, what Lucretius called the "clinamen", the angle off its path that leads to a "...proliferation of new forms."[120] For Serres and other modern commentators on Epicurean science it is turbulence and fluidity rather than matter itself that is the key to nature and to being, with the solid bodies we see and from which we derive meaning being really just "...exceptionally slow moving fluids.... Form itself is never static and local order, which may from within give the appearance of stability, is a minimally open system and will in time return to the global flow from which it arises."[121]

This was the understanding of nature from which Epicurus and Lucretius constructed a philosophical path to happiness in a world of constant contingency. The essential lesson derived from this science was that in a world of chance and chaos the existence of the clinamen, the swerve, implied an element of will and choice, albeit much more limited than the often grandiose wills of secular reformers or spiritual transcenders.[122] One could in the shelter of a private world of self, family and friends overcome the rule of chance and chaos and attain both bodily and inner peace.

The task, then, becomes one of overcoming false beliefs that are based on trusting appearances and believing in human transcendence. In a very real sense this implies a return to the pre-social, uncorrupted child, "...healthy in body *and* mind, not yet exposed to external forces that take hold of it."[123] The child here could, in light of what we have said before, be seen in an anthropological sense as the primitive, the uncorrupted Rousseauean *Emile*, the romantic noble savage. But the Epicurean avoids false beliefs not through child-like ignorance, but by understanding the way nature is:

> If we can learn to imagine things as they are, atoms in their void, endlessly dispersing and reconfiguring, our guilts, our fears, our false desires for unreal cures and unreal pleasures, our illusions, will disappear.....we will then be free to enjoy ourselves, to see accurately and to delight (briefly perhaps but richly) in the ephemeral configurations of atoms that we call ourselves and the world.[124]

The central false belief is the fear of death which has over the millennia persuaded humans to create gods and construct an afterlife, with the result that life, being, becomes one long struggle to attain access to an imaginary afterlife. All this, Lucretius argues, is based on a fear of death that is at its base irrational. Given Epicurean science, death is actually a form of birth—entropy, the inevitable loss of energy, is a condition of being and leads always to creation. Given this, any notion of mastery of the problem of life and death is futile. "We and the world we live in are shifting aggregations of imperceptibly minute atoms falling endlessly through empty space, hence for each of us death is the end, as our component atoms move on to form new alliances."[125]

The problem of the pain that can accompany death proves more difficult, applied logic often meeting its match in combat with a mere toothache. Both Epicurus and Lucretius insist that a proper mental attitude, an understanding of the atomic processes that are occurring, and a mental concentration on fond memories of home and friends can overcome the misery of the pain.

While some modern psychiatrists interpret Lucretius's apparent welcoming attitude toward death as symptoms of a "suicidal personality," the issue is central to his case.[126] If the secular dreams of the public sphere with their statues and arches are to be abandoned along with the spiritual promise of immortality, then the ultimate fate of the private self must be celebrated along with the daily pleasures of life in the restricted realm of the Epicurean garden.

The argument made by Lucretius must centre then on the limitation of desire as the central factor in attaining individual contentment while living and with the attainment of resignation and repose while dying. There is no design, only atoms and void, infinity and chance. All "being" is merely a temporary coming together of compounds of atoms and all such compounds will eventually disintegrate and then reform in new shapes and qualities. There is, then, no intrinsic superiority of one form of being over another and no eternal beings, compounds or qualities, only eternal atoms. Hence there is no sense to any notion of progress or development, no need or ability to conquer the natural, and no ability to affect the future.

There is, in the language of the modern Enlightenment, no individual authenticity to be achieved or realized in struggling for liberty, selfhood or justice. What does exist is being and the conscious drive for pleasure within that state. There is, then, no need to have the public sphere make sense, be progressive, prosperous or victorious. There is only the need to maximize the opportunities for pleasure, and because our particular atomic formation has resulted in consciousness, that pleasure must be both sensual and cognitive since "...mental pleasures depend on a right attitude to bodily feelings", including in this case both death and annihilation.[127]

CONCLUSION

The long era of the "primitive" is over. There remain what the Greeks and Romans will call barbarians, but they exist only to be absorbed into and help develop and diversify a culture that will be based on language, markets, reason, religion and cities. Parallel cultures will prosper in other centres of human settlement that will use, in unique forms, similar tools and attributes.

WHAT ARE WE BY CHOICE, CHANCE AND CONTEXT? | 99

Already with the hegemony of Platonic, Aristotelean and Christian thought we can see the beginnings of the sustainability trouble we 21st century moderns find ourselves in.[128] A radical separation has taken hold in Western culture with humanity imagining itself as qualitatively unique from (and at the apex of) the natural world around it. This separation is grounded in a privileging of consciousness and belief in a soul that is eternally human. It manifests itself in the world through language (naming the objects of nature), increasingly complex tools, and an anthropocentric spirituality. An ever-expanding mastery of nature and elaboration of the tools of domination become the means of proving this uniqueness. And, of course, mastery does not stop at nature's door but is also a sectarian fetish within humanity itself, as the drive to hierarchy, to establishing the domination of human groups over each other, becomes the stuff of history.

But, as the Epicureans would insist, there are not inevitabilities in either processes or resolutions, only tendencies, choices, fluids and chance. The idea, belief, and determination that humans deserve to be—or are ordained to be—set apart from the rest of creation thanks to their reason and/or their soul is the crucial heritage from this first post-primitive era. We know from the remnants of the hunter-gatherer cultures that remain with us and from the shards of evidence from the distant past that this is a dramatic shift, a shift from integration to separation, from an egalitarian world to a world of hierarchy, from a world of leisure to a life of labour.

The degree to which individual humans, or even groups of humans, "chose" settlement over pastoralism, mastery of nature over stewardship, or the polis over the garden is a moot point. Choices were made and by the 4th century BCE Homo sapiens was on a clearly marked path leading to a gradual separation from and ascendancy over the rest of the natural world. But we have identified three alternative paths of human development that were present from the start and remain, to varying degrees, viable in the 21st century.

First, under the generic label primitive we can still find the remnants of the hunter-gatherer and the pastoral ways of life that have survived now for some three to four thousand years of relentless pressure to settle, first on farms and later in urban centres. Decimated by the diseases generated at sites of dense human settlement and exported to the wilderness regions, by wars of extermination and by the tightening of the noose around wilderness itself, the hunter-gatherer and pastoral path remains vibrant in many so-called indigenous groups in the Americas, Asia, Africa and the Arctic. Indeed, as we will explore later, in the light of a potential ecological catastrophe that some argue has been brought on by the mastery agenda of modernity, the ways of life of these indigenous peoples may have important lessons for us all.

100 | THE FALL FROM GRACE

Secondly, the idea of stewardship that can be located within the Christian tradition and even more centrally in the traditions of neighbouring cultures has become the salient objective of a wide range of environmental groups. Ecofeminists, advocates for alternative technologies, Greenpeace activists, animal rights groups and many others take as their starting point the moral obligation of humans to care for and respect the ecosystem as a whole. This will be examined in greater detail in subsequent chapters, but it important to note here its deep roots in Western culture.

Finally, even at the centre of an ancient Greek culture built on the power of human reason and self-consciousness we have an approach to reason, the Epicurean and Stoic, that has the potential to re-direct modernity from mastery to toleration. The Epicurean system is a mechanistic and materialist world view, but without the progressivist and exploitative trappings that usually accompany such a world view. Lucretius is quite clear about the importance of reason even in the Epicurean Garden, where the mind needs more than flowers or gentle streams to free itself from worry:

> ...his power rests with reason alone....dread and darkness of the mind cannot be dispelled by the sunbeams, the shining shafts of day, but only by an understanding of the outward form and inner workings of nature.[129]

Thus it is potentially compatible with our modern understanding of reality but stands in opposition to the prescriptions that followed from those understandings. An Epicurean philosophic approach might afford modernity a path toward salvaging mechanism as a means of understanding reality without adopting its more pathological prescriptive attributes, persuading us perhaps to learn to love simplicity instead of just becoming resigned to it; to see limiting our desires as a source of happiness instead of bitterness.

And once again we can end close to where we began. What seems to link these three early alternatives is a certain attention to what we have earlier identified as the sentiment of existence or "being". Just recently I ran into a very contemporary account of the power of this sentiment to transform. The novelist Ian McEwan (*The Cement Garden, Amsterdam* and most recently *Saturday*) recounts the following:

> It was the Mediterranean spring and I had the day to myself...and I had one of those little epiphanies of *I'm me*, and at the same time thinking, well, everyone must feel this. Everyone must think, *I'm me*. It's a terrifying idea... yet that sense that other people exist is the basis for our morality. You cannot be cruel to someone, I think, if you are fully aware of what it's like to be them. In other words, you could see cruelty as a failure of the imagination, as a failure of empathy.[130]

WHAT ARE WE BY CHOICE, CHANCE AND CONTEXT? | 101

A bit banal? Or commonplace? Perhaps, but also a step toward a deeper understanding of the connection of one to all. As Barry Commoner warned, "everything is connected to everything else" and the fundamental relationships of our world are interdependent.[131] And McEwan is only confirming a long-standing human insight. We have seen Rousseau express these feelings, and Mary Wollstonecraft on her journeys in Norway and Sweden likewise reflects on "...some involuntary sympathetic emotion, like the attraction of adhesion, [which] made me feel that I was still a part of a mighty whole, from which I could not sever myself."[132]

Two final affirmations of this sentiment of being are worth our time and meditative thoughts. In the *Ruined Cottage* Wordsworth's shepherd had in vain turned to science to explain what he could not understand about life, but....

> From Nature and her overflowing soul
> He had received so much, that all his thoughts
> Were steeped in feeling. He was only then
> Contented, when, with bliss ineffable
> He felt the sentiment of being, spread
> O'er all that moves, and all that seemeth still,
> O'er all which, lost beyond the reach of thought,
> And human knowledge, to the human eye
> Invisible, yet liveth to the heart,
> O'er all that leaps, and runs, and shouts, and sings,
> Or beats the gladsome air, o'er all that glides
> Beneath the wave, yea in the wave itself
> And mighty depth of waters.[133]

And in the very last lines of his monumental treatise on the primitive, *Tristes Tropique*, the anthropologist Claude Levi-Strauss reserves for mankind the "privilege" to grasp "....during the brief intervals in which our species can bring itself to interrupt its hive-like activity, the essence of what it was and continues to be, below the threshold of thought and over and above society: in the contemplation of a mineral more beautiful than all our creations; in the scent that can be smelt at the heart of a lily and is more imbued with learning than all our books; or in the brief glance, heavy with patience, serenity and mutual forgiveness, that, through some involuntary understanding, one can sometimes exchange with a cat."[134]

102 | THE FALL FROM GRACE

NOTES

1   Niccolo Machiavelli, *The Prince* (London: Penguin, 1999), pp 79–80.

2   Stephen Quilley and Steven Loyal, "Towards a central theory: the scope and relevance of the sociology of Norbert Elias" (pp 1–22) in Quilley & Loyal, *The Sociology of Norbert Elias* (Cambridge: Cambridge University Press, 2004), p 5.

3   Robert Frost, "The Road Not Taken", *Mountain Interval*, 1920 (www.bartleby.com/119/1.html)

4   Fernandes-Armesto suggests that the most successful human cultures may be those that have been the most stable, successfully resisting the stresses and turmoil of change and progress. p 30.

5   Rousseau, Jean-Jacques, "Letter to Voltaire on Optimism", in David Wooten, ed., *Candide and Related Texts* (Indianapolis: Hackett, 2000). p 112. It is no accident that the communal celebrations of the hippies of the 1960's and 1970's were called 'Be-Ins' and that they featured a kind of calculated primitivism.

6   Rousseau, Jean-Jacques, *Reveries of A Solitary Walker*, p 27.

7   Paul Shepherd may be the one exception since he stopped his notion of an ideal world at the hunt and rejected the village.

8   Jean-Jacques Rousseau, *Discourse on the Origins of Inequality* (Indianapolis: Hackett Publishing, 1987), p 65.

9   Weatherford, Jack, *Savages and Civilization: Who Will Survive?* (New York: Crown Publishers, 1994), p 46.

10   Berman (2000), p 62.

11   Hayden, Brian and Gargett, Robert, "Big Man, Big Heart? A Mesoamerican view of the emergence of complex society", *Ancient Mesoamerica*, vol 1, 1990 (pp 3–20). Citied in Berman, p 75. Fernandez-Armesto says the "..." reasons why industrialization was accepted with all its warts provide a clue to why mass agriculture was tolerated in societies which espoused it. It happened—so to speak—by stealth. Its earliest stages were benign and only when a certain momentum of change had built up were the health and happiness of its victims swept away." p 208.

12   Rousseau, Jean-Jacques, *Discourse on the Origins of Inequality*, p 137.

13   There is an argument that this sentiment of existence is hard-wired in the sense of having a neurological foundation. D. Lewis-Williams and D. Pearce, *Inside the Neolithic Mind* (London: Thames and Hudson, 2005), p 26.

14   Serres, Michel, *The Birth of Physics* (Manchester: Clinamen Press Ltd, 2000), p 177.

15   Meier, Heinrich, "The Discourse on the Origins and Foundations of Inequality Among Men", *Interpretation*, v.16:2, 1988 (pp 211–227), p 219.

16   Cook, Michael, *A Brief History of the Human Race* (New York: Norton. 2003), p 7.

17   "There was no single 'Neolithic Revolution', but rather a series of independent, albeit relatively rare, events that developed separately in a few parts of the world and then diffused to other regions." Alan Simmons, *The Neolithic Revolution in the Near East* (Tucson: The University of Arizona Press, 2007), p 4.

18   Finlayson, p 126.

19   Tobias, p 34.

20   Jack Goody argues that "Just as there were many important inventions well before the Neolithic (speech, tools, cooking, weapons) so too there were many between the Neolithic and the modern periods (metallurgy, writing, the wheel). In recent writings on prehistory, the idea of a sudden revolution produced by the domestication of plants and animals has been replaced by a more gradual progression of events which take one back to the last interglacial.", p 8.

21   "If we...compare the Sumerians with the hunters and gatherers that preceded them or have lived since, we see that the contrast between these dawn people of civilization and any Stone Age people is greater than the contrast between the Sumerians and ourselves.

WHAT ARE WE BY CHOICE, CHANCE AND CONTEXT? | 103

In examining hunters and gatherers we are looking at people who are profoundly other. In looking at Sumerians and other early civilized peoples of the Middle East....we are looking into a very old, very dusty mirror." Alfred Crosby, *Ecological Imperialism* (Cambridge: Cambridge University Press, 1986), p 22.

22 David Lewis-Williams and David Pearce argue that belief in descent into a tunnel and flight to a higher realm "...are both sensations wired into the human brain and are activated in altered states of consciousness......Beliefs in magical flight and vortex travel seem to be inextricably linked to beliefs about a tiered cosmos...". p 69.

23 Alan Simmons, p 4.

24 Boserup (in *Population and Technological Change* 1981) argued that population pressures made a shift to food production necessary. "Agrarian populations were more productive than foragers—in food per unit land rather than in food per unit labour effort." Cited in Bert deVries and R.A. Marchant, "Environment and the Great Transition: Agrarianization", in Goudsblom and deVries, p 103.

25 See James Woodburn, "Egalitarian Societies" (pp 87–110) in Gowdy, and Tim Ingold, "On the social relations of the hunter-gatherer band" in Richard Lee and Richard Daly.

26 Berman, (2000), p 52.

27 Simmons, p 68–9.

28 Bert DeVries, "Increasing Social Complexity", in Goudsblom and deVries, p 185

29 Whittle, Alasdair, "The First Farmers" (pp 136–166) in Barry Cunliffe, *The Oxford Illustrated Prehistory of Europe* (New York: Oxford University Press, 1994), p 136.

30 Simmons, p 118.

31 Rousseau, *Discourse on theOrigins of Inequality*, p 115.

32 Levi-Strauss (1978), p 391.

33 Shepard, Paul, "A Post-Historic Primitivism", in Max Oelschlaeger, ed. *The Wilderness Condition* (San Francisco: Sierra Books, 1992), p 58.

34 Sherratt, Andrew, "The Transformation of Early Agrarian Europe: The Later Neolithic and Copper Ages 4500–2500 BC" in Cunliffe.

35 David Lewis-Williams and David Pearce, p 167.

36 Robert Harrison, p 16. Some see the figure Enkidu in the epic as akin to Rousseau's 'natural man', bridging "...the human/animal opposition" and illustrating the stages of a process of "Neolithicization". Lewis-Williams and Pearce, p 157.

37 Zerzan, John, "Why Primitivism?", *Telos*, No. 124, 2002 (pp 166–172), p 171.

38 Horowitz, Asher, "Laws and Customs Thrust Us Back into Infancy: Rousseau's Historical Anthropology", *Review of Politics* v. 52:2, 1990 (pp 215–241), p 224.

39 Gowdy, John, "Hunter-Gatherers and the Mythology of the Market" in Lee and Daly, p 392.

40 Kingdon, p 314.

41 Stephen Quilley, "Ecology, 'human nature' and civilizing processes: biology and sociology in the work of Norbert Elias" (pp 420–58) in Quilley & Loyal, p 53.

42 in *De re rustica*, cited in Lovejoy and George, p 369.

43 Finlayson, p 203. Bert deVries and R.A. Merchant speculate that "...it may have been a gradual intensification of the relationship between groups of humans, their environment and each other. Clearing plots of land, usually by fire; having animals around, becoming part of their habitat; gathering roots and tubers, gardens emerging. These could have been the slow changes in various places that led to 'agriculture' and its associated phenomena such as sedentarization, domestication and urbanization." Bert deVries and R.A. Marchant, "Environment and the Great Transition: Agrarianization", in Goudsblom and deVries,, p 98.

44 Kingdon, p 147.

45 Harrison, p 197.

46 Tobias, p 207.

47 Stephen Quilley, "Ecology, 'human nature' and civilizing processes: biology and sociology in the work of Norbert Elias" (pp 420–58) in Quilley & Loyal, p 49.

104 | THE FALL FROM GRACE

48 Norbert Elias, "The Sciences" (pp 152–165) in Johan Goudsblom and Stephen Mennell, eds., *The Norbert Elias Reader: A Biographical Selection* (Oxford: Blackwell, 1998), p 156.

49 Harrison, p 198.

50 Edward Wilson, p 143.

51 Weatherford, p 67.

52 Shepard (1991), p 53.

53 Merchant (2003), p 40.

54 Berman, who sees the sedentary life as something akin to a disaster for humans and for the planet, sees pastoralism as "...a reassertion of movement and a deep rejection of sedentary life along with its values." Morris Berman (2000), p 153.

55 See Ronald Wright's review of Brody's *The Other Side of Eden: Hunters, Farmers and the Shaping of the World* in the Toronto *Globe and Mail*, 11 November 2000, p D13.

56 Barkow, Jerome, Leda Cosmides and John Tooby, "Introduction: Evolutionary Psychology & Conceptual Integration", in Barkow, Cosmides and Tooby, *The Adapted Mind* (New York: Oxford University Press, 1992), p 5.

57 Shepard, Paul, *Nature and Madness* (Athens: University of Georgia Press, 1982), p 3

58 Robert Wright is perhaps the most outspoken proponent of this idea: "We live in cities and suburbs and watch TV and drink beer, all the while being pushed and pulled by feelings designed to propagate our genes in a small hunter-gatherer population." p 191. See also Boyden, p 40.

59 Blackburn, Simon, "A not-so common good", *Times Literary Supplement* 5 May 2000, p 31. A review of Alasdair McIntyre's *Dependent Rational Animal*. Rousseau recognized this aspect of the human condition, agreeing with Aristotle that "...society was made necessary by our weakness: "born incomplete, dying incomplete, always prey to the need for other...". In Tzvetan Todorov, *Imperfect Garden: The Legacy of Humanism* (Princeton: Princeton University Press, 2002) p 89.

60 Peter Wilson, p 103.

61 Rousseau, skating dangerously close to a "Noble Savage" construction, refers in *Emile* to the impact of this rural routine on the character of humans: "There are two sort of men whose bodies are in constant activity, and who both surely think little of cultivating their souls—that is, peasants and savages. The former are crude, heavy, maladroit; the latter, known for their good sense, are also known for their subtlety of mind. To put it generally, nothing is duller than a peasant and nothing sharper than a savage. What is the source of this difference? It is that the former, doing always what is ordered or what he saw his father do or what he has himself done since his youth, works only by routine; and in his life, almost an automaton's, constantly busy with the same labors, habit and obedience take the place of reason for him.

For the savage, it is another story. Attached to no place, without prescribed task, obeying no one, with no other law than his will, he is forced to reason in each action of his life. He does not make a movement, not a step, without having beforehand envisaged the consequences. Thus, the more his body is exercised, the more his mind is enlightened; his strength and his reason grow together, and one is extended by the other." Jean-Jacques Rousseau, *Emile*, p118.

62 Clastres, p 193.

63 Harlan, Jack, *Crops and Man* (Madison, WI: American Society of Agronomy, 1992), p 8.

64 Elizabeth Kolbert, "Why Work: A Hundred Years of The Protestant Ethic", *New Yorker*, 29 Nov 2004 (pp 155–160), p 156.

65 Weatherford, p 55. Shepard notes that we "...have more osteoporosis, lung disease and deafness and our average height is smaller than Cro-Magnon hunters living 25,000 years ago." Paul Shepard (1998), p 99.

66 Edward Wilson, p 99.

67 Freeman Dyson, p 4.

68 Leopold, p 29.

69 Kingdon, p 147.

70 Shepard (1967), p 101.

71 Peter Wilson, p 107.

72 Engels argued that the shift to agriculture "...subverted the former collective economy, transforming women's work from "...public production to private household service. The critical development that triggered the change was the specialization of labor that increasingly replaced the production of goods for use by the production of commodities for exchange...Instead of carrying out public responsibilities in the band or village collective within which goods were distributed, women became dependent on men as producers of commercially relevant goods.". in Eleanor Leacock, in Gowdy, p 160.

73 Wokler, Robert "Perfectible Apes in Decadent Cultures: Rousseau's Anthropology Revisited", *Daedalus*, v. 107, 1978 (pp 107–133), p 118.

74 Rousseau, *Discourse on the Origins of Inequality*, pp 113–116.

75 Michael Ignatieff, *The Needs of Strangers* (London: Penguin, 1984), p 107.

76 Gerhard Von Rad, *Genesis: A Commentary* (Westminster; John Knox Press, 1972), p 17.

77 Joseph Meeker, "The Assisi Connection", *Wilderness*, Spring 1988, p 61.

78 Andrew Collier, *Being and Worth* (London: Routledge, 1999), p 64.

79 Roger Sorrell, *St.Francis of Assisi and Nature* (New York: Oxford University Press, 1988), p 141.

80 Von Rad, p 131.

81 Robin Attfield, "Christian Attitudes to Nature", *Journal of the History of Ideas* (vol. 44:3, 1983, pp 369–386), p 374.

82 Anthony Pagden, *The Fall of Natural Man: The American Indian and the Origins of Comparative Ethnology*, (Cambridge University Press, 1982). Outside the city there is no basis for friendships (the highest of the purely human virtues) Nor is it possible to acquire knowledge of the world since "...knowledge depends on consensus and this can only be achieved when men live in close and structured proximity with one another." p 69.

83 Lovejoy and Boas, p 201

84 "The first point is that nothing comes into being from what is not." Epicurus, *The Epicurus Reader* trs and edited by Brad Inwood (Indianapolis: Hackett Publishing, 1994), p 6.

85 It is worth remembering that traditional tribal governance was very 'democratic'. Pierre Clastres observes that tribal societies "...are distinguished by their sense of democracy and taste for equality...that the most notable characteristic of the ...chief consists of his almost complete lack of authority"., p 28.

86 Norbert Elias in Goudsblom and Mennell, p 182.

87 The *Oresteia* by Aeschylus (first performed in 458 BC in Athens) is regarded as the founding myth for the Greek sense of justice via the rule of law instead of the blood feud.

88 Linda Nicholson, *Gender and History: The Limits of Social Theory in the Age of the Family*, (New York: Columbia University Press, 1986), p 106.

89 Sophocles, *Antigone*, trs by Paul Woodruff (Indianapolis: Hackett Publishing, 2001), p4.

90 Sophocles, p 29

91 Fernandez-Armesto, p 45.

92 Aristotle, *Politics* (New York: Oxford University Press, 1962), p 21.

93 Pagden, p 18. From here Aristotle developed his idea of "natural slavery", humans whose intellect has failed to achieve mastery over passions and instincts.

94 See Marianna Torgovnick's *Gone Primitive: Savage Intellects, Modern Lives* for a full discussion of this tradition.

95 Pierre Hadot, *The Veil of Isis* (Cambridge: Harvard University Press, 2006), p 92.

96 Plato, *Phaedrus and Letters VII and VIII* (Penguin, 1973), p 15–16.

97 Plato, *Timaeus and Critias* (London: Penguin, 1977), p 42.

98 Hefner, p 192.

99 Peter Nicholson, "Aristotle: Ideals and Realities", in Brian Redhead, *Political Thought from*

THE FALL FROM GRACE

Plato to NATO (Chicago: Dorsey Press, 1984), p 34.

100 Plato, *Timaeus*, p 98.

101 Charles Taylor, *Sources of the Self*, (Cambridge: Harvard University Press, 1992), p 21.

102 Harrison, p 38.

103 Kirkpatrick Sale, "Foreward" to *Turtle Talk: Voices for a Sustainable Future*, ed. by Christopher Plant and Judith Plant (Santa Cruz: New Catalyst, 1990), p viii.

104 From the *Laws*, cited by W.R. Johnson, *Lucretius and the Modern World*, (London: Duckworth, 2000), p 48.

105 D.R. Dudley, "Introduction" to *Lucretius* (New York: Basic Books, 1965), p 3.

106 Bruno Latour interviewing Michel Serres, *Conversations on Science, Culture and Time*, (Ann Arbor: University. of Michigan Press, 1995), p 59.

107 Martha Nussbaum, *The Therapy of Desire* (Princeton University Press, 1994).

108 Epicurus died painfully after a two-week bout of kidney failure and kidney stones, but he apparently died cheerfully because he "...kept in mind the memory of his friends and the agreeable experiences and conversations they had had together." D. S. Hutchinson, "Introduction", *The Epicurus Reader* (Indianapolis: Hackett Publishing), 1994, p viii.

109 Marshall Sahlins, *Stone Age Economics* (New York: Aldine de Gruyter, 1972); Richard Lee, "What Hunters Do for a Living, or, How to Make Out on Scarce Resources" in Gowdy.

110 M. Hossenfelder. "Epicurus—hedonist malgre lui", in M. Schofield and G. Striker, eds., *The Norms of Nature: Studies in Hellenistic Ethics*. (Cambridge: Cambridge University Press, 1986) , p 247.

111 Nussbaum, p 10.

112 The Roman Stoic Seneca insisted that "Nature suffices for all she asks of us. Luxury has turned her back on nature, daily urging herself on and growing through all the centuries, pressing men's intelligence into the development of the vices...Because the bounds of nature, which set a limit to man's wants by relieving them only where there is necessity for such relief, have been lost sight of; to want simply what is enough nowadays suggests to people primitiveness and squalor." Seneca, *Letters from a Stoic* (Penguin, 2004), p 168.

113 Lucretius, *The Nature of Things* (New York: Norton, 1977), Book 5:1350 p 144.

114 A.C. Crombie. *Medieval and Early Modern Science*. Vol II, (New York: Doubleday, 1959), p 38.

115 Carolyn Merchant. *Death of Nature: Women, Ecology, and the Scientific Revolution*. (New York: Harper, 1980) p 200.

116 Hutchinson, p x.

117 Eric Charles White. "Negentropy, noise and emancipatory thought" in N. Hayles, *Chaos and Order: Complex Dynamics in Literature and Science* (Chicago: University of Chicago Press, 1991), p 263.

118 Serres (2000), p 38. Christian theology agreed that at creation there was chaos, but this was not soothing. Thus Genesis 1.2, "The earth was without form and void, and darkness was upon the face of the deep" Gerhard Von Rad notes that Man has always suspected that behind all creation lies the abyss of formlessness; that all creation is always ready to sink into the abyss of the formless; that the chaos, therefore, signifies simply the threat to everything created.....Faith in creation must stand this test." Von Rad, p 50.

119 John Masson. *Lucretius: Epicurean and Poet* (London: John Murray, 1909), p 68.

120 David Webb, "Introduction" to Michel Serres (2000), p xi.

121 Webb, p xi.

122 This activist dimension of Epicurean thought was what Karl Marx found so attractive. Thus Marx argued that for Epicurus the atom was "...not only the material basis of the world of phenomena, but also the symbol of the isolated individual, the formal principle of abstract individual self-consciousness." The world, the realm of reality, was in a perpetual process of dissolution and rebirth which opened up all kinds of possibilities—whatever is could be otherwise. Above all, for Marx the power of Epicurus' theory lay in its rejection of religion and its creation/discovery of an "energizing principle". Like Epicurus, Marx saw philosophy

and religion as being in "radical opposition." Franz Mehring. *Karl Marx* (Ann Arbor: University of Michigan Press, 1959), p 29.

123 Nussbaum, p 109.

124 W.R. Johnson, *Lucretius and the Modern World* (London: Duckworth, 2000) p 12.

125 David West. *The Imagery and Poetry of Lucretius* (Bristol: Classical Press, 1969), p vii.

126 Charles Segal. *Lucretius on Death and Anxiety* (Princeton: Princeton University Press, 1990), p 9.

127 J.M. Rist. *Epicurus: An Introduction* (Cambridge: Cambridge University Press, 1972), p 105. Here the Stoics part company with the Epicurerans. For the Stoics, reason was the key to virtue and virtue the key to happiness. And this focus on reason set humans apart from animals, making humans "...incomparably higher" on any scale of value. M. Nussbaum, 1994, p 325.

128 We have not reviewed in any detail the contribution of Christian thought to this world view, but Carolyn Merchant's conclusion will suffice for now: From an ecological perspective, the separation of God from nature constitutes a rupture with nature. God is not nature or of nature. God is unchanging, nature is changing and inconstant. The human relationship to nature was not one of **I** to **Thou**, not one of subject to subject, nor of a human being to a nature alive with gods and spirits....The separation of God from nature legitimates humanity's separation from nature and sets up the possibility of human domination and control over nature." Merchant (2003), p 29.

129 Lucretius, *On the Nature of the Universe* (London: Penguin, 1994), p 39.

130 John Freeman, "Pleasure and Pain", an interview with Ian McEwan, *Vancouver Sun* 5 February 2005, p D1.

131 Leslie Thiele (1999) p 257.

132 Wollstonecraft, p 69.

133 Wordsworth, from *Home At Grasmere*, ed, by Colette Clark, (London: Penguin Books), 1960.

134 Levi-Strauss (1978), pp 414–415.

PART TWO

COMING TO SENSE AND ENCOUNTERING SENSIBILITY

CHAPTER THREE
THINKING THE WORLD TO PIECES

"I cannot discover this 'oceanic' feeling in myself"   FREUD

Lest we forget the meta-question of this book, the poet Shelley reminds us of the importance and the complexity of preserving aspects of the past in the present:

> The whole of human science is comprised in one question: How can the advantages of intellect and civilization, be reconciled with the liberty and pure pleasures of natural life? How can we take the benefits and reject the evils of the system which is now interwoven with all the fibres of our being?[1]

A romantic poet, but also a keen realist, Shelley knew that the modern, the substance of civilization was indeed irrevocably "interwoven with all the fibres of our being"—there was to be no going back to some pre-modern past, whether genuine or idealized. But as we have seen, Shelley's mentor Rousseau argued that integral to achieving any connection with this "pure pleasures of natural life" is a pre-rational, even primitive sensibility, what Rousseau and others called an apprehension of the *sentiment of existence.*

And the case being made here is that the preservation and nurturing of access to this sentiment is a crucial component for the successful creation of ecologically sustainable, healthy and creative human cultures. So before beginning this chapter's exploration of reason, science and mastery in modernity, let us return briefly to an image that inspired Shelley and so many other Romantics: Rousseau floating on Lake Bienne off the Isle St.Pierre.

At one point, were we to be eavesdropping from the shore, we would hear him cry out, "O Nature! O my Mother! I am here under your sole protection."[2]

He is, in moments like these, celebrating not only a heightened awareness of his own existence but also a sense of connection with the meta-existence of nature *per se*. And in calmer moments he insists that in these ecstatic unions with nature during which he feels "self-sufficient like God", a kind of truth or clarity is achieved, a clarity that indeed can only be achieved, according to him, without the presence of thought or reason.[3]

A few years later William Wordsworth at his own rural retreat at Grasmere in the English Lake District will cry out in a similar manner in verse:

Embrace me then, ye Hills, and close me in;
Now in the clear and open day I feel
Your guardianship; I take it to my heart;
'Tis like the solemn shelter of the night.[4]

Pierre Hadot in his wide-ranging discussion of how we think about nature cites both Schopenhauer and Goethe as echoing this romantic anxiety and discomfort with reason.[5] And Ralph Waldo Emerson (1803–1882) will be even more expansive in his transcending of reason and mere ego when, in one of his transcendent moments, his head uplifted to infinite space, he becomes "...a transparent eyeball: I am nothing: I see all, the currents of Universal Being circulate through me: I am part and parcel of God."[6] The German neo-Romantic philosopher Wilhelm Dilthey (1833–1911) called phenomena like this *Erlebnis* or "inner experience", one based not on sense data or empirical observation but rather stemming from experiencing the "...concrete flow of life which is temporally and logically prior to all sensate experience and rational reflection."[7] Environmentalist and Sierra Club founder John Muir (1838–1914), who we will encounter at some length later, claimed that in the midst of the wild and remote Sierras "...you *lose consciousness of your separate existence*; you blend with the landscape and become part and parcel of Nature."[8] More recently Elaine Scarry in referring to our experience of unexpected (or unconventional) beauty calls it an "instance of somatic pleasure" which has "sentient immediacy" and is based on a "transcendence of ordinary fact".[9] Charles Taylor sees these experiences as examples of authenticity, of being "... true to ourselves", which we achieve by being connected to a wider whole.[10]

As we will see in the following chapter, for the Romantics of the 19th century and their followers ever since, these somatic moments, epiphanies or ecstatic experiences are linked closely to childhood, seen as a naïve and uncorrupted period prior to reason and context taking hold of and dominating the understanding of experience. What are we to make of this persistent inclination in human cultures to seek and then find a sense of oneness with all creation, with nature? And how might such a pursuit mesh with the conflicting but equally human drive to master nature through the use of force and reason? As we saw, Rousseau responded to this tension by advocating *far niente*, doing nothing, as the only way to sustain the feelings of personal happiness and contentment that flowed from being at one with one's natural surroundings. Henry David Thoreau and then John Muir both retreated to woodland isolation. What fate modernity if such a counsel of retreat and passivity were to be followed?

But enough for now of these romantic views. Here we need to talk of reason and science and hear some new voices if we are to understand the mastery agenda that has played such a powerful role in shaping our modernity. Modernity has no shortage of reasoners, scientists and technocrats from whom to choose, but for a start Sigmund Freud will assume centre stage. Why Freud? For a start because he saw himself as a scientist, someone who based his understandings of human nature and human behaviour on evidence and employed a theory built from that evidence. But just as important, he was also sympathetic to the romantic alternative, even if finally unable to accept it as more than wishful dreaming.

But Freud is also much more than a scientist. Shelley suggested that poets were a culture's tuning forks, sensing the future well before it was clear to other less sensitized folk.[11] Seen in this way, Rousseau and Freud were both poets, sensing trends and events that were to shape decisively the future of subsequent generations. Responding to the violent upheavals he saw coming in 18th century Europe, Rousseau's initial response had been to propose an enlightened, rational social contract. But when faced with rejection and the realization that his intervention was too late, he retreated into introspective reveries. Freud, on the other hand, responded to the looming violence in 20th century Europe by raising the alarm in the name of reason and psychoanalysis.

OCEANIC FEELINGS IN VIENNA — 1927–1931

What must it have been like for Freud to be in Vienna in the 1930s? His academic wars were largely over and psychoanalysis finally established as a discipline and a therapy. His major scientific work had been done, his rivals dispatched and psychoanalysis, while still controversial, increasingly respected. But there was mortality to confront, the painful cancer of the jaw to deal with, and the rumblings of war, racism and fascism to worry about.

Stripped of its vast pre-war multicultural empire, Vienna in the late 1920s was a fading imperial city, now the capital of a mere province, a rump state increasingly fearful of being devoured by its expansive German and Italian neighbours. The post-war "sturm und drang" that surfaced in the 1930s must have felt a clear and present danger to Freud sitting as he was in the cockpit of the troubles. And for Freud not just a danger personally, but also a danger to modernity itself, to reason, science and faith in progress. The mankind as "Creatures of Reason" that had emerged so powerfully from the Enlightenment now seemed to be destined for a darker realm of passion, violence and chaos. Primitivist Balkan nationalisms, Italian fascism clothed in Roman imperial pretensions, and Russian messianic agendas were the issues of the day, while Hitler and the Nazis with their terrible agenda remained just below the surface.[12] But Freud read the tea leaves more clearly than most.

What did he see? More, one suspects, than just the ephemera of current politics. His 1927 book on religion, *The Future of an Illusion*, gives us a clear idea of how a modern scientist he employed reason to diagnose the ills of human culture, that diagnosis then more formally outlined in Freud's 1933 book, *Civilization and Its Discontents*. For Freud the illnesses and violence which he saw as so often associated with the inner life of individuals was too often explained away via delusions or myths, from witchcraft to stubborn denial. Instead, he insisted, reason and science, employed in this case by the craft or "technology" of psychoanalysis, was the proper path for dealing with these issues. But could such a tool be used to diagnose and treat social neuroses more effectively than the religious or repressive means used in human cultures?

In *The Future of an Illusion* Freud opens with the by then commonplace assertion that the function of human civilization is to control the forces of nature, extract from that nature what humans need to thrive, and devise rules to govern the distribution of the wealth derived from nature. But in the tradition of Thomas Hobbes, William Golding, Joseph Conrad and others, Freud saw this civilizational drive in constant conflict—both overt and covert—with deeply embedded human aggressive instincts (cannibalism, incest and killing being the most primal). In what seems an abrupt turn away from conventional progressivist thinking about civilization in general and modernity in particular, Freud makes a case for civilization being inherently coercive, necessarily forcing individuals to renounce or repress these aggressive and egoistic instincts:

> One has, I think, to reckon with the fact that there are present in all men destructive, and therefore anti-social and anti-cultural, trends and that in a great number of people these are strong enough to determine their behavior in human society.[13]

And, of course, he was seeing clear evidence of this in the theatrical posturing of Mussolini and the primitivism displayed in Nazi torchlight parades and mass rallies.

With Vienna itself as a reminder of the ravages of war and tribal nationalisms, with Italy and Spain militarized, with Germany sliding into an overt primitivism, and Russia imposing a violent collectivism, Freud was seeking not just an explanation for these civilizational catastrophes, but rather understand the basis for any resistance to the natural drives or instincts that seemed to inspire them. How do we learn to love the coercion of civilization? How do we renounce or repress any dissatisfaction with endemic social inequality or with the social frustration of our sexual desires? How do we come to accept, and even desire, repression in these and other areas of our lives? For Freud the answer was fear rather than some innate virtue or acquired enlightenment. In particular two powerful fears: fear of our primitive, aggressive "natures", and fear of nature itself. It was nature that doomed us to death. Hence for Freud "...the principal task of civilization...is to defend us against nature".[14]

In *The Future of An Illusion* Freud focused on religion as the means used by humans to respond to these fears. At first aspects of nature itself become gods. These "Nature Gods" somehow exorcize the terrors of nature, reconcile humans to the cruelty of death, or in some way create something which compensates the individual for the suffering which both nature and culture have imposed.[15] Eventually this becomes a singular God (or Allah, or Buddha) who is then elevated above nature. For Freud it was clear that religion was a phenomenon created by humans in civilization and its supernatural pretensions were only illusions.

Hence it could not in the long run serve as a sufficient bulwark against the destructive forces within humans and in nature. Instead, "scientific work is the only road which can lead us to a knowledge of reality outside ourselves. It is once again merely an illusion to expect anything from intuition and introspection; they can give us nothing but particulars about our own mental life."[16] Freud was showing his Enlightenment credentials; there was a real world out there to be studied and reason and science were the ways to find it, study it, understand it, and then master it.

By 1933, when Freud published *Civilization and Its Discontents*, life in Vienna was becoming even more precarious and symptomatic of the problems he had been seeking to address in the Future of an Illusion. And his confidence was waning. While not prone to reveries, the opening pages of *Civilization and Its Discontents*, reads much like one, Peter Gay describing it as a "meditation on belief.[17] Freud, always insisting on the strictly scientific basis of his analyses and understandings, seeks not to dismiss or demean the romantic and sensual yearnings for oneness, but rather to understand them clinically and in the process to at once affirm their reality and undermine their validity.

116 | COMING TO SENSE AND ENCOUNTERING SENSIBILITY

Freud had become friends with the French novelist Romain Rolland who had won the Nobel Prize for Literature in 1915 for his sweeping novel Jean-Christophe. Rolland was an idealist, a musician, biographer, novelist, pacifist and critic whose work had increasingly been inspired by a desire to discover fundamental truths about the human condition. His correspondence with Freud was part of this process and in 1927 Freud had sent Rolland a copy of his critique of religion, *The Future of An Illusion*.

"I had sent him my small book that treats religion as an illusion, and he answered that he entirely agreed with my judgment upon religion, but that he was sorry I had not properly appreciated the true source of religious sentiments. This, he says, consists in a peculiar feeling, which he himself is never without, which he finds confirmed by many others, and which he may suppose is present in millions of people. It is a feeling which he would like to call a sensation of eternity, a feeling as of something limitless, unbounded—as it were oceanic. This feeling, he adds, is a purely subjective fact, not an article of faith; it brings with it no assurance of personal immortality, but it is the source of the religious energy which is seized upon by the various Churches and religious systems, directed by them into particular channels, and doubtless also exhausted by them. One may, he thinks, rightly call oneself religious on the ground of this oceanic feeling alone, even if one rejects every belief and every illusion.

The views expressed by my friend whom I so much honour, and who himself once praised the magic of illusion in a poem, caused me no small difficulty. I cannot discover this oceanic feeling in myself. It is not easy to deal scientifically with feelings."[18]

Freud cannot locate the oceanic feeling, feel it, examine it, prove it and so must claim that it does not formally exist, that Rolland is deluded as are all the others who report such feelings. And this would make sense because for the scientist relies on seeing or experiencing a phenomenon to verify its existence. And so here is the modern dilemma. Feelings are undoubtedly real; they exist but they cannot be seen, touched, or parsed. Freud knew that even if they could not be seen or even replicated, feelings like Rolland's sense of oceanic oneness were nonetheless real and with the proper tools could be analyzed, understood, interpreted and their magic or their curse—their distraction as it were from the task of civilization—thereby defused or negated. Borrowing an insight from the Romantics' own world view, Freud linked Rolland's oceanic feeling to childhood, to a pre-rational time in individual development in which the primitive "pleasure principle" was dominant, memories of which never really disappeared but came back in dreams, in trauma, and eventually in oceanic feelings. While a Rousseau might yearn for these pleasures, Freud saw the only hope for humankind being instead an affirmation of the "reality principle" which served to establish a human individual identity separate from nature and from being awash in oceanic oneness. Thus, for Freud and for most moderns, "....the oceanic with its absence of boundaries and divisions is something we need to be protected from if we are to take our place in the 'mature' culture of the West."[19]

Freud provides this rationale for a Western culture already fully committed to a scientific world view that was based on a radical separation of culture and nature, of the self from all that which surrounds it. Freud in *Civilization and Its Discontents* offers an *ex post facto* explanation of why such a separation is necessary, grounding the primitive (and for him historical) origins of human nature in an innate violence and aggression that requires mechanisms of repression (i.e. civilization) to ensure the survival and prosperity of humanity. Belief in some kind of oceanic reciprocity among humans and between humans and the other is for Freud and other modern "realists" a dangerous illusion. Instead the task of culture is to master the self, the other, and nature. There is a somewhat bittersweet quality to Freud's conclusion. He acknowledges (as did Rousseau before him) that humankind was better off in a primitive condition of knowing no restrictions on instinct, but insisted that "...his prospects of enjoying this happiness for any length of time were very slender. Civilized man has exchanged a portion of his possibilities of happiness for a portion of security."[20]

In a clever sleight of hand theorists of the modern manage to shift the issue toward a dualist frame; our choice is either the risky and inevitably contingent "happiness" of the primitive or the mastery (or perhaps just management) of contingency and risk via the rational tools of modernity. But what is left in the shadows of the story is the millennium we call the Medieval era in Western history, a time in which humankind appears to have been much more embedded within the natural world in a great Chain of Being, neither slave nor master. This natural world (which included humans) was seen as "...a plurality of beings each possessed of its particular function and purpose in maintaining the whole"—very much like contemporary ecological theory.[21] But in our modern literature all this was forgotten and this potential path to a reconciliation of humanity with nature disappeared under the shadow of the label *Dark Ages*.

This is not the place to attempt a full analysis of the various approaches to relations with the natural world that could be found in Medieval culture. We can see in the works of St. Augustine, Thomas Aquinas, St. Francis and others a full range of positions familiar to Greek and Roman times and our own. On the one hand the neo-Aristotelean philosophy that dominated Christian

The modern is one of those slippery notions. Some historians use it to describe the era in which we live, placing its origin in Western culture in the mid-17th century. Others push the date back to the 1400s, seeing the Italian Renaissance as its start. Others, like Horkheimer and Adorno in their influential *Dialectic of Enlightenment* stretch the time frame even further, from about 400 BCE to the present, thereby including the Classical, Medieval, Renaissance and contemporary era. For them it is the existence of a market economy and institutions designed to support it that becomes the hallmark of the modern.

This view has some appeal for my task. For Horkheimer and Adorno, Odysseus is the first "bourgeois enlightened figure," the wily solitary who is "...already homo economicus, for whom all reasonable things are alike...from the standpoint of the developed exchange society and its individuals, the adventures of Odysseus are an exact representation of the risks which mark the roads to success."[28] Thus—they argue—reason began to contend with instinct and passion well before the 17th century, and many centuries before its full triumph in the modernized, instrumental form we encounter in the 21st century.

We can imagine a first phase of humankind as the "long duree" of the Paleolithic, the millennia of seemingly unchanging hunter-gatherer existence that was described in Chapter One. A second phase, shorter but still over 10,000 years, we call the Neolithic, the millennia of human settlement. In some parts of the world a version of the Neolithic lives on. The Third Phase, a mere 3,000 years old, is the modern era of reason, repression and dramatic material progress. To make this elongated notion of the modern work, I will refer to our contemporary, post 18th century era as "modernity", an increasingly globalized phase of a modern era that has perhaps reached its zenith in our time.

thought was fundamentally dualist in regard to nature—there was a clear separation for St. Augustine between things material (nature) and things spiritual.[22] For many Christians, including Aquinas, this led to a general disregard for nature. Since God had given man dominion it was a matter of indifference to Aquinas whether animals suffered but, much like modern attitudes from Kant to the present, he held that a man who shows compassion for animals is more likely to show compassion for fellow humans.[23] For St. Francis on the other hand, humans were just one life form among many, leading to a philosophy of mutuality.[24]

It is to Carolyn Merchant's *Death of Nature* that we owe a clear vision of the Medieval era as one that offered to Western culture a more holistic view of human relations with the natural world. Building on the ideas of St. Francis, the communal, still primarily rural basis of Medieval society, the preference for stability over change and a culture in which women had a more equal standing, Merchant's critique of modernity is based on its subverting of a potentially more organic culture growing out of Medieval communalism.[25]

While from at least the Neolithic on there were many who aspired to mitigate or master certain aspects of the natural world—one thinks of drought, famine, fire, etc.—these aspirations were not yet a system nor were they sustained by an ideology, religious or secular. But that was to change dramatically as European philosophers, political leaders and mechanics of all types began to lose their inhibitions and expand their vision of mastery.

MASTERY

Mastery. A word central to the modernity we all live within. No more oneness with, subservience to or even resting content with a simple appreciation of the rest of creation, but instead disenchantment, exploitation, control, and domination. The mastery of nature paradigm is, depending on the speaker, either the source of alienation and ecological catastrophe or the key to humanity reaching its full potential as (a) God's Creation or (b) a species. And here, approaching the mid-point of our exploration, we must account for the triumph in Western culture of this modernity of reason, science and mastery over the more chronologically pervasive "oceanic", respectful and reciprocal alternative.[26] The path to this modernity was one constructed from several interlocking components, including: a growing conviction that humans were unique and qualitatively separate from the rest of the natural world; an increasing faith and reliance on reason and a corresponding mistrust of feeling; the development of powerful sciences and technologies based on an increasingly instrumentalized reason; and an accompanying ideology of progress.[27] To understand the complexity and power of the mastery agenda we will need to examine each of these components.

How then, have we in modernity built a case for humans being not only unique (species, after all, are by their nature unique), but more importantly special, superior, and in some way final? Religion, particularly Christianity, has played a significant part. While many religions and cults placed humans firmly within the animal or natural world, Judaism, Christianity and then Islam just as firmly placed humans above or outside the rest of the natural world, thereby separating humanity from "...its embeddedness in the larger order that sustains it."[29] This became in effect a primary predisposition in cultures influenced by these religious traditions and in effect means that the now dominant form of the modern has been decisively shaped by spiritual and cultural traditions that stem from the West.

Like all cultural beliefs, this assertion of human exceptionalism was never uncontested. From Pythagoras (569–475 BCE) to PETA (People for the Ethical Treatment of Animals) there has been a consistent argument within Western culture in favour of a more equalitarian stance toward other species and toward nature as a whole. The Roman commentator Pliny recounts in a tone highly critical of his peers that a group of elephants in an arena about to be killed "..entreated the crowd, trying to win their compassion with indescribable gestures, bewailing their plight in a sort in lamentation."[30] Montaigne (1533–1592) writing at the very cusp of modernity insisted that all supposedly unique human qualities could be seen as well at work in the behaviour of animals.[31] Rousseau saw animals as equally as sentient as humans and Jeremy Bentham (1748–1832) insisted that even without reason or language evidence of suffering was enough to demand equality of treatment:

> It may come one day to be recognized that the number of legs, the villosity of the skin, or the termination of the os sacrum, are reasons insufficient for abandoning a sensitive being to the caprice of a tormentor. What else is it that should trace the insuperable line? Is it the faculty of reason, or perhaps the faculty of discourse? But a full grown horse or dog is beyond comparison a more rational, as well as a more conversable animal, than an infant if a day, a week, or even a month old. But suppose the case were otherwise, what would it avail? The question is not, Can they reason? Nor, can they speak? But, can they *suffer?*[32]

Bentham was part of a long chain of European writers, philosophers and political figures who deplored the pain that humans inflicted upon animals, including Pythagoras, Empedocles, Erasmus, Thomas More, Thomas Tryon, Montaigne and Percy Shelley.[33] And in the 21st century the cutting edge position occupied by organizations like PETA is that animals exist to "...occupy themselves and for no other reason. That humans take advantage of other animals in any way, simply because we are stronger or smarter, PETA sees as the abiding moral outrage of our time."[34]

But these early equalitarian or holistic arguments were blunted by a growing acceptance by the 17th century version of *Natural Law* which provided an increasingly secular rationale for human exceptionalism. For both the pagan Aristotle and the Christian Thomas Aquinas nature was purposive, dynamic and teleological, striving toward some defined end. Existence was neither arbitrary nor over-determined by accidents and the particularities and varieties of phenomena observed by humans were subsumed by a universal drive. Nature—including humankind—aimed at perfections. Taking their lead from Plato's theory of forms, natural law theorists held that each species in nature had a specific perfection to which it aspired. One can imagine here a fully developed redwood tree, a fully developed horse or a truly virtuous person.[35] As we will see, it was relatively easy for advocates of a more secular modernity to use this natural law tradition to eventually limit to humanity alone this natural drive to perfection, to make teleology or purpose, as it were, conscious and anthropocentric. By the 18th century, Mary Wollstonecraft could conclude that "The world requires, I see, the hand of man to perfect it."[36]

This is a crucial shift, from an early natural law theory which gave a kind of agency to all living forms trying to reach their particular perfection, to a natural law restricted to humans. Val Plumwood sees this shift as *the* crucial development leading to our specific version of modernity. Nature now "...lacks all goals and purposes of its own...Any goals or direction present are imposed from *outside* by human consciousness....Nature is neutral, indifferent and meaningless, with no interests or significance of its own, a mere endless hurrying of particles; any significance or value it might have for humans is an arbitrary product of human consciousness."[37] The utopian dreams of 18th century *philosophes* and revolutionaries, the progressive confidence of Victorian industrial culture and the transformational beliefs of socialists, communists and capitalists are all manifestations of this perfectionist anthropocentric vision. But is was modern science that really made the case for truth and for its accessibility, insisting that it was successful because its theories about the working of nature were true — not just logically or rationally, but actually.[38]

In the 20th century, many came to see belief in human exceptionalism and its accompanying teleology as dangerous delusions. For post-modern ironists like Richard Rorty all phenomena are hopelessly contingent, stemming as they must from events that could have been different. For the anthropologist Clifford Geertz humans are "by nature" just incomplete animals and dependent for all their powers not on any "nature" or teleology but rather on the disparate cultures they create. Rorty agrees, insisting that "...socialization, and thus historical circumstance, goes all the way down...there is nothing beneath socialization or prior to history which is definatory of the human."[39] And Darwinism, taken seriously as opposed to crudely, focused as we have seen on the contingent dimension of natural selection and rejected any idea of perfection or a direction to evolution. The unresolved tensions between these two fundamental views of nature and humanity — one progressive the other contingent — preoccupied the 20th century and have been passed on to us in the 21st.

But despite these very contemporary doubts about the possibility of any over-arching meaning or purpose to being, the dominant tendency in modernity has been to retain a commitment to some variant of this revised natural law tradition. In its purest form, belief in natural law implied a certain conservatism, a justification for the status quo. Thus for Aristotle some humans were slaves *by nature* and Greeks were superior to barbarians *by nature*. The gods had chosen to create a world that was rigidly hierarchical and segmented, with each component charged with striving to reach the ideal assigned to it *by nature*. But — and here is the key to all subsequent thinking about natural law and human nature — the gods had given humans the potential for reason, for rational thought. Barbarians, like other animals, still acted only on instinct in following natural law's dictates but those humans with reason were able to choose to acknowledge (or not) the wisdom of nature's plan.[40] Later,

COMING TO SENSE AND ENCOUNTERING SENSIBILITY

for Christians, the natural order or Great Chain of Being was seen as evidence of God's goodness and it was incumbent upon humans to choose to obey the laws of that system or, if they chose otherwise, to suffer damnation.

Well into the early modern era then, natural law continued to dominate understandings of the key human issues concerning the purpose and meaning of life, in effect affirming the presence of both. But with the increasing importance being attached to human reasoning powers as opposed to heavenly manipulations, the determinism and teleology implicit in a God-devised natural law began to wane. Humans were now being clearly seen as a separate case, unique among created things. And as we will see in the next chapter, serious debates began to rage on the issue of human nature itself, especially as it might have been revealed in a so-called state of nature, seen here as the prolonged era of the hunter-gatherers that we explored in the first chapter.

The "path taken" now begins to transform itself into a highway. The English Philosopher Thomas Hobbes (1588–1679) sets the tone in thus famous excerpt from the *Leviathan*:

> ...during the time when men live without a common power to keep them all in awe, they are in that condition which is called war; and such a war, as is of every man, against every man....In such condition, there is no place for industry; because the fruit thereof is uncertain; and consequently no culture of the earth; no navigation, nor use of the commodities that they be imported by sea; no commodious building; no instruments of moving....no knowledge of the face of the earth; no account of time; no arts; no letters; no society; and which is worst of all, continual fear, and danger of violent death; and the life of man, solitary, poor, nasty, brutish, and short.[41]

Homo homini lupus — man is wolf to man. These were "natural urges" reflecting the Christian commitment to the truth of an original sin, a Fall from Grace. For Hobbes and his contemporary Francis Bacon (1561–1626) this was a fundamental starting point in understanding the human condition, with sympathetic or beneficent (e.g. unnatural) behaviour being either a result of remarkable piety or a mastery of one's own nature by rational self-interest. Freud would provide a more secular explanation of the same understanding in the 20th century.[42]

Writing in the midst of a brutal civil war in England, Hobbes and his contemporaries were easily persuaded that human nature was less than serene and virtuous. All was not lost, however, and there was no need to lapse into despair, nihilism or a belief in "final days". Despite being driven by pride, selfishness and instinctual aggression, humans were also the only animals with speech, reason, foresight and a seemingly limitless curiosity — all key aspects of the human nature that was to replace the more benign human of ancient natural law theory.[43] Culture and Nature were to be "at war" within the human, with reason pitted against instinct, mind against body, and progress against chaos. These 17th century theories were only repeating the reason/emotion/appetite struggles in humans that Plato described in dialogues like the *Phaedrus*. There he used the metaphor of the human charioteer trying to control the contending drives of a black and a white horse. Plato's complex trio of reason, emotion and appetite was simplified by the Hobbesians with emotion and appetite being collapsed into one, thus preserving a dualist world view.

It should be clear by now what a fundamental challenge this dualist view is to my objective of finding crucial value in our deep historical past, indeed of possibly reconciling key aspects of that past with the modern. A dualist framework implicitly works against reconciliation—even against reciprocity—driving one instead toward resolution of the dualism by victory or, in the Hegelian sense, by a synthesis which gives rise to its antithesis—and hence a new dualism. It is possible of course to use the dualist frame to argue on the one hand that "nature" comprises what is universal, inherent and thus good and that the "...really universal elements in human nature are to be seen in their simplicity and purity only in savages or in primeval mankind."[44] Or, from the other more post-modern position, that nature has no moral content and issues of the good or the virtuous are always the negotiable and contingent products of human culture.

There can, however, be only a speculative dualism concerning the relationship between modern humans and their primitive ancestors. We still do not know for sure whether Neanderthals could speak and have no clear idea whether curiosity and a sense of the future, both attributes of a heightened consciousness, drove Homo sapiens onto the Eurasian steppe or if the movement was merely a response to climatic stimuli. But by the 17th century there was a more assured starting point concerning the claim for a unique human nature; the observed differences between humans and all other animals. With cognitive factors such as reasoning, speech and creativity increasingly seen as more defining than physical strength or spiritual claims about the presence or absence of souls, a solid case for human exceptionalism could more easily be constructed.

Rene Descartes (1596–1650), notorious now among environmentalists and advocates for animal rights for his insistence that animals were more akin to mere machines (biomechanisms) than to humans, led the charge in making the case for humans being both unique and the pre-eminent life form. In doing so, he acknowledged from the start the radical nature of his claim by rejecting the arguments of "certain of the Ancients" (e.g., Pythagoras) who held

COMING TO SENSE AND ENCOUNTERING SENSIBILITY

that animals did indeed have language (and hence thought) but that we were simply unable to understand it. This ancient claim is now, of course, very avant garde but Descartes in the 17th century would have none of it.[45] Instead he insisted that animals, even if they displayed signs of certain skills and abilities, in fact did not have "mind", only instinctual, mechanical responses: "... it is nature which acts in them according to the disposition of their organs."[46] Scholars have rightly focused on this assertion by Descartes as signaling a truly decisive shift in human relations with the rest of the natural world, first rejecting all claims for animals having souls and then dismissing mind and conscious feeling as well.

As Carolyn Merchant and others have pointed out, the work of Descartes, Bacon and their colleagues transformed the way we perceive nature—indeed the universe as a whole—from what had been a vast collection of living organisms into a machine that could be probed, taken apart, improved or disposed of at will. It is possible then for the non-human realm to become merely a collection of mindless, affect-less objects. This in turn cleared the way for a dramatic era of progress in science and civilization on the one hand and on the other ecological exploitation and eventual degradation.[47]

Despite their still deep religiosity, 17th century figures like Descartes and Bacon were beginning to anticipate the more secular version of this special status for humans. Bacon insisted that humans were the end of development, the final being following a long chain of creation. But he argued that humans were not created thus as in the Biblical story—rather the human animal only becomes human with the acquisition of reason and the resulting arts and technologies. Thus, "Just as it is inappropriate to charge carnivorous animals with murder, or blame dogs for incest, virtue and vice do not appropriately refer to man until he is both distinct from and higher than animals....and it is the arts and reason...which will help man to shoulder the responsibility of being the final cause of the world."[48] And this responsibility is to be governed, according to Bacon, by convention, by political efficacy, by will and—now consigned to the end of the chain of authority—by religious belief.

But what does all this mean for our understanding of the relationship between humans and the natural world? Descartes was not just an explorer of the nature of mind and soul but also a serial vivisectionist, exploring all aspects of body and brain through the dissection of animals. He knew through viewing the bodies of the animals he cut open that they looked very much like the bodies of opened human cadavers. He writes to the Marquess of Newcastle:

> Since the organs of their body are not very different from ours, it may be conjectured that there is attached to those organs some thoughts such as

THINKING THE WORLD TO PIECES | 125

we experience in ourselves, but of a much less perfect kind. To which I have nothing to reply except that if they thought as we do, they would have an immortal soul like us. This is unlikely, because there is no reason to believe it of some animals without believing it of all, and many of them such as oysters and sponges are too imperfect for this to be credible.[49]

So what are we to make of this? The hard logic of the Cartesean method shows us the way. Thinking, consciousness, and pain would all seem to require soul and since a sponge cannot reasonably be said to have a soul, it can have none of the other attributes either. And since there are no clear grounds to differentiate a sponge from any other living being except humans (who can reason and speak), then no animals have souls or mind.

Yet Descartes could not ignore the physical similarities between the bodies of many animals and those of humans. He avoided a crude and hard to sustain rigid dualism by arguing that such animals did not experience pain "in the strict sense", so that while pain may be present, it was abstract, not "consciously experienced." The pain such animals feel will not be perceived or thought of by them to be "part of them."[50]

A very reasonable exercise in reason one might say, akin to David Hume's skeptical comment that it was not contrary to reason to prefer the destruction of the whole world to the scratching of one's finger.[51] Starting from the unreasonableness of a sponge having soul and the difficulty of parsing the animal world between souled and soul-less, it was much easier to limit soul to humans who obviously possessed it. And since the evidence for soul was mind (I think, therefore I am), it must be reasonable that only humans have minds. But animals clearly had brains and nerve endings, so genuine pain must be a phenomenon of mind and hence not be present in animals, even though a pain-like sensation might remain but not be felt since feeling was a function of mind, not brain.

While I am not at this point going to contest this perspective, it may be useful to remind us briefly of the opposite view. One of the most "unreasonable" of commentators on the human-animal connection was the farmer-poet John Clare (1793–1864). We read earlier the ending of his poem on the Badger in which all distinctions are blurred.

> They get a forked stick to bear him down
> And clap the dogs and take him to the town,
> And bait him all the day with many dogs,
> And laugh and shout and fright the scampering hogs.
> He runs along and bites at all he meets:
> They shout and hollo down the noisy streets.

126 | COMING TO SENSE AND ENCOUNTERING SENSIBILITY

All feel pain and all are animals and all are thinking, feeling and in some fundamental way soul-less. Clare is careful not to anthropomorphize too much, the badger "grinning" at one point but otherwise behaving much as on might expect an animal to behave in such circumstances. But he was cruelly done by in this rural scene of what Plumwood calls radical exclusion. Not unlike Pliney's elephants at the Roman Circus, the fate of Clare's badger remains a powerful indictment of human culture.

The 17th century advocates for animal rights and vegetarianism had relied on winning the debate over whether animals had souls. But their case collapsed in the face of the instrumental reason of Descartes and Bacon and the scientific advances based on the use of that reason that followed. Later activists like Shelley would rally to Clare's (and the badger's) defence in the 19th century by arguing that the violence we inflict on animals (and on other persons) whets our appetite for violence against each other. "By choosing to sustain ourselves on animal food we have chosen to wage a kind of war on nature, *torture* and exploit animal life, and to introduce physically and spiritually corrupting influences into our lives."[52]

Shelley's critics argued that the badger's alertness and instincts for self-preservation are only the primitive building blocks of a much more complex nature. Humanists from the ancient world to the present reject such anthropomorphizing as Clare's badger (or for that matter, the penguins in the recent popular film *The March of the Penguins*), insisting that for animals "...there is no reflection or anticipation, but only an immediate awareness of the environment and a reaction to it."[53] Echoing the Roman philosopher Cicero, the modern social ecologist Murray Bookchin argues that the qualitative "...ontological divide between the non-human and the human is *very* real....animals...adapt, while human beings...innovate."[54]

The assertion in the 17th century that reason made humans truly unique was only the first step toward a sense of mastery over nature, a mastery that remained limited by the restraints implicit in the still pervasive religiosity or spiritualism of the times. Descartes' celebration of reason had assumed that "I think, therefore I am", but it also took as a given the existence of God. The third of his *Meditations*, "Of God; That He Exists", seems to many modern readers an odd and even illogical accretion to an otherwise aggressively rational and secular argument. But in the 1630's there was a more powerful political logic that persuaded Descartes to make this connection, writing as he was in the midst of the great Catholic Counter-Reformation.

In England, Descartes' contemporary Francis Bacon had more room to maneuver and it was here, starting with his *Novum Organum* (new Instrument or Method) that we can see the real beginning of instrumental reason's

THINKING THE WORLD TO PIECES | 127

triumph over both revelation and rationalism. Bacon's formulation of reason — now manifested in the form of science — served to prove the validity of the Biblical granting to humankind dominion over the Earth. Bacon in effect used a theological argument to support the uninhibited pursuit of secular science, insisting that a new applied and experiment-driven science "... could help to restore a dominion over Nature that had been intended for humankind but which had been lost through the Fall."[55] Leiss describes Bacon's argument as "irresistible", with scientific success becoming "signs" that humankind was once again Right with God.[56]

We can pause and look closely at the significance of these shifts, especially at the work of Francis Bacon whose contribution often seems to be overshadowed by the Cartesean *cogito*. Descartes, sitting alone by the fire in his Dutch refuge from Catholic censorship, had an epiphany, a kind of oceanic experience in which he saw the truth of the *cogito* in all its wonderful simplicity:

> I resolved to pretend that nothing which had ever entered my mind was any more true than the illusions of my dreams. But immediately afterwards I became aware that, while I decided thus to think that everything was false, it followed necessarily that I who thought thus must be something; and observing that this truth: *I think therefore I am*, was so certain and so evident that all the most extravagant suppositions of the skeptics were not capable of shaking it, I judged that I could accept it without scruple as the first principle of the philosophy I was seeking.[57]

While he was in fact a practicing scientist, even more so than Bacon, Descartes remained more a rationalist than an empiricist in the Baconian sense. Bacon's work, on the other hand, clearly pointed the way toward empirical research, a defined scientific method and a stress on experimentation, an approach designed to "command nature in action"[58] in order to enlarge "...the bounds of Human Empire, to the effecting of all things possible."[59]

To make "all things possible" required a new world view, one based not merely on observations or sense experience, but rather on theories, patterns or schemata imposed on sense experience. These theories were in turn derived not from visionary insights or tradition-bound concepts but from careful observation coupled with experimentation. The observations of Copernicus (1473–1543), confirmed by Galileo (1564–1642) via the telescope had in effect made a mockery of mere sense perception. The heavens did not move in the way that our eyes or our authorities told us they did.

And so there is a strange reversal in the way we discover the truth about the real world. In the ancient and medieval cultures of the West and the Middle East to know the truth was to "see" it, to grasp it via either sense perception or by rational intuition.[60] With the work of the great 17th century trio of Descartes, Galileo and Bacon the senses and intuition became suspect and prone to error, seen as being linked too closely to the bias of the observer and the shifting traditions of the age. The focus of the new science was on a more intense, objective observation and on the imposition of Platonic-like models or forms that are seen to exist in mathematical formulations behind the observed phenomenon. Thus "...a realm of eternally unchanging objects (the objects of mathematical physics) exists *behind* the fluctuating and deceptive realm of sense experience."[61] The world we actually (or seem to) live in, the world of the senses, must therefore be subordinate to this 'real' world of true being that lies behind it, a world made palpable not through logic, religion or vision, but through science.[62]

So reason takes a strange turn here. A rational person might have concluded that the information being received from senses and from seemingly logical deductions was the best way to judge the truth of something. But this kind of reason (or rationalism) was no longer to suffice. Instead, this fluid realm of direct experience of 'living nature' was seen to be a secondary, derivative dimension, a "...mere consequence of events unfolding in the *realer* world of quantifiable and measurable scientific *facts*."[63] In his famous dialogue *The Starry Messenger*, Galileo ridiculed the reasoning of Simplicius, his advocate for an Earth-centered universe, and insisted instead on the "irreducible and stubborn facts" revealed by his telescope.[64]

To accomplish this understanding of the reality behind the senses a scientific method was required, one which purged the truth-seekers of all subject-relative properties such as beauty, utility, goodness, colour, and smell, "...in order to isolate the properties that are essential to the thing as it is in itself—namely the features of a thing that can be quantified, such as mass, velocity, and position."[65] What remains, then, is essentially mathematics, the one tool that could fulfill Descartes' need for the "clear and distinct" ideas necessary to perceive reality.[66]

This scientific way of seeing would soon infuse others areas of inquiry in which truths were sought. The 17th century political philosopher John Locke (1632–1704) is described by his biographer as particularly unimaginative in his observations, insisting on rejecting both traditions and private visionary insights in order to discover only the "...plain, measurable, publicly verifiable facts."[67] No mystery now how Charles Dickens finds his Thomas Gradgrind, the teacher in *Hard Times* who drilled his students with facts, "...nothing but Facts. Facts alone are wanted in life. Plant nothing else and root out everything else. You can only form the minds of reasoning animals upon Facts."[68] For Locke it was clear that nature had no intrinsic value, and acquired value only when it became property useful to humans.[69]

The issue being confronted here was Freud's issue in his rejection of Rolland's oceanic feeling. Despite attempts from Plato to Aquinas to find an objective basis for fundamentally subjective aesthetic ideas like beauty, for the 17th century advocates of a new, instrumental reason such qualities remained too dependent on the psychological state of the observer to be within the purview of science or reason. Feelings, such as those derived from an aesthetic response to nature, were seen as passive, not active, grounded in the "animal" part of us, not in the reasoning that both separates us from and elevates us above other animals. Such feelings are of the body, not the mind.

The romantic poet/philosopher/scientist/writer Johann Wolfgang von Goethe (1749–1832) provides us with a classic illustration of the perils and/or reality of this subjectivity in his classic novel *The Sorrows of Young Werther*. He has his young protagonist Werther respond directly to nature in two distinct passages, the first when he is happily in love with his Lotte, the second four months later when that love has been thwarted by her decision to marry another. In the first passage (Letter of 10 May), his feelings for nature are reminiscent of Rousseau's ecstatic feelings floating on Lake Bienne:

> When the beloved valley steams around me, and the lofty sun rests on the surface of the impenetrable darkness of my forest, with only single rays stealing into the inner sanctuary, then I lie in the tall grass beside the murmuring brook, while on the earth near me a thousand varied grasses strike me as significant; when I feel the swarming life of the little world between the grass blades, the innumerable, unfathomable shapes of the tiny worms and flies, closer to my heart, and feel the presence of the Almighty, who created us in his image.....I succumb to the overpowering glory of what I behold.

Then, only three months later when his mood or mental state has changed to a deep depression, nature appears to him in quite a different guise:

> There is no moment which does not consume you and yours with you, no moment when you are not, and of necessity, a destroyer; the most innocent stroll costs a thousand crawlers their life, as one step destroys the laborious structures of the ants, trampling a little world into an ignominious grave. Ha! it is not the great, rare catastrophes in the world, the floods that wash away your villages, the earthquakes that engulf your cities, which touch me; what undermines my heart is the consuming power which lies hidden in the whole of nature; power which has formed nothing that does not destroy its neighbour, destroy its own self. And so I stagger on in terror! Heaven and

earth and their interplaying forces all around me! I see nothing but a monster which, eternally swallowing, chews its eternal cud.[70]

Hence the perils in allowing mere observation or the senses to come to conclusions about reality. The nature Werther observes has not significantly changed from May to August, but the changes within the observer have made it seem so. Goethe was linked to an early 19th century natural philosophy movement that took issue with the cold and restrictive reason of Francis Bacon. Most popular in Germany, this approach attempted a blending of facts and feeling, objectivity and subjectivity in understanding natural phenomena. Thus despite the weaknesses inherent in Werther's emotion-driven observations, his opposite in the story, the steady and rational Albert, is no better an observer. But this attempt to blend or allow a co-existence of feeling with reason proved poor competition to the ability of a more instrumentalized reason to dominate the new modernity. For Bacon and Descartes, only a scientific method grounded in a very instrumental notion of reason could provide a safe path through this maze of observer bias, outdated traditions and religious restrictions.[71]

But what is this reason that Bacon, Descartes, Galileo, Machiavelli and the other 17th century moderns were privileging? We need to reflect in a more systematic manner on this human talent for reason and its role in creating our contemporary modernity. Humans, after all, have been thinking and reasoning for millennia but with widely varying conceptions of reason's relative weight or value in constructing and defining cultures and societies. We need to have a clear idea of what the modern world came to understand as reason. Specifically, we need to think about how the reasoning capacity of humans became in the modern West this kind of "instrumental reason", a focusing of mind on the "...calculation of efficient means to given ends..." which has in turn spawned unprecedented increases in knowledge, power, prediction and control while consigning issues of ultimate purpose, the good and the beautiful to precarious subjectivities.[72]

It is the means taking precedence over ends that is most characteristic of this new, instrumental reason. No longer are there universal goals of the good or the beautiful toward which reason strides, no broad framework of values guiding scientific exploration. Instead means such as efficiency, cost-effectiveness and utility become ends. Rationality is recast as "...the calculation of efficient means to given ends. The narrowing of focus has meant an unprecedented increase in knowledge and power, prediction and control, but it has also meant that ultimate purposes, the good and the beautiful generally, have become a matter of only subjective conviction."[73] Stanley Diamond, who we heard from earlier in reference to the "primitive", sees this as a reduction of reason to a functional rationality, a privileging of achievement which subsequently comes to characterize all that is important in human existence.[74] The rational, in the words of William Leiss, hence becomes limited to "...that which is serviceable for human interests."[75]

So we return to reason and the nature of modernity. Rousseau was a "Man of Reason" but it was severely tempered by his savouring of reverie and insistence on an essential subjectivity to truth. For the more powerful and iconic voice of reason we turn to Immanuel Kant (1724–1804). Bacon, Descartes, Newton, Galileo and other heroic figures in the modern pantheon broke the back of the medieval commitment to sustaining the core traditions of Greek and Roman thought. It was the work done in the 18th century Enlightenment that completed the transition to a new instrumentalized, empirical and pragmatic reason. For Kant, as for his ancient predecessors and enlightened contemporaries, reason was "...man's release from the womb of nature."[76]

So here we are again! There was a womb, a pre-birthing era in the human story — pre-reason and in a sense pre-being. And after this birth humans came to see themselves as the true *end* of nature and that nothing "...that lives on earth can compete with him in this regard [and] his nature privileged and raised him above all animals. And from then on he looked upon them, no longer as fellow creatures, but as mere means and tools to whatever ends he pleased."[77] Potentially humankind knows no limits once the chains of empathy and holism are loosed and religious proscriptions voided. The other-than-human world becomes completely value-free and nature becomes entirely a means to human ends, leading inexorably to a "fundamental alternation to the life of man in Western Civilization."[78]

So the first step is taken: humans are qualitatively unique. The philosophers of the early modern period in Western history have asserted that humans are the final end of species development, the final objective of creation. By what criteria can such a claim be made? Kant, like so many of his predecessors, started with reason. The debates of the 17th century had been sidetracked by the issue of soul and could not overcome the difficulty of proving or disproving the existence of soul in beast as well as human.

Reason, though, held out more promise. Descartes had already claimed that "sponges feel not" and "beasts reason not" followed on closely. But what did it mean to reason or to possess reason? For Kant, reasoning beings "...stand out against the background of nature, just in that they are free and self-determining...Everything else in nature...conforms to laws blindly. Only rational beings conform to laws that they themselves formulate."[79] It is our will, not mere inclination, that shall drive human action. This is a powerful

132 | COMING TO SENSE AND ENCOUNTERING SENSIBILITY

dualism, an alexandrine sword slashing at the knot of species continuity, with individual self-determination and its accompanying notion of freedom the decisive measure of value.

While this places humans, in Kant's words, "...infinitely above all other things on earth...", it was no license for abuse.[80] If for no other reason than to improve our own moral worth, we are advised by Kant to treat animals humanely and with consideration. This is an important codicil to the rule of humans over nature, reminding us of the instructions in Genesis to be stewards of nature. Earlier humanists like Spinoza argued that despite the likelihood of animal sentience, the privilege of reason condoned us "...making use of them as we please."[81] But for Kant, Rousseau, Shelley and some other modern thinkers, compassion for sentient animals was one of the components for embedding compassion within human cultures. Indeed Kant even speculated that while God ("nature" really) may consider man more important than the lion, He/it would still not likely prefer one man over the entire species of lions.

But this sensitivity to humaneness must not distract us from the real import of Kant's privileging of reason — of thinking, consciousness, and will. Bacon had asserted that man's intellect, his reason, would be a force for the liberation of humans from nature: "...by art and the hand of man she [nature] is forced out of her natural state, and squeezed and moulded."[82] Building on this idea, Kant two hundred years later proclaimed the possibility of a new human nature, one in which reason not just represses instinct, but in fact replaces it with a new, second nature.[83] So the development of reason means, in the Enlightenment tradition inspired by Kant, the growth of human faculties beyond the desires that we share with the other animals. This "ultimate moral end of the human species" will require, Kant argues, human domination of nature by science and domination of the self by reason.[84]

This perspective on the relationship between humans and nature — both external and internal — is based on what Mary Midgely has called the "strong anthropic principle", the idea that "..the universe does indeed have a central purpose, namely the exaltation of man."[85] Kant's ultimate moral end for humanity, then, is to use their powers of reason to pursue progress in all aspects of life in order to pursue the end toward which civilization is moving.[86]

But Kant was as much a moral philosopher as he was the founder of theories of practical reason. We have seen his concern that human uniqueness not be used as a rationale for cruelty and we know from his other works that he thought a great deal about the importance of beauty and the sublime. He regarded Rousseau's understanding of the importance of feeling, morality and reverie as crucial components of the enlightened modernity he hoped would develop in the 18th century. And he remains in university departments of philosophy *the* central spokesman for reason and enlightenment. But reason applied, scientific reason, moved in quite different directions under the more instrumental influence of the mechanics, inventors, scientists and entrepreneurs who followed Bacon and Descartes.

Instrumental reason or what is sometimes referred to as "formal rationality" or "subjective reason" and manifested in praxis as scientism and positivism, moved some distance from Kant's philosophic ideals. The issue of the ends to which actions might lead, absolutely central to Kant and other Enlightenment thinkers, took second place to means. The new scientism or instrumental reason, critical theorist Max Horkheimer argued, "...does not ask whether ends are intrinsically rational, only how means are fashioned to achieve whatever ends may be selected."[87] Advancements in knowledge via instrumental reason are thus concerned solely with questions of practical use. Bacon's demand that only facts be used became the hallmark of modern science, with only that which could be measured being considered facts.

The new scientific instrumental reason that emerged in full bloom in the 19th century was concerned with two dimensions of reality; on the one hand the other-than-human natural world and on the other the cultural and personal world of humans themselves. Most commonly the impact of instrumental reason in our relations with the natural world has been seen in terms of "disenchantment", an idea first suggested by Max Weber and more recently popularized by the work of Carolyn Merchant. Charles Taylor in his *Sources of the Self* calls instrumental reason a "neutralizer of the cosmos", nature ceasing to serve as an arbiter of the good since it is now seen as a mere mechanism — a means rather than an end.[88] Reason becomes the tool that humans can use to control and dominate a nature that is increasingly seen as inherently hostile to humanity — a mechanism with no moral sense, only chaos or self-interest.

As a machine, or a collection of machines, nature quickly comes to be seen as potential property, its value no longer intrinsic but rather linked to its potential use; mechanisms that need to be tamed and brought into the realm of human culture.[89] Zygmunt Bauman describes this using a garden metaphor, the objective of science being to use its awesome power to first control and then improve on nature, to construct a garden.[90] And the garden idea had a powerful grounding in religion as well, with Christianity in particular determined to break the primitive and pagan flirtations with "sacred groves" and holistic world views in general. Instead, a tamed nature existed to serve the needs of enlightened humans. John Locke expressed the connection clearly:

> God gave the world to men in common; but since he gave it them for their
> benefit and the greatest conveniences of life they were capable to draw from

134 | COMING TO SENSE AND ENCOUNTERING SENSIBILITY

it, it cannot be supposed he meant it should always remain common and uncultivated. He gave it to the use of the industrious and rational—and labor was to be his title to it—not to the fancy or covetousness of the quarrelsome and contentious.....hence subduing or cultivating the earth and having dominion, we see, are joined together. The one gave title to the other.[91]

And so nature becomes property, an object of mastery and possession. And why not? Anticipating Darwin, Charles Lyell in his *Principles of Geology* (1833) built on Locke's argument in reasoning that in "...obtaining possession of the earth by conquest, and defending our acquisitions by force, we exercise no exclusive prerogative." All species, Lyell insists, engage in such struggle and have "slaughtered their thousands" as they spread their influence.[92] Ralph Waldo Emerson, who we encountered earlier transcending the world of being and will encounter later trying to enjoy wild Yosemite with John Muir, waxes eloquent in urging the transformation of wild America into a garden:

> This great savage country should be furrowed by the plough, and combed by the harrow; these rough Alleganies should know their master; these foaming torrents should be bestridden by proud arches of stone; these wild prairies should be loaded with wheat; the swamps with rice; the hill-tops should pasture innumerable sheep and cattle....How much better when the whole land is a garden, and the people have grown up in the bowers of a paradise.[93]

Even modern critics of human pretensions to dominion over nature acknowledge that the attempt to dominate one's environment is a "necessary process and one which man shares with all the more highly developed forms of organic life".[94] As we will see, not all critics are as forgiving as Leiss on this matter, but it does seem clear that at some level human aspirations to control their natural environment are a matter of just "doing what comes naturally".

But the transformation of consciousness and language into reason and from there into instrumental reason with its cadres of scientists has resulted in a degree of human domination that has created a possibly counter-productive 'culture of mastery'. It is now asserted that we humans are "...as powerful as the moon", our installed power plants having a capacity comparable to tidal power.[95] E.O. Wilson argues that we have achieved the dubious distinction of becoming a "geophysical force" unequalled by any previous species, a force able to alter the atmosphere itself.[96]

Tzvetan Todorov worries that after mastering the nature we know, the instrumental reason and its scientific-technical apparatus that so dominates contemporary modernity will begin to "...envisage that another reality, better adapted to our needs might emerge...", perhaps in the form of gene therapy, artificial selection and plastic trees.[97] So despite myriad speculations about utopian destinations made possible by this mastery, surely we know now that an unfettered instrumental reason has created as many problems as it has solutions and in the opinion of many is the direct cause of what we see now as an ecological crisis.

Michel Serres, scientist and cultural critic, sets us a new task for 21st century reason. With modernity, he says, we have resolved the issue of how to dominate the world. The key question we face now is "...how can we dominate our domination; how can we master our own mastery?"

> Being masters imposes crushing responsibilities, suddenly driving us far from the independence we so recently believed would henceforth be the bed of roses of our new powers. From now on we are steering things that, in the past, we didn't steer. In dominating the planet, we become accountable for it. In manipulating death, life, reproduction, the normal and the pathological, we become responsible for them. *We are going to have to decide about everything....* about the physical and thermodynamic future, about Darwinian evolution, about life, about the Earth and about time......[98]

If Serres' analysis is correct, then we need to turn our attention to the human side of the equation, because this is in fact the central issue being raised in this book: has modern culture equipped humankind with the appropriate perspectives, values and tools needed to address the ecological challenges provoked by this new modernity? Can we find within modernity the mental as well as material resources — the cultural tools — to create ecologically sustainable human cultures? Or, to borrow a phrase from Isaiah Berlin, is the human timber present after nearly 300 years of a culture built on the dominance of instrumental reason, scientism and technology too crooked to rely upon?[99]

Many would say yes to that question, arguing that the "unbound Promethean" dimension is at the core of modernity, enabling humans to create a material culture that "...continually enhances our capacity to transform and manipulate non-human nature" and as a by-product threatens an ecological catastrophe.[100]

While it is true that we have "...lost the contact with the earth and its rhythms that our ancestors had...", it seems equally true that the "...possibility of a radically different ethic or a new value system separate from and independent of the Enlightenment mentality is neither realistic nor authentic."[101] Terry Hoy insists, along with Jonathan Bate, Tu Weiming, Charles Taylor and others that we need instead to find ways to broaden the Enlightenment Project, to critique its values and find a "..corrective to the distortion in the instrumental-utilitarian tradition".[102] After all, along with its technocratic

136 | COMING TO SENSE AND ENCOUNTERING SENSIBILITY

and industrial distortions, the Enlightenment also bequeathed to us progressive ideas of justice, political liberty, altruism, and a focus on the dignity of the individual. But where should we start with the corrective?

We can start with Homo Economicus, the version of human nature that lies at the heart of modernity as defined by economists, technocrats, media, politicians and, not surprisingly, most participants in that modernity. This perspective sees humankind as:

> ...naturally acquisitive, competitive, rational, and calculating and is forever looking for ways to improve his material well-being. He rations his time from an early age to get the training needed to earn an income, and he carefully allocates this income among the dizzying array of goods and services available in the marketplace......We believe that to want more things is a natural human attribute. We value the individual above society. Competition and expansion, not cooperation and stability, describe the rules by which our world operates....Not only do individuals prefer more to less, but those with more possessions are accorded higher status in market societies..... One's value as an individual is largely a function of economic success, of accumulating (and consuming) more wealth than does one's neighbor.[103]

Add to this Freud's influential argument that we are naturally (instinctively) aggressive—homo homini lupus—and we end up with a very problematic animal, at once atomized, alienated, acquisitive and aggressive.

It is easy to see, given this model of human nature, why a highly instrumentalized and focused approach to reason has proven so popular in a culture dedicated to expansion and achievement. Achievement of these "economic" ends does, however, come at a price. Since mobility is one of the means by which these ends are realized, community itself becomes an instrument instead of a "home place", returning to a kind of hunter-gatherer relationship with place. Nature as well, once a home in a more holistic sense but now shorn of any spiritual or inherent value, becomes simply a collection of material objects upon which to act in order to achieve human ends. This version of human nature, based on a drive to achieve at all costs, is then in an antagonistic relationship to the human nature we explored in the previous chapters and with our concern for ecological sustainability. A culture constructed around the ideal of Homo Economicus, despite its claims to be an engine of progress, can in the end only constrict the range of human possibilities. By transforming nature into a realm of objects we risk becoming objects as well, an "...alienated *thing* in a world of other, equally meaningless things."[104]

But of course nothing in a phenomenon as complex as a culture is ever as clear as this. There are countervailing pressures in modernity as we have seen and will explore in more detail in subsequent chapters. But this idea of progress, of science and instrumental reason being the key to civility, to assuring the triumph of the civil over the barbaric, peace over terror, or truth over superstition remains the most powerful intellectual and cultural "story" of our era. And central to this story is the idea that these progressive accomplishments are the result of the "...gradual yet relentless substitution of human mastery over nature for the mastery of Nature over man".[105] Whatever problems remain are simply the result of not enough civilization. Thus in our time the defeat of Terrorism requires more cameras in public places and more sophisticated intelligence services, just as bringing modern democracy to Baghdad requires more bombs and soldiers.

In his impressively summative book on *Civilizations*, Felipe Fernandez-Armesto senses a "loss of nerve" within this version of modernity, a collapse of self-confidence in Western superiority. The current conflict (civil war?) in Iraq, the scatological terror ranging from Timothy McVeigh to the Tamil Tigers, the IRA and al-Qayeda all seem to bear out his very post-modern insistence that the rational, instrumental planning so central to modernity faces in the 'real world' a "...genuinely chaotic system:"

> In reality, planning almost always went wrong. Science proved more efficient in equipping evil than in serving good. 'Scientifically' constructed societies turned into totalitarian nightmares....Even the successes of genuine science were equivocal. The motor car and the contraceptive pill did wonders for individual freedom, but they also threatened health and challenged morals. Industrial pollution could choke the planet to death. Nuclear power could save the world or destroy it. Medical advance has encumbered us with imbalanced and unsustainable populations, while disease-bearing organisms evolve immunity to our antidotes.[106]

A jaundiced view to be sure and vigorously disputed by those who would defend the progressive developments that characterize much of modernity.

We are left with two questions for our interlocutors Rousseau and Freud. For the latter, we need to wonder whether the science and reason he so brilliantly espoused and furthered really must expunge from human experience the "oceanic feelings" that his friend reported sensing. These deeply subjective "feelings" for the spiritual and for nature in its most material landscape sense seem too integral to human nature itself to be abandoned so readily. For Rousseau we need to explore his seeming primitivism, his call for a return to nature. Was not Rousseau perhaps "...presenting the model of an earlier and a happier time in an effort to humanize the society in which he lived."[107]

COMING TO SENSE AND ENCOUNTERING SENSIBILITY

NOTES

1 Shelley, "Vindication of Natural Diet", in David Lee Clark, p 83.

2 Jean-Jacques Rousseau, *Confessions* (London: Penguin, 1953) p 594.

3 Rousseau, *Reveries*, p 89.

4 Wordsworth, from *The Recluse, Part I*, in Colette Clark, p 27.

5 Hadot, p 215

6 Cartmill, p 117.

7 Paul Mendes-Flohr, *Martin Buber: A Contemporary Perspective* (Syracuse: Syracuse University Press, 2002), p 10.

8 From the conclusion to Muir's Essay "Twenty Hill Hollow", cited in Michael Cohen, *The Pathless Way: John Muir and the American Wilderness* (Madison: University of Wisconsin Press, 1984), p 212.

9 Elaine Scarry, *On Beauty and Being Just*, reviewed by Stuart Hampshire, *New York Review of Books*, 18 November 1999, pp 42–45), p 42.

10 Taylor, (1991), p 91.

11 "Poets are the hierophants of an unapprehended inspiration; the mirrors of the gigantic shadows which futurity casts upon the present...". Percy Shelley, "A Defence of Poetry" in D.L. Clark, p 297. Note: he was quite liberal in his definition of poets.

12 Berman says that Nazism was the only organized political attempt to replace the Enlightenment model, but unfortunately was a "satanic version of holism". Morris Berman, *Coming To Our Senses* (New York: Bantam, 1989) p 298.

13 Sigmund Freud, *The Future of an Illusion* (trs and edited by James Strachey, New York: Norton, 1961), p 8.

14 ibid, p 19.

15 ibid, p 22.

16 ibid, p 40.

17 Peter Gay, *Freud: A Life for Our Time* (New York: Norton, 1988), p 544.

18 Sigmund Freud, *Civilization and Its Discontents* (New York: Norton, 1961), p 10–11.

19 Torgovnick, p 207.

20 Freud, *Civilization and Its Discontents*, p 73.

21 Kate Soper, *What Is Nature?* (Oxford: Blackwell, 1995), p 25.

22 Augustine's early adherence to Manichean doctrine, of course, helped in establishing his dualist position.

23 John Passmore, "The Treatment of Animals", *Journal of the History of Ideas*, v.36:2, 1975 (pp 195–218), p 199.

24 John Mizzoni, "St. Francis, Paul Taylor and Franciscan Biocentrism", *Environmental Ethics* v. 26:1, 2004 (pp 41–56), p 47.

25 Merchant's interpretation of the Medieval era is controversial. See, for instance, Brian Vickers, "Francis Bacon, Feminist Historiography and the Dominion of Nature" in *Journal of the History of Ideas*, v. 69:1, 2008 (pp 117–141).

26 "...our remote ancestors seem to have accepted without question that they were part of the great animal continuum. For the earliest creatures we might reasonably classify as human, this was actually an observable fact, since for most of the hominid past they co-existed with other, similar species. Primitive wisdom, moreover, was bound to defer to animals bigger, stronger, or faster than humans. Animals who were enemies were treated with awe, those who were allies with admiration.....They (mesolithic hunters) accepted their dogs as full

THINKING THE WORLD TO PIECES | 139

members of society, burying them with the spoils due to prowess and, in some cases, with more signs of honour, such as gifts of rare, blood-red ochre and spoils of the hunt, than are found in the graves of men. For most of the past, people adopted totemic ancestors, worshipped animals or zoomorphic gods, and clad their shaman-elites in animal disguises." Fernandez-Armesto, p 38. See also, Nurit Bird-David in Gowdy.

27 Charles Taylor defines instrumental reason as "...the kind of rationality we draw on when we calculate the most economical application of means to a given end. Maximum efficiency, the best cost-output ration, is its measure of success." Taylor (1991), p 5.

28 Horkheimer and Adorno, p 61.

29 Leiss, (1994/1972), p xi.

30 Martha Nussbaum, *Upheavals of Thought: The Intelligence of Emotions* (Cambridge: Cambridge University Press, 2001) p 89.

31 Cartmill. He notes that Montaigne even saw religion at work with animals, ants reverently bringing home the bodies of their dead and elephants saluting the rising sun.

32 Bentham, Jeremy. *Introduction to the Principles of Morals and Legislation*, in Burns, J.H. and Hart, H.L.A. (eds.) *The Collected Works of Jeremy Bentham.* (New York: Oxford University Press, 1996), p 283.

33 See Tristram Stuart, *The Bloodless Revolution: A Cultural History of Vegetarianism from 1600 to Modern Times* (New York: Norton, 2006).

34 Michael Specter, "The Extremist", *The New Yorker*, 14 April 2003 (pp 52–67), p 54. The philosopher and animal rights supporter Peter Singer is perhaps the best known voice in this cause.

35 Mathew Illathuparampil, "Normativity of Nature" (pp 225–235) in Drees, p 228. Even Aristotle, who rejected the idea of a perfect 'form' for each phenomenon in nature, insisted that each individual member of a species contained within itself a potential for perfection.

36 Wollstonecraft, p 122.

37 Plumwood, (1993), p 110.

38 Stephen Clark, "Have Biologists Wrapped Up Philosophy?", *Inquiry* , v. 43, 2000 (pp 143–66), p 153.

39 Richard Rorty, *Contingency, Irony, and Solidarity* (Cambridge: Cambridge University Press, 1989), p xiii.

40 Erazim Kohak, *The Embers and the Stars* (Chicago: University of Chicago Press, 1984), p9.

41 from *Leviathan*, cited in Richard Ashcraft, *Hobbes's Natural Man: A Study in Ideology Formation*, *Journal of Politics*, v.33, 1971, p 1096.

42 "....men are not gentle creatures who want to be loved, and who at the most can defend themselves if they are attacked; they are, on the contrary, creatures among whose instinctual endowments is to be reckoned a powerful share of aggressiveness. As a result, their neighbour is for them not only a potential helper or sexual object, but also someone who tempts them to satisfy their aggressiveness on him, to exploit his capacity for work without compensation, to use him sexually without his consent, to seize his possessions, to humiliate him, to cause him pain, to torture and to kill him." Sigmund Freud, *Civilization and Its Discontents*, p 68.

43 John Scott, "The Theodicy of the Second Discourse: The Pure State of Nature and Rousseau's Political Thought", *American Political Science Review*, 86:3, 1992 (pp 696–711); John Sheriff, *The Good-Natured Man: The Evolution of a Moral Ideal, 1660-1800*, (Montgomery: University of Alabama Press, 1982).

44 Lovejoy and Boas, p 112.

140 | COMING TO SENSE AND ENCOUNTERING SENSIBILITY

45   See for instance, Alasdair MacIntyre, *Dependent Rational Animals* (Chicago: Open Court, 1999); Joanne Lauck, *The Voice of the Infinite in the Small: Re-Visioning the Insect-Human Connection* (Boston: Shambala Press, 2002), and Kohak.. Felipe Fernandez-Armesto has made this a central issue in his books on humankind: "...the animal rights movement has been remarkably successful in challenging us to identify what, if anything, entitles us to privileged treatment, compared with other animals. Beyond human diversity lies an indistinct frontier between human and animal realms which were once thought to be mutually exclusive. In modern times, the problem of distinguishing humans from animals has inspired some radical solutions. According to Descartes, animals resembled machines, whereas in the human machine there was a ghost. His followers went further: the cry of a beaten dog was no more proof of pain than the sound of an organ when its keys were struck.....In the present state of knowledge of the continuities between humans and other animals, it is impossible to sustain any account of human nature as sharply differentiated as Descartes". p 2–3.

46   Rene Descartes, *Discourse on Method* (London: Penguin, 1968), p 75. For Descartes, there were only two kinds of material in existence, thinking matter (*reg cogitans*) and extended matter (*reg extensa*) and animals were in the latter category.

47   Roderick Nash, *The Rights of Nature*,(Madison: University of Wisconsin Press, 1989) p 18. See also Carolyn Merchant, *The Death of Nature*.

48   Heidi Studer, "Strange Fire at the Altar of the Lord: Francis Bacon on Human Nature", *The Review of Politics*, v. 65:2, 2003 (pp 209–235), p 221. Rousseau will later agree, insisting that primitive humans in pre-social conditions might have been 'happy', buy could not yet be virtuous.

49   Letter to the Marquess of Newcastle, http://pubpages.unh.edu/~jel/descartes.html. Cavendish was a prominent opponent of anthropocentrism, arguing that animals had a full range of passions and their own reason and language. Keith Thomas, p 128. Recent research has concluded that it is unlikely, for instance, that lobsters or other invertebrates feel pain despite their unseemly thrashing about when put in boiling water. *Vancouver Sun*, 16 February 2005, p A13.

50   John Cottingham, "A Brute to the Brute? Descartes' Treatment of Animals", *Philosophy* (v. 63, 1988, pp 175–183). And Tom, Regan, *The Case for Animal Rights* (Berkeley: University of California Press, 1983).

51   David Hume, Treatise of Human Nature, Book 2, Section 3 (http://www.class.uidaho.edu/mickelsen/texts/Hume%20Treatise/hume%20treatise2.htm#PART%20III) Hume was not, of course, advocating such an egoistic stance, since in his next line he insisted that "Tis not contrary to reason for me to chuse my total ruin, to prevent the least uneasiness of an Indian or person wholly unknown to me."

52   Onno Oerlemans, *Romanticism and the Materiality of Nature*, (Toronto: University of Toronto Press, 2002), p 115. This was, of course, not a new argument. Shelley is borrowing from Pythagoras who wrote in the 6th century BC.

53   Morris Berman (2000), p 37.

54   Murray Bookchin, (1995), p 16. See also Lovejoy and Boas, p 250 for reference to Cicero.

55   John Hedley Brooke, "Improvable Nature?" in Drees, p 160.

56   Leiss, (1990), p 78.

57   Descartes, p 53.

58   Francis Bacon, *The Great Instauration* in J. Weinberger, ed., Francis Bacon, *The Great Instauration and New Atlantis* (Arlington Heights: Crofts Classics, 1980), p 21.

THINKING THE WORLD TO PIECES | 141

59 Bacon, p 70.

60 Kohak, p 15.

61 Leiss, (1994/1972), p 128.

62 Of course, in another part of the world a very different realm lay behind that revealed by the senses. For Hindus, Buddhists and Taoists "…the world is conceived in terms of movement, flow and change…an inseparable web whose interconnections are dynamic and not static. The cosmic web is alive; it moves, grows and changes continually. Modern physics, too, has come to conceive of the universe as such a web of relations and…has recognized that this web is intrinsically dynamic." Fritjof Capra, *The Tao of Physics* (Boston: Shambhala, 1991), p 192.

63 David Abrams, *The Spell of the Sensuous* (New York: Vintage, 1996), p 34.

64 Whitehead cited in Pete Gunter, "The Disembodied Parasite and Other Tragedies; Or: Modern Western Philosophy and How to get out of it", in Oelschlaeger (1992), p 213.

65 Guignon, p 31.

66 "The only clear idea one can have of objects is that they are extended in height, depth and breadth, that is to say the idea of them that can be expressed mathematically." F.E. Sutcliffe, "Introduction", Rene Descartes, *Discourse on Method and the Meditations*, p 15.

67 Maurice Cranston, *John Locke: A Biography* (New York: Oxford University Press, 1985), p 163.

68 Dickens, p 1.

69 Eric Katz, *Nature as Subject: Human Obligation and Natural Community* (New York: Rowman and Littlefield, 1997), p 225.

70 Johann Wolfgang von Goethe, *The Sufferings of Young Werther* (New York: Ungar, 1957), pp 15 and 69.

71 This issue will become the undoing of areas of scientific inquiry such as physics in the 20th century thanks to the work on quantum theory by Werner Heisenberg and others who proved that the observer necessarily become an active participant in the observation and impinges on the result.

72 Charles Larmore, "Romanticism and Modernity", *Inquiry* (v.34, 1991, pp 77-89), p 78. Sometimes these 'subjectivities' can be quite shocking. The following instrumental reasoning from World Bank Vice President Lawrence Summers in 1991, which suggests shifting high pollution industries to developing countries is a case in point: "…health impairing pollution should be done in the country with the lowest cost, which will be the country with the lowest wages…I think the economic logic behind dumping a load of toxic waste in the lowest wage country is impeccable and we should face up to that…I've always thought that under-populated countries in Africa are vastly under-polluted, their air quality is probably vastly inefficiently high compared to Los Angeles or Mexico City….: Cited in Elizabeth Dore, "Debt and Ecological Disaster in Latin America", *Race & Class* (v. 34:1, 1992), p 85.

73 Larmore, p 78. See also John Ely, "Ernst Bloch, Natural Rights, and the Greens" in David Macauley, ed., *Minding Nature: The Philosophers of Ecology*, (New York: The Guilford Press, 1996).

74 Diamond, p 102.

75 Leiss, (1972/1994), p 149.

76 I. Kant, "Conjectures on the Beginning of Human History" in *Kant: Political Writings*, ed. By Hans Reiss (Cambridge: Cambridge University Press, 1991), p226.

77 Kant, p 225.

78 Leiss, (1972/94), p 109.

142 | COMING TO SENSE AND ENCOUNTERING SENSIBILITY

79 Taylor, (1992), p 365.

80 In Kant's "Lectures on Anthropology", cited in Allen Wood, "Kant on Duties Regarding Nonrational Nature", *Proceedings of the Aristotelean Society*, v. 72, 1998, p 4.

81 From Spinoza, *Ethics*, Part IV, Proposition 37, Scholium I, cited in Nussbaum (2001), p 317.

82 Bacon, p 27.

83 Kant, in Reiss, p 63.

84 Frank and Fritzie Manuel, *Utopian Thought in the Western World* (Cambridge: Harvard University Press, 1979), p 519.

85 Midgley, p 19.

86 Ron Vannelli, *Evolutionary Theory and Human Nature* (Boston: Kluwer Academic Publishers, 2001), p 2.

87 Leiss, (1994/1972), p 149.

88 Charles Taylor, (1992) p 148. Taylor notes that while instrumental reason has had many positive outcomes, "...there is also a widespread unease that [it] not only has enlarged its scope but also threatens to take over our lives. The fear is that things that ought to be determined by other criteria will be decided in terms of efficiency or 'cist-benefit' analysis, that the independent ends that ought to be guiding our lives will be eclipsed by the demand to maximize output.....for instance, the ways the demands of economic growth are used to justify very unequal distributions of wealth and income, or the way these demands make us insensitive to the needs of the environment, even to the point of potential disaster." Taylor (1991), p 5.

89 Katz, p 225

90 Bauman, *Modernity and the Holocaust* (19898/2001) p 92. Of course, gardens have weeds, hence Bauman's linking of an instrumental modernity to the Holocaust. "All visions of society-as-garden define parts of the social habitat as human weeds...". p 92.

91 Locke, *Second Treatise of Government*, in Gruen and Jamieson, p 21.

92 In Keith Thomas, *Man and the Natural World* (London: Penguin, 1984), p 242.

93 Merchant (2003), p 105.

94 Leiss, (1994), p 106.

95 Guy Denielou, "Developing a Company Outlook in Universities", *European Journal of Education* (v. 20:1, 1985), p 19.

96 From Wilson (2003) cited in *Scientific American*, Feb. 2002, p 84.

97 Tzvetan Todorov, *Imperfect Garden: The Legacy of Humanism* (Princeton: Princeton University Press, 2002), p 22.

98 Latour interviewing Michel Serres, p 173.

99 See Isaiah Berlin, *The Crooked Timber of Humanity* (Princeton: Princeton University Press, 1991).

100 Stephen Quilley, "Ecology, 'human nature' and civilizing processes: biology and sociology in the work of Norbert Elias" (pp 420–58) in Quilley & Loyal, *The Sociology of Norbert Elias* (Cambridge: Cambridge University Press, 2004), p 53.

101 Tu Weiming, "Beyond the Enlightenment Mentality", (pp 3–21) in Mary Tucker and John Berthrong, eds., *Confucianism and Ecology: The Interrelation of Heaven, Earth, and Humans* (Cambridge: Harvard University Press, 1998), p 4.

102 Terry Hoy, *Towards a Naturalistic Political Theory: Aristotle, Hume, Dewey, Evolutionary Biology, and Deep Ecology* (Westport, Conn: Praeger, 2000), p ix.

103 Gowdy, p xvi.

104 Morris Berman, *The Reenchantment of the World* (New York: Bantam, 1981), p 3.

105 Bauman, p 96.
106 Fernandez-Armesto, p 545.
107 Geoffrey Symcox, "The Wild Man's Return: The Enclosed Vision of Rousseau's Discourses", in Edward Dudley and Maximllian Novak, *The Wild Man Within: An Image in Western Thought from the Renaissance to Romanticism*, (Pittsburgh:University of Pittsburgh Press, 1972) p. 243.

CHAPTER FOUR

RECURRENT SENSIBILITY: ON GOLDEN AGES, NOBLE SAVAGES AND A MODERN PRIMITIVISM

"I must love and admire with warmth, or I sink into sadness"
MARY WOLLSTONECRAFT

And so the Cartesean trap is set. The long history of either/or dualism in Western culture achieves a kind of apotheosis with Descartes' definitive division of everything into thinking matter and extended matter: *I think, therefore I am; it thinks not, therefore it is not.* In philosophy, politics and daily life the dualist beat goes on — we're either part of the solution or part of the problem, we're male or female, we can have jobs or a vibrant environment. The actual lived world of humans, however, looks much different leaving room as it does for yes and no, but also maybe and always, never or sometimes. And since we now accept that we can also be male or female or a bit of both, perhaps we can also learn to sustain other seeming paradoxes such as having both jobs and a sustainable environment.

In beginning to posit alternatives to the mastery agenda of modernity our task is to avoid walking into this dualist trap and to illustrate instead the complexity of the combinations and options we encounter and that we can create in working our way toward an ecologically sustainable culture. The problem, of course, is the lure of clarity. True to its Cartesean roots, modern dualism insists on matters being "clear and distinct" in order to be credible but with alternative paradigms such clarity is often not attainable nor even desirable.

146 | COMING TO SENSE AND ENCOUNTERING SENSIBILITY

The case for complexity, for holism and even for paradox within modernity arrives toward the end of the 18th century with the emergence of Romanticism as both a component of and challenge to the rationalist agenda of the Enlightenment. The Romanticism I am going to focus on strays beyond the confines of the famous poetry of the era, including as well particular approaches to politics, science and social issues as well as literature and the arts.[1] It is appropriate, therefore, that we gain our initial insights into this alternative paradigm by turning to Mary Wollstonecraft, novelist, educator, the founding voice of modern feminism, and a revolutionary democrat in the finest tradition of the 18th century European Enlightenment. She is all that and more. But the prevarications and seeming contradictions that swirl about her responses to and assessments of both nature and culture in her *Short Residence in Sweden, Norway and Denmark* (1796) illustrate the difficulties faced then and now in demanding clear and distinct categories.

OCEANIC FEELINGS CONTESTED AGAIN

"How frequently has melancholy and even misanthropy taken possession of me, when the world has disgusted me, and friends have proven unkind. I have then considered myself as a particle broken off from the grand mass of mankind—I was alone, till some involuntary sympathetic emotion like the attraction of adhesion, made me feel that I was still part of a mighty whole from which I could not sever myself."[2] So Mary Wollstonecraft is at once a Cartesean particle and a romantic unity; an aspect of the human condition that is displayed in vivid detail throughout these twenty-five Letters from her Scandinavian travels.

And she had much to be disgusted with, hurt by and enthralled with on her travels North during the Summer of 1795. Raised in urban poverty in London, she had been a teacher, writer and editor and had achieved considerable notoriety as the author of *A Vindication of the Rights of Women* (1792). She had left an increasingly repressive England in 1793 and traveled to France just in time to miss the idealistic early days of the French Revolution, but in time to witness the start of a bloody Reign of Terror beginning with the brutal execution of the king and queen (1793) and ending with the equally brutal execution of the executioners in 1794. Despite her democratic credentials she was in danger herself as the Revolution began to consume it's own. She was, after all, English in a time of "Scarlet Pimpernels", English-inspired economic destabilization plots, and English-funded émigré armies threatening invasion.

In the meantime, her lover/companion and father of her child, the American entrepreneur and smuggler Gilbert Imlay, had turned away from

RECURRENT SENSIBILITY: ON GOLDEN AGES, NOBLE SAVAGES AND A MODERN PRIMITIVISM | 147

her, triggering the first of two suicide attempts just prior to her departure for Scandinavia. So the world was not a happy place for Mary, and like others who relied too completely on the steadfastness of lovers or whose hopes for true political change had been dashed by the Reign of Terror in Paris, she spun into depression and dark thoughts.

But she was rescued from sinking into what her contemporary Samuel Taylor Coleridge described as a "...an almost epicurean selfishness..." on the part of his peers who were disillusioned with politics and philosophy. Reconciled briefly with Imlay, she set out with her infant daughter and a servant on a trek through Scandinavia looking for one of Imlay's cargo ships that had gone missing during one of his smuggling ventures. On this extraordinary mission through the rough northern frontier of European culture she missed the cargo but discovered instead images of a golden age of "...independence and virtue; affluence without vice; cultivation of mind, without depravity of heart, with "ever smiling liberty" the nymph of the mountain. I want faith! My imagination hurries me forward to seek an asylum in such a retreat from all the disappointments I am threatened with."[3] Earlier, in her novel of 1787, *Mary, A Fiction*, she described such sensibility as the "...most exquisite feeling of which the human soul is susceptible...", with the passions tempered by reason so that impulses of the heart need no correction.[4] Watching the daily life of fishermen and farmers in this rural and undeveloped part of Europe, she professed at first to see only harmony and tranquility.

But then, a few lines later in a kind of modernist schizophrenia she warns of the dangers of falling into such imaginative reveries when the world is "...in such an imperfect state of existence" and counsels that the world "...requires the hand of man to perfect it," that the virtue of a culture is necessarily linked to its scientific improvements and that the evidence she sees of human industry and progress means that it is impossible for humans to remain "...in Rousseau's golden age of stupidity."[5] She tires quickly of the simplicity of the rural tranquility she so recently admired as "...reason drags me back, whispering that the world is still the world, and man the same compound of weakness and folly, who must occasionally excite love and disgust."[6]

There is a kind of journey within her letters from the North, one that starts with a desperately depressed and distressed woman who "...opened her bosom to the embraces of nature," who relished the simplicity and sublimity of the nature and the cultures she observed on the northern frontier of modernity and who dreamed of trees being "philosophers...ever musing."[7]

But then she tires of simplicity and begins to notice and celebrate signs of progress, labour, invention and ambition and insists she must emerge from the woods in order "...that I may not lose sight of the wisdom and virtue which exalts my nature."[8] But then, democrat and reformer that she is, when she reaches the bustling bourgeois world of Copenhagen it repels her, convinced as she is that one cannot "chase after wealth" and love humanity at the same time. For the progress-minded bourgeois, family, virtue, "greatness of mind," patriotism are all easily sacrificed for business: "...a greedy enjoyment of pleasure without sentiment embrutes them till they term all virtue, of an heroic cast, romantic attempts at something above our nature."[9]

Mary was 15 when Goethe's story of Werther, the young man who killed himself from thwarted love, appeared and her easy immersion into the realm of deep — and dangerous — sensibility reflects the influence of that emotionally torrid 1774 best seller. Like Werther, she asserts her own psychological vulnerability: "For years I have endeavoured to calm an impetuous tide — labouring to make my feelings take an orderly course — I was striving against the stream. I must love and admire with warmth, or I sink into sadness."[10] She was 17 when the American Declaration of Independence arrived on the political scene and 19 when Rousseau died, 23 when his Confessions and then his *Reveries of the Solitary Walker* appeared in print. Irritated by the anti-feminism and excessive emotionality in Rousseau's early works, Mary found in these later, more introspective texts a fellow "heroic misfit."[11] Her four major books appeared in quick succession from 1786 to 1792 and she was in Paris and pregnant at age 34, before her journey North on her mission for the feckless Imlay.

What do these thoughts and the biography of an 18th century woman have to tell us about "recurrent sensibility" or about alternatives to Cartesean certainty? Wollstonecraft embodies in a particularly fulsome way both the paths of modernity that I am attempting to identify: the one reasoned, progressive and scientific and the other more reciprocal, sensitive and holistic. She seems unable or incapable of choosing between them, one moment longing for immersion in the vast sublimity of nature and single human desires, the other demanding to enter the lists to ensure that politics serves the needs of the many in the march into a progressive future.

Wollstonecraft's indecision could be the result of her particular history or, to be even more mundane, symptomatic of the manic-depressive strain in her family.[12] But I want to make a different claim for her and others like her who seem to display this kind of cultural schizophrenia. And here we enter the realm of the paradox, a realm already deeply mined by Rousseau, where positions are maintained that are essentially self-contradictory, though based on a valid deduction from acceptable premises.[13] Morris Berman sees this ability to sustain a paradox as akin to a "genetic memory", a throwback to primitive times, a "mature ambiguity...sustaining the tension of this conflict so that a deeper reality can emerge."[14] Wollstonecraft and others in what I will argue is a "tradition" in Western culture celebrate rather than repress this ambiguity and the paradoxes it necessitates because they sense that something more profound than Cartesean certainties are needed to answer deeply felt human questions and the even more pervasive needs of the natural world.[15] Mary Wollstonecraft survives across time, therefore, not just because she happens to connect with a contemporary social issue (feminism), but because the issues she raises and the manner in which she speaks to them are of fundamental importance and go to the heart of the crisis of modernity.

In the last chapter we explored the origins and nature of what has come to be the dominant world view in a now globalized modernity; the use of an instrumentalized reason and its twin prosthetics, science and technology, to master the natural world along with the social and the psychological worlds of humankind. This "mastery agenda" has deep roots in Western culture and while some of this depth was referred to, the focus of the exploration was on the modern post 16th century era. That same focus will occupy us in this look at a counter or alternative tradition within that same Western culture.

In devoting significant attention to what appears to be a "failed" or perhaps hopelessly backward-looking tradition, I am following the lead of E.P. Thompson, the historian whose work has had a tremendous influence on the evolution of my own thought. He opens his monumental *The Making of the English Working Class* with a plea to reject telling the "story" of mankind as one of winners and losers. His case is worth citing in full as an introduction to what follows:

> I am seeking to rescue the poor stockinger, the Luddite cropper, the "obsolete" hand-loom weaver, the "utopian" artisan.....from the enormous condescension of posterity Their crafts and traditions may have been dying. Their hostility to the new industrialism may have been backward-looking. Their communitarian ideals may have been fantasies. Their insurrectionary conspiracies may have been foolhardy. But they lived through these times of acute social disturbance, and we did not....Our only criterion of judgment should not be whether or not a man's actions are justified in the light of subsequent evolution. After all, we are not at the end of social evolution ourselves. In some of the lost causes of the people of the Industrial Revolution we may discover insights into social evils which we have yet to cure.[16]

Wollstonecraft's pitching back and forth between celebrating progress and longing for rural simplicity, awkwardly trying to occupy some middle ground between reason and feeling, culture and nature, the beautiful and the sublime give us some important clues about the difficulties involved in trying to negotiate the space between apparent winners and losers. From a 21st century perspective her position as a woman, an "outsider" from the dominant intellectual tradition and an outspoken "radical feminist and democrat" gives her perspective on the tensions of her time added weight.

To shed light on these tensions and especially on the "losing option" within modernity—the romantic, holistic, small-is-beautiful version that was to be overwhelmed in the 19th century—we need to start at the beginning once again, back to the issues raised in the first two chapters, that is to the debates about human nature. Four issues in particular will occupy our initial discussion: the so-called state of nature; the existence (or not) of an innate moral sense; the myth of a Golden Age; and the idea of the Noble Savage. Each of these issues was central to the eventual emergence of the first coherent contemporary critique of modernity by advocates of what became known as "Romanticism".

This becomes then the starting point for conceptualizing a counter-tradition or alternative means of organizing human relations with nature, the other and the self.

A DIVERSION — ON GOING BACK
OR GOING FORWARD BY LOOKING BACK

When Rousseau argued in his essay on the origins of inequality that humans were "happier" in more primitive times, he was parodied by his arch-rival Voltaire who said in a letter: "I have received, Monsieur, your new book against the human race, and I thank you. No one has employed so much intelligence to turn us men into beasts. One starts wanting to walk on all fours after reading your book. However, in more than sixty years I have lost the habit".[17]

This, of course, is simply a version of the old canard, *If you don't like it here, why don't you go back where you came from?* so often used in response to immigrants or outsiders who offer critiques of the status quo. No doubt Voltaire would have preferred that Rousseau at least go back to Switzerland, if not to a state of nature. Of course Rousseau had never suggested a "return" to the primitive or to a state of nature, but instead urged us to look back upon this distant past in order to arrive at a different vision of the human possibility and to show us how far our 2,000 year focus on reason and intellect has taken us from that possibility.[18]

The human possibility that Rousseau, and later the Romantic writers of the 19th century, caught a glimpse of when they looked back upon an imagined primitive era was a possibility of an existence built on a fundamental equality among humankind and a humanity living with nature, not above it. A fundamental social equality ensured an absence of the alienation from self

RECURRENT SENSIBILITY: ON GOLDEN AGES, NOBLE SAVAGES AND A MODERN PRIMITIVISM | 151

and other that they saw as pervasive in modernity, and the more intimate connection with nature put an end to dualism and mastery and hence eliminated humanity's second great alienation, that from nature.

From Rousseau through to 19th century Romantics and 20th century environmentalists the accusation that these speculations were simply naïve nostalgia for an unattainable (and even imaginary) past has proved a powerful weapon in the hands of the defenders of the form of modernity we have created over the past 300 years.[19] But except for a few true primitivists, the accusation is clearly a red herring. The central argument of the position we will be exploring here is, as Rousseau first argued, that "...enlightened social man can attempt to recover some of the advantages of the primitive independent state, without losing the benefits of the social state."[20]

History may be the "problem" since it was within the process of historical time that we allowed reflection to overwhelm intuition, pride to swallow self-esteem and imagination to rule desire, but the solution must involve going forward, not back and in doing so we need to "...attack not the instrument [reason], but the hand that guided it."[21] The objective is not to return to nature, but to grow through an understanding of nature and our place within it in order to rediscover "...in a new (political and moral) form our original (natural and animal) wholeness."[22] Paul Shepard, one of the more outspoken modern advocates for the relevance of the primitive for our time, put it this way:

> In advocating the *primitive* we seem to be asking someone to give up everything, or to sacrifice something: sophistication, technology, the lessons and gains of History, personal freedom, and so on. But some of these are not *gains* so much as universal possessions, reified by a culture which denies its deeper heritage. *Going back* seems to require that a society reconstruct itself totally, especially that it strip its modern economy and re-engage in village agriculture or foraging, hence is judged to be functionally impossible. But that assumptions misconstrues the true mosaic of both society and nature, which are composed of elements that are eminently dissectible, portable through time and space, and available. You can go back to a culture even if its peoples have vanished, to retrieve a mosaic component, just as you can transfer a species that has been regionally extirpated, or graft healthy skin to a burned spot from a healthy one.[23]

It is not mere nostalgia or naïve idealism, then, that motivates a backward glance by those troubled by our present, but rather a sense of "functionality", of the "use" of the past not just to understand the present, but to inspire visions of alternative futures. This ambition was captured eloquently by T.S. Eliot in his famous lines:

And the end of all our exploring
Will be to arrive where we started
And to know the place for the first time.[24]

We need to "know" our distant past with the depth and intuitive sense that Eliot implies if we are to be able to select those aspects of it that might contribute to a more sustainable humanity.

THINKING ABOUT A *STATE OF NATURE*

In the 18th century, the issues of human nature and human origins were central components of the various positions being taken in the general re-thinking about the human possibility that characterized the times. The Christian version of a Garden of Eden ruined by human curiosity and redeemable only by obedience to the rules of an established church was no longer tenable in a culture now more interested in scientific and secular understandings. Just as important, 250 years of discoveries of cultures beyond Europe, some highly developed and others strikingly primitive but all definitely different had called into question the singularity of human development. Finally, the revival of interest in classical learning and the idealization of Greek and Roman antiquity gave 18th century philosophers and political leaders a new (or old) set of ideas to bring into play.

And so we start again with the issues raised in Chapter One — what were the roots of human nature? The idea of "stages" or "development" was already a popular explanatory tool in the 18th century and so "human development" came to be the model for understanding human nature. We were not simply "created", but in a still pre-Darwinian sense we "evolved".

Two very distinct ancient or primitive stages were posited by most attempts to describe this development: a state of nature in which primitive humans were hardly differentiated from other animals, struggling daily in a dangerous world; or an earlier arcadian Golden Age in which climate and soil yielded food without work or strife and where leisure fostered art and innocent pleasures.[25] We will explore each of these stages in turn.

In imagining a state of nature as a pre-social, pre-civilized era of human existence, 18th century theorists like Rousseau, Locke, Buffon, Smith, Kant, Shaftesbury and others took their lead from ideas they encountered in classical texts and ancient myths. Mary Wollstonecraft's daughter, Mary Shelley, gives us a clear insight into one powerful ancient source with her story of *Frankenstein, Or the Modern Prometheus*, with Victor Frankenstein breathing knowledge into a created but primitive and child-like human. Her husband Percy Shelley was likewise fascinated by *Prometheus Bound*, the tragedy by Aeschylus and composed his own version, *Prometheus Unbound*. For Percy Shelley, as Mary recounts in her notes to his poem, Prometheus was a regenerator of humankind, who "...unable to bring mankind back to primitive innocence, used knowledge as a weapon to defeat evil, by leading mankind, beyond the state wherein they are sinless through ignorance to that in which they are virtuous through wisdom."[26] As we will see, Percy Shelley, the quintessential modern romantic, is here tapping into the ancient triad of simplicity coupled with reason leading to virtue.

What the Shelley's each encountered in Aeschylus' Prometheus story was a grim set of images of humanity in a state of nature:

> ....first of all, seeing, they saw in vain, hearing, they heard not; but, like the shapes of dreams, through all their weary life they confounded all things aimlessly, and they knew neither houses woven of brick to face the warmth, nor work of wood, but they dwelt underground, like swarming ants, in sunless depths of caves. No token had they of winter nor of flowery spring nor of fruitful summer wherein to trust, but wholly without judgment did they act, until I showed them the risings of the stars and their perplexing settings.[27]

In the Greek myth it was Prometheus who, disobeying the gods, gave these humans fire and knowledge of agriculture, enabling them to escape their grim state of nature. Plato accepted this harsh view of humanity's original state, though he suggested there might have been an even earlier Golden Age when "...men were without art or invention." But for Plato that era had passed and until the intervention of Prometheus, humans were in a pitiable state.[28] Interestingly, the human story in Plato's version is one of decline and renewal, one in which development occurs but seems inherently unstable.

While the Romans were generally more interested in Golden Ages than in the primitive, speculations about the state of nature by Seneca and especially Lucretius subsequently had a great influence on their 18th century readers. Obsessed by the corruption of Roman civic culture at the turn of the millennium, the rectitude these writers portrayed in their speculations on the state of nature made sense to Enlightenment critics of their own absolutisms, which they saw as rife with a similar corruption. Unlike the troll-like pre-fire humans imagined by Aeschylus, the 1st century Roman Stoic Seneca imagined a time of equality, peace and an absence of corrosive desire, an a-historical era in which men "...who found shelter from the sun in some thick wood, and protection from cold and rain in some lowly refuge beneath the leaves, passed tranquil nights without a sigh....To them by day as well as by night the prospect of this most lovely dwelling place of ours lay open."[29]

But for Seneca these barely social creatures were neither wise nor virtuous for "...it is an art to become good" and these humans contained only the stuff of virtue, not virtue itself. This will be a recurring theme, the natural containing the potential for the good, but lacking the means to transform that potential into virtue. For Seneca, the ideal would be a return to the innocence of material simplicity but supplemented by the conscious, rational, and disciplined virtue of the Stoic. One has a sense here of how the Stoic ideal would be attractive to many contemporary environmentalists.

By far the most thorough classical work on the state of nature is to be found in the *Nature of the Universe* by the 1st century epicurean poet Lucretius. Here humans in the state of nature were:

- tougher and larger than men of his day
- mobile rather than sedentary
- content to accept whatever nature delivered
- living without fire in caves and forests
- unaware of concepts of the common good, morals or laws
- carnivorous
- long lived [30]

Being closer in time to the era of hunters and gatherers than us, observers of the human condition like Lucretius had more vivid examples to refer to and were in more intimate contact with the myths and stories that told of their own ancient times. They imagined an "artless" time, with no property, no social groups beyond perhaps the family, no government and no moral codes. Lives were governed by instinct, intuition and opportunity with an edenic absence of the idea of sin. And it was these images with their attendant qualities of freedom, peacefulness and contentment forming the very base of human nature that would later prove attractive to the Romantics and others in their attempts to define a position for humans within nature that did not rest on domination by force or privilege.

Despite these earlier myths and traditions, it is testimony to the robust strength of modernity that the 17th century formulation of Thomas Hobbes concerning the nasty, short and brutish lives of humans in the state of nature remains one of the more powerful ideological pillars supporting the structure of modern societies. Alongside it is the equally strong core belief that humans are innately aggressive, that, as Freud argued, aggression is an "...instinctual disposition in man...the hostility of each against all and of all against each."[31] If Homo homini lupus—man is wolf to man—then one can only imagine what such a human is to nature!

The third pillar of modern belief, that humans avoid the perils of nature by employing their aggressive instincts in on-going drives to advance at all costs their material well-being—Homo economicus—completes a rather savage notion of a fearful, aggressive and avaricious humankind. As we have seen in Chapter Three, with this set of assumptions the world becomes one in which these aggressive, self-aggrandizing humans pit their material and mental skills against a too often malevolent nature in order to bend it to their will. This is the dominant paradigm of human nature that we have inherited from the beginnings of modernity and which the turmoil and trauma of the 20th century so heartily reinforced.

But we know from the discussion in Chapter One of the research done by modern anthropologists and archaeologists that there is persuasive evidence that these assumptions about core beliefs rest on a weak foundation, evidence that humans in our deep past—an era at least akin to a state of nature—were cooperative, peaceful, at relative peace with the nature around them, and content with a steady state material culture. The early proponents of an alternative paradigm for modernity, while not yet privy to this anthropological evidence, nonetheless started from the very similar assumptions.

The two great social contract theorists of the 18th century—Locke and Rousseau—led the way in calling the Hobbesian position into question, Locke first by softening the "brutish" quality of the state of nature and Rousseau then by rejecting altogether the dispositional package made necessary by that supposed brutish nature. Locke achieved this softening by asserting that despite what he insisted was a "blank slate" in terms of specific principles and virtues, humans were "naturally good" because even prior to language and society they were possessed of an innate reasoning capacity. This innate reason was the source of the "natural law" that governed humans even in this very primitive stage of development, a natural law that decreed all to be equal and independent. These primitive humans needed only to consult their reason and thereby be assured that "...being all equal and independent, no one ought to harm another in life, health, liberty or possession."[32]

But the other two core beliefs of modernity prospered under Locke's purview. There remained a powerful spiritual or religious quality to this pre-positioning of reason, what Lucien Colletti calls a "transcendental investiture", stemming no doubt from Locke's deep religious faith.[33] And this faith found easy alliance with the anthropocentrism of Bacon and the idea of Homo economicus. For Locke "in the beginning all the world was America", a blank natural environment awaiting cultivation by free, unitary subjects with powers of reasoning and a sense of justice.[34] Having no reason, nature could have no claim on human benevolence or kinship and since nature existed to be used

COMING TO SENSE AND ENCOUNTERING SENSIBILITY

by humans to advance their interests, those humans who chose not to employ their reason in such a task—e.g., indigenous peoples—also lost much of their claim to benevolence or kinship.

If Locke softened Hobbes, it was Rousseau who turned him on his head and in doing so laid the foundation for a radical critique of modernity and advanced a more complex picture of humans in nature. In his speculations on humans in a pristine state of nature Rousseau opens with a fiction, a Robinson Crusoe-like situation of *man alone* in nature. Or perhaps Tarzan would be a more familiar example. In this fictional or philosophical landscape borrowed freely from Lucretius the human is a mere animal among other animals and within nature: no reason, no language, no conscience, no self-awareness. These humans were peaceful unless provoked, mostly vegetarian and "...oriented wholly toward present existence."[35] They were randomly nomadic and apart from sexual reproduction and perhaps a rudimentary family affiliation—and Rousseau stubbornly resists even this nod to sociality—"...their lives are essentially solitary and indolent."[36]

> His imagination paints no pictures; his heart yearns for nothing; his modest needs are readily supplied at hand; and he is so far from having enough knowledge for him to desire to acquire more knowledge, that he can have neither foresight nor curiosity. The prospect of the natural world leaves him indifferent just because it has become familiar. It is always the same pattern, always the same rotation....His soul, which nothing disturbs, dwells only in the sensation of its present existence, without any idea of the future, however close that might be, and his projects, as limited as his horizons, hardly extend to the end of the day.[37]

Indifferent to his own and nature's future, indeed indifferent to the very idea of past or future, mortality or meaning, Rousseau's *savage man* was "...healthy, vegetarian, physically strong, well coordinated, stupid and solitary;" the perfect model for Mary Shelley's Creature in *Frankenstein*.[38] But like her Creature, Rousseau's savage man was no blank slate. There were two deep predispositions present from the start: an instinct for self-preservation and a feeling of sympathy for other sentient creatures in distress. This was the core of the *wildness within*, the twin building blocks of all that would follow in human development; from the instinct for self-preservation will flow a sense of self-esteem and from sympathy will flow virtue—but only after these proto-humans develop a social world within which these two domains of moral virtue could develop.

Rousseau's savage is clearly no human we know. But then neither are the millennia of Homo sapiens who lived in the long era of the Paleolithic that we encountered in Chapter One. As we know from that discussion, there can be no fixed time of origin for Homo sapiens, no actual state of nature from which humans emerged. As Tim Megarry reminds us, "...the only original human condition is one of endless change and development."[39] And Rousseau's image of the lone human remains problematic since it is likely that our evolutionary ancestors were already highly socialized in families and small groups long before the emergence of Homo sapiens as a distinct species.[40]

Still, Rousseau's "mental construct", as Todorov calls it, is a useful way to help us imagine what we might find at the core of humanity when layer upon layer of cultural accretions are peeled away. It may help us get at some "real" facts about us as opposed to facts that are relative to time or place and simply comparable to other facts.[41] Rousseau proposes that these beings, our Homo sapiens ancestors, were "happy" if not yet virtuous, and that in their complete membership in or oneness with the natural environment within which they lived — animal, mineral and vegetable — they experienced to the full the "sentiment of existence" that he himself had glimpsed while floating on the water off the Isle St. Pierre. This pre-social, pre-linguistic human lived within the order of the natural world and had no impulse to change it. And there was no regret at the lack of a means to do so.[42] It is our great handicap and impediment to happiness, Rousseau might argue, that thanks to the development of consciousness in our more sophisticated brain we DO have the means and hence the ambition and the regret.

For Rousseau these early humans were not "thinking" about their position in nature, or about their relations with other humans. Instead, exercising their innate sense of sympathy with all about them, they were "feeling" existence. It was to rediscover or rehabilitate this realm of feeling that Rousseau was after in creating his model of savage man in a state of nature, a being who feels and who shares the sentiment of self-love with other animals.

Self-love and sympathy as human core qualities or perhaps instinctual dispositions, signal a very different human from that suggested by Hobbes or Freud. When coupled with a similar predisposition toward cooperation and sharing which has been proposed by several anthropologists, we begin to see a human nature that runs counter to the prescriptive images that emerged from the work of Bacon, Descartes and Hobbes. We also see more clearly now that humans, by their very nature, seem caught somewhere within a continuum between the need to bond on the one hand and the need for autonomy on the other. Instead of seeing individuals moving deftly along this terrain between equally powerful drives, in Western culture this continuum is most often transformed into a duality in which individuals experience often pathological tension or retreat to ambivalence as they are forced to choose one side

COMING TO SENSE AND ENCOUNTERING SENSIBILITY

and repress the other.

Rousseau in his own life may seem to lean toward the autonomy side of the continuum, toward an assertion of a natural individualism, yet he clearly realized that reason, language, conscience, self-awareness and forethought were essential qualities for long-term human well-being and that they could only develop within a social milieu. To that extent Rousseau remains within the circle of political theorists who support the social contract as an essential step in assuring the stability of that milieu. But for our purpose of exploring new possibilities for human relationships with the other-than-human natural world, Rousseau adds this deeply subjective personal, holistic and primitive awareness of a connection with the whole as yet another human predisposition side by side with the well-established qualities of sociality and autonomy. He fears, though, that the necessary bonding with other humans via language and reason will so privilege these latter tools that the more holistic and sensual bonding with nature will be put at risk. Hence our crucial need to nurture this deeply subjective disposition to identify with the whole of nature by tempering the social contract with what Michel Serres has called a *natural contract*.

Holistic—what might this often abused word mean for our purposes? For the Romantics of the 19th century, taking their lead from 18th century speculations about the state of nature which, in turn, took as their inspiration 1st century and earlier versions of that prehistoric era, holistic or holism implies a deep continuum bridging life and death, past and future, humans and nature. It borders on mystical and has strong connections with what we now refer to as "Eastern" and pre-modern thought. Here there is a focus on an underlying life force—variously called *mana, wakan* or *dharma*—"...anterior to the individuality of persons or objects."[43] Here the self is an "extended self", part of a much larger but still essentially organic or material whole and often awed "...before the forces that lie outside human control."[44]

Minimizing the importance of death in the continuum of existence was an on-going task. Epicureans and Stoics led the way, Seneca noting that: "there is nothing to very great about living—all your slaves and animals do it."[45] One of the more moving and impressive attempts to convey this re-positioning of individuals existence is Percy Shelley's poetic reverie on first encountering Mont Blanc in Switzerland. Even an excerpt conveys Shelley's sense of the seamlessness of life:

> *Far, far above, piercing the infinite sky,*
> *Mont Blanc appears,—still, snowy, and serene -*
> *Its subject mountains their unearthly forms*
> *Pile around it, ice and rock; broad vales between*

Of frozen floods, unfathomable deeps,
Blue as the overhanging heaven, that spread
And wind among the accumulated steeps;

.....

The wilderness has a mysterious tongue
Which teaches awful doubt, or faith so mild,
So solemn, so serene, that man may be,
But for such faith, with nature reconciled;
Thou hast a voice, great Mountain, to repeal
Large codes of fraud and woe; not understood
By all, but which the wise, and great, and good
Interpret, or make felt, or deeply feel.

.....

All things that move and breathe with toil and sound
Are born and die; revolve, subside, and swell.
Power dwells apart in its tranquility,
Remote, serene, and inaccessible:
And this, the naked countenance of earth,
On which I gaze, even these primaeval mountains
Teach the adverting mind.[46]

The scene of chaos and destruction Shelley confronted in 1816 as he viewed the work of Mont Blanc's glaciers, avalanches and storms produced, according to his wife Mary Shelley, "deep and powerful feelings" in the poet.

Like his deceased mother-in-law Mary Wollstonecraft in her contemplation of wild nature in Scandinavian fjords, Shelley is struck by the sublime emotions of fear and awe when standing before Mont Blanc. Wollstonecraft prevaricates, at first responding by sensing her membership in "a mighty whole", but later siding with the more modern agenda of overcoming the sublime by the mechanisms of progress.[47] Shelley, however, has no desire to climb Mont Blanc and thereby assert human power over sublimity, but instead wants to learn, to accept, and to "..make felt and deeply feel" the truth that "...the material world is infinitely larger and more complex than we are, and is indifferent towards our existence."[48]

Shelley parts company here in important ways with Kant and the use of the sublime in Enlightenment thought. Shelley is content to let the power of Mont Blanc, it's sublime quality, simply be and in some way merge with it. For Kant, the existence of feeling of sublimity when confronted by such a phenomenon was potentially overwhelming and thus had to be "defeated" by reason. These sublime moments, then, were for Kant occasions to demonstrate

COMING TO SENSE AND ENCOUNTERING SENSIBILITY

the power of reason over nature.[49] It is this distinction, captured by Shelley's response to Mont Blanc, that makes Shelley part of the shadow side of modernity. He would no doubt agree with his fellow poet/philosopher John Clare that he was, in the end, "...just another inquisitive mammal going about its daily activities."[50] Just another mammal! This idea—obvious on the one hand, potentially revolutionary on the other—provides a key entry point for an alternative paradigm for modernity, one resting not on a ratio-linguistic anthropocentrism but rather on life itself being considered a "universal state."[51] Keith Thomas calls this idea that the world does not exist for man alone "... one of the great revolutions in modern Western thought."[52] Humans came to be seen as just one link in the great Chain of Being, in which:

> Each hated toad, each crawling worm we see,
> Is needful to the whole as well as he.[5]

Onno Oerlemans calls this an early example of "biocentric imagination" which collapses the distinction between animal and human.[54] Much earlier, St. Francis had argued, in what must have seen at the time as a kind of blasphemy, that just as humans may from time to time need to kill animals, wild animals in turn had God's permission to kill humans, a kind of early version of modern mutualism.[55] Responding to Voltaire's assertion of human precedence in nature, Rousseau insisted that while it may be no true compensation for death that "...the corpse of a man nourishes worms, wolves or plants...in the system of the universe, it is necessary to the preservation of the human species that there should be a circulation of substance between men, animals and vegetables."[56] This "best of all possible worlds" defence of nature clearly embeds humans within a much larger system, one looking much more like a State of Nature than the anthropocentric modernity of Bacon, Descartes or Kant. We come close, in fact, to a vision of the world as the 4th century BC Greek philosopher Epicurus understood it, one in which humans belong to a world of finite living things, without immortality, where the "...soul decays as the limbs rot and fall away, scattered."[57]

This massive de-centering from the human was a retreat toward the holism that Rousseau and others imagined had characterized humanity in its most sustainable—both materially and emotionally—era. This was not the imagined brutal state of nature, but rather a Golden Age" a time in both myth and anthropology when humans "got it right".

ON GOLDEN AGES

A peaceful, safe, idyllic life, so sweet,
The people led in that old former age;
It was sufficient that the food to eat
Was yielded by fields by custom at every stage;
Excess was never a cause for any outrage;
The grindstone was unknown, and the mill too;
Such food as mast and haw-berries were the rage,
And water from the cold well was their due.
Then was the ground still untilled by the plough,
Corn springing up, unsown by any hand,
They rubbed for flour, and ate just anyhow.
No man had yet driven furrows through his land;
No man found fire from flint at his command;
Untrimmed and unprepared still lay the vine;
No man had yet crushed spices on demand
To clear, nor to make sauce of galantyne.
GEOFFREY CHAUCER, "THE FORMER AGE"[58]

If theories about a state of nature are attempts to imagine a prehistoric, pre-social, pre-linguistic but nonetheless historical era of human existence, the parallel track of stories, myths and legends of a Golden Age are speculations of a different sort. What Nussbaum calls infantile and Kant merely futile, empty yearnings, are nonetheless powerful myths that persist in virtually all cultures.[59] And some are not myths at all but rather are interpretations of an era in ancient times that is seen to retain the holism of the state of nature without the brutishness.

The classic Golden Age myths are located in the deep past, in edenic times before any notion of a Fall, and express the "...revulsion of men from the contingencies, the cares and the corrupt compromises of a degrading present."[60] Human history is thus a progressive deterioration, but with the myth of a better time in the past holding out hope that a return is possible, that the human nature present then is present still within. The Christian Eden story is built on an initial harmony and perfection in which humans live at peace with all other divine creations. But it also posits the "...idea of human separability from, and potential mastery over, nature. Finally, there is a *Fall* from grace following Adam and Eve's refusal to obey the divine mandate not to eat of the Tree of Knowledge. The story concludes with the exile from the garden, the subsequent necessity of labor and its challenge of their mastery over nature."[61]

162 | COMING TO SENSE AND ENCOUNTERING SENSIBILITY

One cannot help but see elements of the human progression from childhood through adolescence to adulthood in this story, with death (by nature!) being the only possible end, barring some return to the essence of the childlike condition of peace with self and with nature.[62] This attraction to the child metaphor, to the idea that we can and must attempt to nurture the intuitive powers of understanding that we had in our earliest youth is particularly powerful in Romanticism. And so the idea of a Golden Age provides one more argument, slender reed though it is in our modern times, for the idea that the future reconciliation of humankind with itself and with nature involves going back as much as going forward.

The Greek poet Hesiod (8th century BCE) provides the clearest imagery of the Golden Age passed on to Western culture, supplemented by the Roman poet Ovid (1st century BCE). Hesiod tells us of five ages in human history, beginning with a Golden Age when the gods created a race of golden but mortal humans. They lived "...with hearts free from sorrow and remote from toil and grief...all good things were theirs...the fruitful earth spontaneously bore them abundant fruit...and they lived in ease and peace upon their lands."[63] Sadly, at some point a sleep-like death ended their pleasant sojourn on earth. They were replaced by a race of lesser, silver humans who behaved foolishly and were destroyed by Zeus. The third age of bronze were too warlike and ended up destroying themselves. A fourth age of more heroic characters ended up becoming gods and were followed by the age we (and Hesiod) live in, the Iron Age in which we are "...forced to work in sorrow and waste away and die."[64]

Ovid modifies or expands upon Hesiod in some important ways. The humans of the Golden Age are vegetarians, all metals remain buried deep in the earth, a natural goodness prevails eliminating any need for laws, and a general satisfaction with place removed any need for travel.

> *This age understood and obeyed*
> *What had created it.*
> *Listening deeply, man kept faith with*
> *The source*[65]

And central to this keeping faith with "the source" was the idea of balance, that in these arcadian settings the "design" rested on a balance between humans and nature. For Ovid the Iron Age embodies all the various defects of the human condition—crime, war, greed, property, mining, etc.—and he has Jupiter destroy it all with a flood, letting only one virtuous couple survive. It is an easy transition from this myth to the general acceptance of Genesis and the idea that the assigned role for these new humans will be to maintain that balance by being stewards of the Earth.

The Iron Age—which still persists—stands condemned for its abandonment of a pastoral ideal held to be central to human well-being and—a modern environmentalist might argue—planetary well-being as well. Instead of being harbingers of progress, mining, metallurgy, manufacturing and the constant movement of people were seen as pressing at long-standing taboos and common sense restrictions, prying too audaciously into nature's secrets.[66] Thus as in contemporary debates about atomic energy, recombinant DNA or genetically modified food, the idea of "dangerous knowledge" has a long history, as does the yearning for stability, simplicity and peace of mind.

The second variation on the Golden Age story addresses these latter yearnings more directly and also has its origin in ancient mythology. Based in Stoic and Epicurean philosophy and carried forward through Lucretius, Seneca, early Christian theology, arcadian utopias, and Rousseauean celebrations of rural simplicity, this tradition assumes many forms but always retains a focus on certain key attributes, including:

- matching desires with means
- a focus on simplicity and frugality
- stewardship and vegetarianism
- high value on leisure

What largely disappears in this approach are the supernatural aspects found in Hesiod and Ovid. The humans in these Golden Age scenarios deliberately choose their Arcadia, living a life that theoretically could therefore be lived again if only the will to do so were revived.

For our purposes there are two key elements of this Golden Age story that are important for a process of reformulating human relationships with the natural world. The first is the issue of desires, needs, wants and means, or to put it in its more controversial modern guise, the issue of self-imposed limits on human progress, development and even imagination in order to end the culture-driven imbalance between our desires and our faculties.[67] The second, clearly linked to the first, is the desirability of a more simple, self-sufficient life-style in order not just to better sustain the environment but also to enhance human well-being materially, socially and psychologically.

To grapple with desire we turn once again to Epicurus and his Roman disciple Lucretius, since their formulation of the argument for limits remains as relevant today as it was 2,500 years ago. As Nussbaum points out, there is no mystical or supernatural dimension to the materialist analysis of Epicurus. Our desires, and our determination to pursue them, are purely cognitive in origin:

164 | COMING TO SENSE AND ENCOUNTERING SENSIBILITY

> ...the central cause of human misery is the disturbance produced by the seemingly 'boundless' demands of desire, which will not let us have any rest or stable satisfaction. But, fortunately for us, the very same desires that cause anxiety, frenetic activity, and all sorts of distress ...are also the desires that are thoroughly dependent on false beliefs, in such a way that the removal of the belief will effectively remove the desire...[68]

The *foundation* of everything, then, is simply being, an awareness of and enjoyment of an existence which is to some extent fortuitous and certainly temporary. Epicureans, Cynics and Stoics alike all preached that we are "...unhappy and irrational in proportion to the multiplicity of [our] desires...[and that]..the good is to be reached...by being able to dispense with everything extraneous to oneself, while maintaining an inner state of untroubled satisfaction," what Rousseau would later come to see as a sentiment of existence.[69] Over two millennia ago Greek philosophers made the point that pleasure is not cumulative, that more is not better in the case of true pleasure. In the 18th century Montesquieu suggested it should be possible to adopt modern needs without modern desires and Rousseau at the end of his life counseled that "...the source of true happiness is within us..", in being fully conscious of our "inner tranquility" not in our possessions.[70] And we cannot, of course, not be struck with the cultural persuasiveness of this approach to being. It has an even more powerful influence in Buddhist and Confucian thought and remains in the mainstream in many cultures. Despite this long tradition, however, many of us seem to remain committed to achieving happiness by trying to satisfy an "endless cycle of desire...that in the end leads to compulsive behavior and finally to all pervasive feelings of emptiness, futility and despair," just as Epicurus said it would.[71]

The second aspect of the Golden Age stories that is relevant to our search for cultural traditions that might support a modernity based on ecological sustainability concerns size, self-sufficiency, and simplicity—and, in a deeper *sense*, community. As is true in so many cases, these issues appear fully formed in the cultural and philosophic debates in ancient Greece, particularly between Epicurus and the more dominant schools of Plato and Aristotle. They are very contemporary debates as well, often characterized as traditional versus modern or conservative versus liberal and they centre on the pressure to abandon family, friends and faith as the arbiter of the good in favor of the organized and codified justice of the city, state and society. Again, the situation is posed in the classic Cartesean language of conflict, tension and dualism.

Because such issues strike at the very core of peoples' lives they are often presented most vividly through literature and art. As we saw in Chapter Three, the *Antigone* by Sophocles centres on this issue of kinship ties versus

RECURRENT SENSIBILITY: ON GOLDEN AGES, NOBLE SAVAGES AND A MODERN PRIMITIVISM | 165

citizenship. For Epicurus, the intimacy of friendship and family were more reliable arbiters of duty and justice than Socratic reasoners from afar. In the 18th century, two millennia later, the rise of the modern city and its attendant bureaucratized culture was offering the same challenge to epicureans like Rousseau, radicals like William Godwin, and philosophic poets like Coleridge, Wordsworth, and Shelley.

Rousseau once again sets the initial context for the debate that will follow for the next 200 years. His Golden Age is not a mythical time at all, though not yet truly "historic" either. It is set somewhere between the Paleolithic and the Neolithic eras, a time of small human settlements in which the household was the dominant unit of social and political life. Rousseau argued that it was only when embedded within the family, self-sufficient and free of jealousy and pride that one could truly *be* whatever one *is*.[72] A world of limited desires in small spaces, gone forever in its Neolithic form but retrievable still in more modern times. In his best-selling novel *Julie; Or the New Heloise*, Rousseau struck a balance between the rigidity of the social contract community and the extreme individualism portrayed in his *Emile*, creating instead a rural ideal, one in which household-based humans live in perfect harmony with nature:

> Picture an ideal world similar to ours, yet altogether different. Nature is the same there as on earth, but its economy is more easily felt, its order more marked, its aspect more admirable. Forms are more elegant, colors more vivid, odors sweeter, all objects more interesting. All nature is so beautiful there that its contemplation, inflaming souls with love for such a touching tableau, inspires in them both the desire to contribute to this beautiful system and the fear of troubling its harmony; and from this comes an exquisite sensitivity which gives those endowed with it immediate enjoyment unknown to hearts that the same contemplations have not aroused.......The inhabitants of the ideal world I am talking about have the good fortune to be maintained by nature, to which they are more attached, in that happy perspective in which nature placed us all, and because of this alone their soul forever maintains its original character. The primitive passions, which all tend directly toward our happiness, focus us only on objects that relate to it, and having only the love of self as a principle, are all loving and gentle in their essence.[73]

And so began the great 19th century flirtation with going "back to the land", from Wordsworth's retreat to the Lake District, to self-sufficient Owenite communities, to the infatuation with the peasantry by the Russian *narodniki*. Here in the countryside necessity, that yoke we must all bear, was seen as rooted in the rhythms of nature instead of the artificial and hopelessly ambitious and capricious conventions of culture.[74]

For many it was a retreat from the obviously superior power of Coketown, the urban, industrial, Baconian modernity that they had attempted to resist, but failed. Wollstonecraft as we have seen remained torn between the two, wanting both the high culture of the city and the cottage in the country and she remains quintessentially modern in her cultural ambivalence. Her husband William Godwin remained in the city, but in his fiction idealized the potential for rural utopia In his novel *St.Leon* the main character lives in the "bosom of nature" and enjoys "...the golden age renewed, the simplicity of the pastoral life without its grossness."[75] Coleridge, in an escapist and pantheistic dream hatched in 1794 with his friends Wordsworth and Southey, yearned for an American utopia they named pantisocracy on the banks of the Susquehanna River in Pennsylvania. Writing to a tethered ass, he offers:

> *And fain would take thee me, in the Dell*
> *Where high-soul'd Pantisocracy shall dwell!*
> *Where Rats shall mess with Terriers hand-in-glove*
> *And Mice with Pussy's whiskers sport in Love...*[76]

It never happened. Wordsworth chose to retreat to the Lakes, Coleridge turned to drugs and Southey became Poet Laureate. But the pastoral impulse, still fueled by golden images of peace and plenty, retained its appeal. E.P. Thompson tells us that the Chartist leader Fergus O'Connor, in the midst of a decade long struggle in the 1830's against the disruptive power of the new modernity, yearned for such a lost Rousseauean paternalist community where "...every man lived by the sweat of his brow...when the weaver worked at his own loom, and stretched his limbs in his own field, when the laws recognized the poor man's right to an abundance of everything."[77]

In the newly established United States, various Owenite and other rural utopian experiments were established throughout the 1800s. Alexis de Tocqueville, in his seminal warnings about the dangers of excessive individualism in *Democracy in America*, bemoaned the erosion of "internalized discipline" without which one "...enters an endlessly anxious quest for desire after desire."[78] One can hear echoes of Epicurus, Lucretius and Seneca here. The Jeffersonian ideal of a decentralized, rural but sophisticated culture in which the family farm is the "...chief repository of the virtues necessary for the republic.." was seen to be at risk.[79] We will return to this theme in the next chapter, but first there is a final component of the Romantic Paradigm we need to establish.

RECURRENT SENSIBILITY: ON GOLDEN AGES, NOBLE SAVAGES AND A MODERN PRIMITIVISM | 167

SYMPATHETIC NOBLE SAVAGES

I am as free as Nature made man,
'Ere the base Laws of Servitude began,
When wild in woods the noble Savage ran.[80]

Striding out from the state of nature into the warm sunlight of the Golden Age, the Noble Savage epitomized for European cultural critics the independent, self-sufficient, child of nature whose seeming health, vigor and goodness seemed an obvious indictment of the flaccid and corrupt civilization they saw themselves inhabiting.

Dryden may have been the first to coin the phrase, but the idea that the primitive was in some inherent way ennobled is almost as old as the notion of the Golden Age. The Roman Tacitus attributed noble virtues to the barbarian Germani, describing them as brave, democratic, hospitable, hardy and mercifully free of the bonds of luxury.[81] Peter Martyr in 1516 described the peoples encountered by the Spanish explorers in the Americas as living in a Golden Age, "..without slanderous judges or books, satisfied with the goods of nature, and without worries for the future."[82] The Spanish priest Bartolome De Las Casas in 1542 described the native peoples of South America as being "...without malice or guile....notions of revenge, rancour and hatred are quite foreign to them...have no urge to acquire material possessions...and are totally uninterested in worldly power and ...have...a natural goodness."[83] William Penn in 1683 noted that in "....Liberality they excel, nothing is too good for their friend....the most merry creatures that live, feast and dance perpetually...if they are ignorant of our pleasures, they are also free from our pains."[84] But it was the Jesuit Fathers in Canada who may have had the greatest influence based on their extensive work with the Hurons, who they saw as having "... uncorrupted virtue and innate nobility."[85]

Of course, we now understand that in the Americas as well as other areas with aboriginal populations the reality of their lives was more complex than these early accounts implied. In the Americas, north and south, most of the population was, as Terry Glavin reminds us, "..town dwelling agriculturalists who had radically transformed the landscape."[86] But Glavin also notes that there were also many groups who were living in equilibrium with other species and the land and had devised complex systems of resource management. This complexity was not invisible to those who wished to either praise or condemn the quality of life observed in the "new worlds."[87]

Rousseau's resounding defence of the primitive era in human evolution in his 1753 *Discourse on the Origins of Inequality* (he never called it "noble"[88])

168 | COMING TO SENSE AND ENCOUNTERING SENSIBILITY

which had so angered Voltaire was the culmination of over 200 years of European encounters in distant lands with what appeared to many to be virtual utopias. For example, in Jonathan Swift's *Gulliver's Travels* (1726), the sources of the Houyhnhnms nobility and virtue was "...the lack of everything prized in civilized society." Whether encountered on the edge of empires, in the newly discovered America, in the South Seas or created in literature and myth, the primitive "other" was ennobled by some, while scorned by others. Their admirers tended to ascribe to these peoples a native wisdom or reason which enabled them to live peaceful and full lives in harmony with the laws of nature without the benefit of European culture or religion. Michael Ignatieff describes this tradition eloquently and links it with the Epicurean and Stoic traditions discussed earlier:

> In all the moral traditions that confronted the coming of capitalist modernity, the man of virtue was the man of few needs: in the Stoic discourse, the man of self-command; in the religious discourses of Calvinism and Jansenism, the saint; in Renaissance civic humanism, the citizen; in the reveries of Scottish Lowland gentry like James Boswell, the austere Highland chieftan. From the moment that European commerce began to press its imperium into the forest, jungle and savannah of the Americas, this judging figure took on the guise of the savage. Chained and beaten though he was in the plantations of his masters, he roamed free in their conscience, judging both their cruelty and the triviality of their desires....The savage's virtue...was premised on economic autarky, on his self-reliance as a hunter-gatherer in the lost paradise before mine and thine...In all these forms—Stoic hero, saint, citizen and savage—the man of few needs served to pass judgment on the modern man of vanity.[89]

No where is this set of images more persuasive than in the mid-19th century works of James Fenimore Cooper. His Indian heroes are "dignified, firm, faultless, wise, graceful, sympathetic, intelligent and of beautiful bodily proportions reminiscent of classical sculpture."[90] For the critics of this perspective, as we saw earlier with Voltaire, "...to express a preference for the savages is ridiculous bombast..." since they were clearly more like children, idiots and illiterates than reasoning humans.[91] By the early 19th century the decades of dislocation and disease that followed upon the encounters with Europeans led observers like C.F. Volney to see them as a fallen people, the slave of their wants and noble only in their ability to endure continued degradation.[92]

For the proponents of the Noble Savage ideal on the other hand, there were several qualities in particular that were singled out as crucial for developing their critique of the new urban, industrial and bureaucratic modernity they were witnessing being born. The Noble Savage was perceived to be living still at the "origin" of humankind and as such was able to provide clues as to the innate qualities humans possessed. This idea of the childhood of man coincided with an intense interest in the 18th and early 19th centuries in the process of human development and the education of children, or in the potential of education to make "better" adults. As well, the lifestyle of the people that Europeans encountered seemed much more focused on leisure and enjoying a "happy mediocrity" (as we saw in Chapter One) than on labor, which called into question the idea of progress and the constant escalation of desire.[93] Finally, and perhaps most importantly for our purpose here, the construction of the Noble savage ideal celebrated the presence of an innate empathy, pity or sympathy within humans who were more in touch with nature and a natural style of existence, a sympathy that Rousseau and others argued had been repressed in more advanced human cultures.

These three themes are intimately connected, with the erosion and then repression of sympathy being the key.[94] This innate quality of sympathy for fellow sentient creatures—a kind of disinterested benevolence—was the foundation for what the Scottish moral philosopher Francis Hutcheson in his *Inquiry into the origin of our ideas of beauty and virtue* (1725) saw as our "moral sense". Since it was seen as innate to the species, Hutcheson and his followers like Rousseau, David Hume and Adam Smith argued that everyone possessed it naturally; it was "as inalienable as the external senses."[95] Moral insensitivity, then, becomes a kind of madness, a radical deviation from what is natural. For Kant, "..an erring conscience" was a "logical impossibility", the problem of evil or insensitivity resting with our refusal or inability to employ our reason in following our conscience.[96] Hume, less attracted to the power of the rational than Kant, made sympathy the foundation of his moral theory and insisted that our "oughts", our judgments about the right thing to do, came directly from our intuitions. Hume's friend Adam Smith, in the classic opening passages in his *Theory of Moral Sentiments* (1759) positing observing a friend being tortured "on the rack", argues that it is our imaginative abilities that allow us to be empathetic, to place ourselves in the situations of others even if they are vastly different from our own.

And this argument for an innate moral sense is not a mere 18th century phenomenon. The great anthropologist Claude Levi-Strauss saw the faculty of pity "...which in nature replaces laws, morals and virtue and without which...there can be no law, no morals and no virtue in society.." as the key to our chance to "...live together and build a harmonious future."[97] The novelist Ian McEwan insisted recently that: "You cannot be cruel to someone, I think, if you are fully aware of what it's like to be them. In other words, you could

COMING TO SENSE AND ENCOUNTERING SENSIBILITY

see cruelty as a failure of the imagination, as a failure of empathy."[98] The scientist James Q. Wilson sees the moral sense as having obvious "adaptive value...if it did not, natural selection would have worked against people who had such useless traits as sympathy, self-control, or a desire for fairness in favor of those with the opposite tendencies."[99] Martha Nussbaum is convinced that we are more Rousseauean than Hobbesean since "..if we are made aware of another person's suffering in the right way, we will go to his or her aid."[100] The philosopher Charles Taylor sees an instinctual reluctance to inflict death or injury on another and an inclination to come to the aid of the injured or endangered and Felipe Fernandez-Armesto notes that we see the same inclinations in chimpanzees, our nearest genetic relative.[101]

CONCLUSION

While the majority opinion amongst contemporary observers of humankind may reject the idea of an innate moral sense in favor of various social constructivist approaches, the consistency over time and its cross cultural credentials (the Confucian scholar Mencius (372–289 BCE) insisted that "all human beings have the mind that cannot bear to see the sufferings of others."[102]) make the innate sympathy argument worth considering as a potentially a powerful component of a new approach to human relations to nature. Even if it remains only a predisposition rather than a fully operational quality, it is a predisposition that is certainly worthy of being nurtured since it would be a valuable tool in our attempts to treat fellow humans and the rest of natural world with more sensitivity and wisdom.[103]

It is this increasing erosion of what the Romantics would call our sympathetic imagination that opens the door to the "expansion of desire problem" outlined earlier, the untempered focus on the self at the expense of an affinity toward the other, the community and the natural world. Each individual, in a sense, becomes instead a means to pursue an endlessly expanding array of material desires and sensual pleasures—and reason transformed into science and then into technologies becomes the irresistible instrument employed to satisfy those desires. For the Romantics, then, the new modernity is built on a foundation of radically deteriorating human relationships caused by alienation from nature, "...the destruction of the old organic and communitarian forms of social life [and] the isolation of individuals in their egoistic selves."[104]

But how do compassion or sympathy, typically reserved for human relations, affect our treatment of animals or the rest of nature? It is perhaps easy to imagine how they might affect our responses to certain mammals, especially those with big brown eyes! Nussbaum points out that the movement for the humane treatment of animals relied heavily on generating a sense of empathy in order to in turn promote compassion for their suffering. She cites the novel *Black Beauty* which, despite excessive anthropomorphizing, showed that "...we are much more likely to have appropriate compassion for the pain of animals if we are able to at least try hard to reconstruct their experience of the bad things we do to them."[105] The movement from empathy with and compassion for a horse or other mammal to a forest or an ecosystem is a long trek which we may only just be starting. But it is, really, just a re-tracing of a previous journey and, as we will see, may be the essential starting point for creating an ecologically healthy human culture.

We return then to the issue posed at the beginning of this book, posed here by Bill Readings in his exploration of the crisis in modern education: "how to move from the state of nature to the state of reason without destroying nature, or, put differently, if we are what we *are* by nature, how can we retain that sense of being while moving into the modern via reason which tends to move us toward the *mechanical*, artificial or non-human?"[106]

I want to try to respond to this in the next chapters, but to conclude this discussion of the contribution made by what I have called the Romantic tradition I want to return to Mary Wollstonecraft, and to her daughter.

Mary Wollstonecraft's life was flanked by two people with whom she was "intimate" but whom she never met—Rousseau on the one hand and her daughter Mary Shelley on the other. In her *Vindication of the Rights of Women* and in various letters and essays she offered trenchant criticisms of Rousseau's analysis of the role of women and of his seeming preference for an original primitivism over modern culture, but she also admitted that "...I have always been half in love with him."[107] She read as much of Rousseau's work as was available, his deeply reflective *Reveries of A Solitary Walker* being one of her favourite books and, according to Richard Holmes, influential in shaping the confessional nature of her book on Scandinavia.[108] Despite her misgivings concerning his ideas about gender, she accepted the prevailing view that Rousseau was the intellectual and spiritual force behind the French Revolution and despite equally trenchant misgivings about certain events in Paris, she *was* a revolutionary and a firm believer in the ideals of 1789.

We have, then, three generations before us, spanning the hundred years that witnessed the birth of the modernity we know today. Rousseau, as we have seen, raised what remain the central questions about that modernity and offered several alternative paths aimed at softening its acceleration of humanity's dramatic turn away from its more organicist social and ecological way of being. Isaiah Berlin, that keen observer of this decisive era in Western culture, reminds us of the central components of this Rousseauean world view:

- religious, conservative and inner directed
- intolerant of a singular focus on reason, centralization, and scientific progress
- personal relationships and inner life more important than the values of the external world
- family the basis of true human existence
- loose texture of life founded on affection, tradition and local values
- minimum of interference by trained experts or remote officials
- submission not struggle is the way to serenity and truth
- learn to be contented with one's own insignificance—to learn to submit is to learn to understand
- genius is individual and incapable of being cultivated by social organizations
- superiority of the individual conscience over institutional authority
- celebration of provincialism, simplicity, innocence and roots in local life
- faith in day to day action, instinct and face-to-face encounters with things and men
- wisdom comes from true participation in life, from action rather than contemplation[109]

While perhaps too prone to push people toward nostalgia, this world view nonetheless embodied a radical critique not only of the absolutism of the previous era but also of the emergent bourgeois culture and its focus on industry, on instrumental reason, and on materialism.

Mary Wollstonecraft is a more transitional figure and in many ways is more representative of our time than either Rousseau, her daughter Mary, or Percy Shelley. Writing in the midst of the revolutionary maelstrom that forged the birth of modernity, Mary Wollstonecraft emerges as much a Kantian as a Rousseauean. Given that Kant felt that Rousseau was the "Newton of the Emotions" and therefore a perfect companion to his own championing of practical reason, Wollstonecraft's middle ground between the two is a quite respectable position. We have seen her romantic side at the start of this chapter. The Kantian side is just as strong.

Walking the centre line of cultural analysis, Wollstonecraft argues that it is in fact reflection/reason coupled with the imaginative powers that it releases or nurtures that allows for both progress and the active memory of a more sensual and primitive past. "The more I see of the world, the more I am convinced that civilization is a blessing not sufficiently estimated by those who have not traced its progress; for it not only refines our enjoyments, but produces a variety which enables us to retain the primitive delicacy of our sensations. Without the aid of the imagination all the pleasures of the senses must sink into grossness.....who will deny that the imagination and understanding have made many, very many discoveries...which seem only harbingers of others still more nobler and beneficial. I never met with much imagination amongst people who had not acquired a habit of reflection."[110] So reason and its power of reflection are what gives us the imagination to put into practice the Baconian agenda. "The increasing population of the earth must necessarily tend to its improvement, as the means of existence are multiplied by invention." A very modern idea that! Adrift in the often seductive wilds of a Rousseauean northland, she finds "....it necessary to emerge again, that I may not lose sight of the wisdom and virtue which exalts my nature."[111] She enjoys dreaming about the "...fables of the golden age" but as we saw earlier, reason drags her back to the present, to a world that needs improvement.

As shown by her scathing denunciation of commerce and the bourgeois values she encountered in her visit to Copenhagen at the end of her travels, Wollstonecraft's optimism was based on her faith in revolutionary change. Her daughter Mary Shelley and her poet husband Percy Shelley had to chart a new critique living as they did in an age of profound political reaction. So with the Shelleys we see the beginning of the great division in assessing modernity, a divide that persists to our time. Rousseau's prescription for a social, political and economic order that was founded on simplicity, sustainability and sympathy was championed in various ways during the era of great possibilities associated with the French Revolution, but was perverted by bonapartism and eventually denounced by the conservative regimes that flourished after 1815. Wollstonecraft's hopes for the kind of political change that would humanize and democratize the new modernity her generation witnessed being born was not to be realized. And it was this reality that the Shelleys and their peers faced in the early decades of the 19th century, the era when modern industrial and scientific culture began it's sustained and devastating drive to master the natural world.

Mary Shelley's most famous written work, *Frankenstein, or the Modern Prometheus*, can be seen as a critique of these mastery pretensions and at the same time a rejection of the optimism implicit in much of her mother's work. Victor Frankenstein is to create a "new man" by combining a grab bag of "holistic" and scientifically primitive ideas about unity and simplicity with the tools and systems of modern science—blending Rousseau and Bacon as it were—in order to create a more perfect world.[112] He yearns in doing so to penetrate the "citadel of nature" in order to "unfold to the world the deepest mysteries of creation".[113] And, of course, such an ill-conceived penetration turns

COMING TO SENSE AND ENCOUNTERING SENSIBILITY

out badly for Victor, for his Creature and for the society that surrounds them. But it was not just "bad science" that Mary Shelley is warning us about, it is the separation of rational, scientific inquiry—the stuff of progress—from humane values and obligations that she sees as the flaw in the modern project. She has Victor reflect that "...if the study to which you apply yourself has a tendency to weaken your affections and to destroy your taste for those simple pleasures in which no alloy can possibly mix, then that study is certainly unlawful, that is to say, not befitting the human mind."[114]

From their exile in Italy after 1817, Mary and Percy Shelley can serve as our transition to the emergence of a vigorous defense of the natural world that would become a global movement in the late 20th century. Percy Shelley, militant vegetarian and eloquent observer of the beautiful and the sublime, saw clearly the dilemma posed at the start of our inquiry, that is, how to save what we want while preserving what we need. Like Rousseau, the poet Shelley found part of the answer in moments of reverie, or in other "abnormal" states in which naïve insights could triumph, even momentarily, over self-consciousness. In a reflection on the qualities of childhood, which Freud would no doubt see as part of the problem, Shelley finds a solution:

> Let us recollect our sensations as children. What a distinct and intense apprehension had we of the world and of ourselves. Many of the circumstances of social life were then important to us which are no longer so. But that is not the point of comparison on which I mean to insist. We less habitually distinguished all that we saw and felt, from ourselves! They seemed, as it were, to constitute one mass. There are some persons who in this respect are always children. Those who are subject to the state called reverie feel as if their nature were dissolved into the surrounding universe, or as if the surrounding universe were absorbed into their being. They are conscious of no distinction. And these are states which precede, or accompany, or follow an unusually intense and vivid apprehension of life. As men grow up this power commonly decays, and they become mechanical and habitual agents.[115]

Victor Frankenstein lived a dualist frame of humans and nature, creator and creature, Mary Shelley's nightmare vision of where the Cartesean system would lead humankind and nature. Percy Shelley begins to chart a return to a modern holistic vision with a return to the naïve insights of children, the primitive and the poet. Aware of the powerful forces arrayed against such a course, in his *Ode to the West Wind*, he calls for an alliance with nature itself to carry the message forward:

RECURRENT SENSIBILITY: ON GOLDEN AGES, NOBLE SAVAGES AND A MODERN PRIMITIVISM | 175

Make me thy lyre, even as the forest is:
What if my leaves are falling like its own!
The tumult of thy mighty harmonies
Will take from both a deep, autumnal tone,
Sweet though in sadness. Be thou, Spirit fierce,
My spirit! Be thou me, impetuous one!
Drive my dead thoughts over the universe
Like withered leaves to quicken a new birth!
And, by the incantation of this verse,
Scatter, as from an unextinguished hearth
Ashes and sparks, my words among mankind!
Be through my lips to unawakened earth
The trumpet of a prophecy! O, Wind,
If Winter comes, can Spring be far behind?[116]

NOTES

1 Here I am following the lead of Michael Lowy and Robert Sayre's work in *Romanticism Against the Tide of Modernity* (Durham: Duke University Press, 2001).

2 Wollstonecraft, p69.

3 Wollstonecraft, p 149. For the Coleridge quote, see E.P. Thompson, *The Making of the English Working Class* (New York: Vintage, 1963), p 176.

4 Cited in William St. Clair, *The Godwins and the Shelleys* (London: Faber and Faber, 1989) p 272.

5 Wollstonecraft, p 122.

6 Wollstonecraft ,p 149.

7 Wollstonecraft, p 119.

8 Wollstonecraft, p 122

9 Wollstonecraft, p 191-3.

10 Wollstonecraft, p 111..

11 Janet Todd, *Mary Wollstonecraft: A Revolutionary Life* (London: Phoenix Press, 2000), p 307.

12 Todd, p 74.

13 *The American Heritage® Dictionary of the English Language*, Fourth Edition (Boston: Houghton Mifflin Company, 2000).

14 Berman, (2000), p 6.

15 For all her realism and materialism, Wollstonecraft could not abandon the need for a soul. "I cannot bear to think of being no more—of losing myself—though existence is often but a painful consciousness of misery; nay, it appears to me impossible that I should cease to exist, or that this active, restless spirit, equally alive to joy and sorrow, should only be organized dust—ready to fly abroad the moment the spring snaps, or the spark goes out, which kept it together. Surely something resides in this heart that is not perishable—and life is more than a dream." p 112.

16 Thompson, p 12–13.

17 Maurice Cranston, *Jean-Jacques: The Early Life and Work of Jean-Jacques Rousseau* (London: Allen Lane, 1983), p 306.

18 Ernst Cassirer, *Rousseau, Kant, Goethe* (Princeton: Princeton University Press, 1945), p 10.

19 John Zerzan refers to these as ideologically derived "official lies" designed to buttress a bankrupt Hobbesian view of human history. in Gowdy (pp 255–280).

20 Peter France, *Politeness and Its Discontents: Problems in French Classical Culture*, (Cambridge: Cambridge University Press, 1992), p 188.

21 W.D. Falk, *Ought, Reason and Morality* (Ithaca: Cornell University Press, 1986), p 269.

22 Jean Starobinski, "The Antidote in the Poison", in *Blessings in Disguise* (Cambridge: Harvard University Press, 1993), p 127.

23 Paul Shepard, "A Post-Historic Primitivism" in Max Oelschlaeger (1992), p 80.

24 In *Little Gidding*, cited by Guignon, p 55.

25 Shepard, (1967), p 76.

26 "Notes on Prometheus Unbound, by Mrs. Shelley" in *The Complete Poetical Works of Percy Bysshe Shelley*, ed., by Thomas Hutchinson (London: Oxford University Press, 1948), p 271.

27 Aeschylus, *Prometheus Bound*, trs by T. G. Tucker (Melbourne: Melbourne University Press), 1935. p 26.

28 Lovejoy and Boas, p 159.

29 Lovejoy and Boas, p 273. See also Seneca, *Letters from A Stoic* (London: Penguin, 1969), pp 162-3.

RECURRENT SENSIBILITY: ON GOLDEN AGES, NOBLE SAVAGES AND A MODERN PRIMITIVISM | 177

30 Lucretius, *On the Nature of the Universe*, (London: Penguin, 1994), p 153.

31 Freud, *Civilization and Its Discontents*, p 77.

32 Cranston, (198 5) p 209.

33 Lucien Colletti, *From Rousseau to Lenin* (New York: Monthly Review Press, 1972), p 150

34 Ronald Meek, *Social Science and the Ignoble Savage* (Cambridge: Cambridge University Press, 1976), p 22; Peter Weston, "The Noble Primitive as Bourgeois Subject", *Literature and History*, v. 10:1, 1984, (pp 59–71).

35 Daryl Rice, "Teaching Rousseau: Natural Man and Present Existence", *Teaching Political Science* v. 16:4, 1989 (pp 148–152), p 150.

36 Wokler, (1978), p 117.

37 Rousseau, *Discourse on Inequality*, p 90.

38 Francis Moran, "Between Primates and Primitives: Natural Man as the Missing Link in Rousseau's Second Discourse", *Journal of the History of Ideas*, 54:1, 1993 (pp 37–58), p 52.

39 Megarry, p 12.

40 Ter Ellingson, *The Myth of the Noble Savage* (Berkeley: University of California Press, 2001), p 93. Some argue that Rousseau knew this and only advanced the solitary human as a philosophic starting point, this despite his many arguments defending his position. The consensus that humans are inherently social is a powerful one, starting with Aristotle's insistence on the family as the human baseline: "....man is by his nature a pairing rather than a social creature, inasmuch as the family is an older and more necessary thing than the state, and procreation is a characteristic more commonly shared with the animals. In the other animals partnership goes no further than this; but human beings cohabit not merely to produce children but to secure the necessities of life." Aristotle, *Ethics*, (Penguin), p 280. Contemporary views cast the net even wider: "People are constituted by their relationships. We come into being in and through relationships and have no identity apart from them. Our dependence on others is not simply for goods and services. How we think and feel, what we want and dislike, our aspirations and fears—in short, who we are—all come into being socially." Daly and Cobb, p 161.

41 Todorov, p 84.

42 Scott, p 706.

43 Guignon, p 19.

44 Guignon, p 19.

45 Seneca p 126

46 Thomas Hutchinson, (1970) p 532.

47 To this degree Wollstonecraft embedded herself firmly within the Kantian tradition of Enlightenment thought. For Kant the sight of nature's immense power does at first overwhelm, but then we remember that though physically weak we possess mind and will which can "...assert themselves mentally over the blind force of nature". In Lesley Sharpe, *Friedrich Schiller* (Cambridge University Press 1991), p 125. Thus we are transcendent, the sublime fear we experience is within us, not in nature. (Kate Soper, "Looking At Landscape", *CNS*, v. 12:2, 2001, pp 132–138)

48 Oerlemans, p 119.

49 Susan Neiman, *Evil in Modern Thought* (Princeton University Press, 2002), p 883. To judge nature to be sublime is to be aware of something surpassing any capacities I ever dreamed.... The sublime is not merely chaotic; it is overwhelming....The awe that accompanies the sublime comes not just from the feeling that I could not have created something as crazy as lightning, but from the thought that on balance, I would not have done so. Practical reason

cannot forget that the sublime is always dangerous."

50 McKusick, (1991), p 236.

51 Sharon Ruston, *Shelley and Vitality* (Basingstoke: Palgrave, 2005), p 3.

52 Thomas, p 166.

53 Henry Baker, "Universe, or a Poem Intended to Restrain the Pride of Man" (1727) cited in Keith Thomas, p 169.

54 Oerlemans, p 77.

55 Mizzoni, p 47.

56 Jean-Jacques Rousseau, "Letter to Voltaire on Optimism", in John Hope Mason, ed. *The Indispensable Rousseau* (London: Quartet Books, 1979), p 115

57 Serres, (2000), p 165.

58 Trs by Peter Dean, 2005, www.brindin.com/povc2008.htm

59 Nussbaum, (2001), p 185; Immanuel Kant, "Conjectural Beginning of Human History" in Lewis Beck, ed., *On History* (Indianapolis: Bobbs-Merrill, 1963), p 67.

60 R.A. Leigh, "Jean-Jacques Rousseau and the Myth of Antiquity in the 18th Century" (pp 155–168), in R.R. Bolgar, ed. *Classical Influences on Western Thought*, (Cambridge: Cambridge University Press, 1979), p 158.

61 Candace Slater, "Amazonia as Edenic Narrative" (pp 114–131) in William Cronon, ed., *Uncommon Ground: Re-Thinking the Human Place in Nature* (New York: Norton, 1996), p 116.

62 Interestingly, this focus on childhood is widely cross-cultural. "A prominent Buddhist theme is to "imitate the child-like mind" or to show one's "original face"—modes of experience "...uncluttered by discursive reflection". In creating elaborate rationalizations to justify meeting needs/desires we "...effectively remove ourselves from our natural setting. The child-like mind is originally intimate with its world , for the child has not yet learned to compartmentalize his or her experiences into socially acceptable places. The emotional ties between the child and the environment are powerful and often uncontrolled. As the child develops, however, it is easy for the original intimacy to be lost. The mystery and splendor that accompanies an emotional attachment to nature can become covered, forcing us , in adult stages, to recultivate our child-like minds." David Shaner, "The Japanese Experience of Nature" (p 163–182) in Callicott and Ames, p 180.

63 From Hesiod, "Works and Days", in Lovejoy and Boas, p 27.

64 Vincent Geoghegan, "A Golden Age: From the Reign of Kronos to the Realm of Freedom", *History of Political Thought*, v. 12:2, 1991 (pp 189–207), p 191.

65 Ted Hughes, *Tales from Ovid* cited in Jonathan Bate, (2000), p 29.

66 Harry Levin, *The Myth of the Golden Age in the Renaissance* (Bloomington: Indiana University Press, 1969), p 22.

67 This is Rousseau's primary lesson to young Emile in *Emile*, and he insists that it is better to decrease desires rather than increase faculties or means: A being endowed with senses whose faculties equaled his desires would be an absolutely happy being. In what, then, consists human wisdom on the road of true happiness? It is not precisely in diminishing our desires, for if they were beneath our power, a part of our faculties would remain idle, and we would not enjoy our whole being. Neither is it in extending our faculties, for if, proportionate to them, our desires were more extended, we would as a result only become unhappier. But it is in diminishing the excess of the desires over the faculties and putting power and will in perfect equality." Rousseau, *Emile* p 80.

68 Nussbaum, (1994) p 105.

69 Lovejoy and Boas, p 120.

RECURRENT SENSIBILITY: ON GOLDEN AGES, NOBLE SAVAGES AND A MODERN PRIMITIVISM | 179

70   Stelio Cro, *The Noble Savage: Allegory of Freedom* (Waterloo: Wilfrid Laurier University Press, 1990), p 122; Rousseau, *Reveries of the Solitary Walker*, p 83. Leon Emery, "Rousseau and the Foundations of human Regeneration", *Yale French Studies* v. 28, 1961 (pp 3–12), p 6.

71   Guignon, p 47.

72   Judith Shklar. "Rousseau's Two Models: Sparta and the Age of Gold", *Political Science Quarterly*, v. 81, 1966 (pp 25–51), p 42.

73   Jean-Jacques Rousseau, *Dialogues* p 9.

74   Mark Cladis, "Rousseau and the Redemptive Mountain Village", *Interpretation* v. 29:1, 2001 (pp 35–54), p 42. Rousseau argued that Nature was not capricious. In his Letter to Voltaire on the Lisbon Earthquake he places the blame for the disastrous consequences for the city on the humans who chose to build tall buildings of brick in a known earthquake zone.

75   In Peter Marshall, *William Godwin* (New Haven: Yale University Press, 1984), p, 208.

76   Rupert Christiansen, *Romantic Affinities* (London: Sphere, 1988), p 152.

77   Thompson, p 230.

78   Alexis deToqueville, *Democracy in America*, (v. 2 p 149) cited in Roger Boesche, "Why Did Toqueville Fear Abundance?", *History of European Ideas*, v.9:1, 1988, p29.

79   Kimberly Smith, (2003) p 15.

80   John Dryden, *The Conquest of Granada by the Spaniards* (1672), cited in Ellingsen, p 8.

81   Hoxie Fairchild, *The Noble Savage: A Study in Romantic Naturalism* (New York: Columbia University Press, 1928), p 5.

82   Cro, p22.

83   Bartolome De Las Casas, *A Short Account of the Destruction of theIndies*, (London: Penguin Books, 1992), p 10.

84   P.J. Marshall & G. Williams, *The Great Map of Mankind* (London: J.M. Dent & Sons, 1982), p 99.

85   Geoffrey Symcox, "The Wild Man's Return: The Enclosed Vision of Rousseau's Discourses" (pp 223–247), in Edward Dudley and Maximllian Novak, *The Wild Man Within: An Image in Western Thought from the Renaissance to Romanticism*, (Pittsburgh: University of Pittsburgh Press, 1972) p 227.

86   Glavin, p 133.

87   For a review of debates about the stance of indigenous peoples in the Americas toward nature, see Shepard Kretch, *The Ecological Indian: Myth and History*, (New York: Norton, 1999) and Michael Harkin and David Lewis, eds., *Native Americans and the Environment: Perspectives on the Ecological Indian* (Lincoln: University of Nebraska Press, 2007).

88   "...in Rousseau's construction, the savage was in some ways happier and more fortunate than civilized man precisely *because* he was not, and could no be, 'Noble': lacking the abstract concepts of good and evil that civilization had invented, he was also spared the practical effects of socioeconomic and moral exaltation and degradation that developed alongside them....His mind was unable to form abstract ideas." Ellingson, p 82.

89   Michael Ignatieff, *The Needs of Strangers* (London: Penguin, 1986), p 94.

90   Ketch, p 19.

91   Eighteenth Century French economist Jacques Turgot, cited in Meek, p 71; John Locke cited in Marshall and Williams, p 192.

92   Ellingsen, p 110–115. citing Constantin Francois Chassebouf, comte de Volney, *On the Indians or Savages of North America* (1803)

93   Christopher Thacker, *The Wildness Pleases* (New York: St. Martin's, 1983), p 162.

94   "Sympathy is both a vivid awareness of the other person's pain and an altruistic urge to end it. In the Greek it means "suffering in unison, suffering incurred by a feeling of affinity with

180 | COMING TO SENSE AND ENCOUNTERING SENSIBILITY

another person who is suffering." Lauren Wispe, *The Psychology of Sympathy*, (New York: Plenum Press, 1991), p 68.

95 Carey Daniel, "Reconsidering Rousseau: sociability, moral sense and the American Indian from Hutcheson to Bartram", *British Journal of Eighteenth Century Studies*, v. 21, 1998 (pp 25–38), p 27.

96 Laurence Lockridge, *The Ethics of Romanticism* (Cambridge: Cambridge University Press, 1989, p 98.

97 Claude Levi-Strauss, "Rousseau: Father of Anthropology", UNESCO *Courier*, March 1963 (pp 10–14), p 14.

98 Interview with Ian McEwan, Vancouver *Sun*, 5 Feb 2005, p D15.

99 James Q. Wilson, p 23

100 Martha Nussbaum, *Frontiers of Justice* (Cambridge: Harvard University Press, 2006) p 412.

101 Taylor,(1992), p 5.; Fernandez-Armesto, p 36.

102 Cited in Alasdair MacIntyre, *Dependent Rational Animals* (Chicago: Open Court Press, 1999), p 123.

103 The social constructivist versus essentialist debate is on-going. Richard Rorty insists that "socialization, and thus historical circumstance, goes all the way down, there is nothing 'beneath' socialization or prior to history which is definatory of the human." (*Contingency, Irony, and Solidarity*, Cambridge University Press, 1989, p xiii). Herbert Marcuse insisted "... there is no such thing as an immutable human nature." (in "Ecology and the Critique of Modern Society", *Capitalism, Nature, Socialism*, v. 3:3, 1992, pp 29–38). William Cronon notes that the idea of a human nature must compress diverse and complex phenomena into a "flat, colorless cartoon" ("Introduction", Cronon, ed., (1996), p 35.) Eric Gander, on the other hand, insists that Human nature is back. The thinking seems to be that a complex and richly detailed human nature does exist, that it is to a large degree scientifically knowable, that it differs markedly between the sexes, that it delimits a set of viable human cultures, and that, because of all this, it makes a big difference when we set out to discuss moral;, ethical, and political questions." *On Our Minds* (Baltimore: Johns Hopkins, 2003, p 3.)

104 Lowy and Sayre, p 42.

105 Nussbaum, (2001), p 333.

106 Bill Readings, *The University in Ruins* (Cambridge: Harvard University Press, 1996), p 63.

107 Letter to Gilbert Imlay, 22 September 1794 cited in Todd, p240.

108 Richard Holmes, ed., Mary Wollstonecraft, *A Short Residence in Sweden...* (London Penguin, 1987), p 282.

109 Isaiah Berlin, "The Magus of the North " (Johann Hamann), *New York Review of Books*, 21 Oct 1993, p 64.

110 Mary Wollstonecraft, *A Short Residence in Sweden*, (London: Penguin, 1987), p 72–3.

111 Wollstonecraft (1987) ,p 122.

112 See Gregory Dart, *Rousseau, Robespierre and English Romanticism* (Cambridge: Cambridge University Press, 1999) p 4–5

113 Mary Shelley, *Frankenstein* (London: Penguin,1985), p 88 and 96.

114 Mary Shelley, (1985), p 103.

115 Percy Shelley, "essay on Life" in David Lee Clark, p 174. Martha Nussbaum notes (citing Freud) that "...although at times the infant's world is a Golden Age world, these times alternate with times when the world is hungry, distressed, and in discomfort." Nussbaum (2001) p 185.

116 Percy Shelley, "Ode to the West Wind", in Thomas Hutchinson, p 579.

PART THREE
RE-THINKING CULTURE AND NATURE

CHAPTER FIVE

FINDING ECOLOGICAL SENSIBILITY IN A MECHANISTIC CULTURE

"Going to the mountains is going home"   JOHN MUIR

Beliefs in Golden Ages or Noble Savages have proved grist for romantic mills and been important in keeping alive cultural alternatives, but do not provide practical solutions to the pressing environmental issues so crucial to achieving ecological sustainability in our time. As if in reaction to these almost dreamy motifs and as testimony to their inadequacy, more pragmatic voices have sought to build bridges between the sensibilities aroused by these mythic and romantic stories and the need for a more substantive grounding for the tasks involved in bringing about change in the real world. In this chapter we will be exploring some of these more contemporary approaches.

The 19th century was in many ways a high water mark for the rational, technocratic and industrial culture envisaged by the likes of Bacon, Descartes, Galileo and Newton and the era of the Enlightenment that followed. The European (and American) system of colonial and imperial domination of markets and resources assured an expansion of industrial production at truly remarkable rates, rapidly transforming Western culture into an urbanized and industrial global powerhouse. Simultaneously, thanks to the cultural, social and environmental stresses produced by the expansion of this global system, the dissident or alternative ideas I have associated with a tradition starting with Epicurus and moving through Lucretius, St. Francis, Rousseau and the Romantics has emerged from the shadows to claim a position as an alternative vision for modernity.

So it is time now to assess the possibility of this counter-tradition offering a challenge to the materialist, consumption-driven and environmentally

RE-THINKING CULTURE AND NATURE

disastrous modernity that has come to characterize our era. Focusing on the ideas of critics such as John Muir, Henry David Thoreau, Carolyn Merchant, Aldo Leopold, Val Plumwood and Michel Serres, we will explore the basis for their contributions to a new ecological sensibility in ideas about evolution, in a renewed biocentric concern for all life forms, and in the systemic/holistic approach of ecocentrism. From these scientific and philosophical underpinnings, the chapter will conclude with examples of some contemporary practical strategies for moving a modernity based on ecological sustainability from the shadows to the mainstream.

The modernity devised in Western culture rests on a sea of particles. These "pieces of matter" in their various combinations make up ourselves and the phenomena we experience. From the triangles of Plato with their connecting hooks and barbs, the falling atoms of Democritus, the swerving atoms of Epicurus, the protons discovered by Rutherford to the now hypothesized quarks, we operate in a world of swirling, falling, combining and disengaging invisible particles. As we have seen, our sciences and technologies are based on this privileging of the part over the whole. Bacon convinced us that understanding began by breaking down objects and problems into their smallest components. Descartes needed "clear and distinct" images and understandings that could only be derived from observation of parts rather than wholes.

And then we operationalize. In the opening pages of *The Wealth of Nations*, Adam Smith offers us the story of the division of labour and modern pin-making. He shows us that while one worker would be fortunate to be able to make one straight-pin a day working on his own, ten workers can produce up to 48,000 pins a day by specializing: "One man draws out the wire, another straights it, a third cuts it, a fourth points it, a fifth grinds it at the top receiving the head."[1] It is a short step from the straight pin to Henry Ford's assembly line in which an automobile is created from a collection of parts assembled by mechanical workers. This is indeed a World View in which:

> The world consists of an array of precisely demarcated individual things or substances, which preserve their identity through time, occupy definite positions in space, have their own essential natures independently of their relations to anything else, and fall into clearly distinct natural kinds. Such a world resembles a warehouse of automobile parts. Each item is standard in character, independent of all other items, in its own place, and ordinarily unchanging in its intrinsic nature.[2]

This atomistic-mechanistic paradigm has an obvious link to our conceptualization of nature, which necessarily becomes a collection of discreet objects or phenomena which may or may not be connected. And if perchance undesirable connections are found (e.g. greenhouse gases and global warming), the mechanistic solution is simply to re-arrange the phenomena.

This is far cry from Shelley's child-like imagining of no distinctions between self and nature, no boundaries, no distinct parts—only wholes, only an overlapping series of fluid relationships. Equally part of modernity, this alternative way of assessing and relating with phenomena outside the self focuses not on the working of atoms or sub-atomic particles, but on the realm of sensate experience. And from Lucretius through the Romantics to contemporary post-Newtonian physics this sensual realm is noted for its fluidity, a state of being particularly difficult to quantify or deconstruct. Michel Serres introduces us to a very different Lucretius, a materialist still, but suddenly not the mechanist he is so often accused of being:

> Lucretius launches us into movement—everything in his work begins with turbulence....if you follow his vortices, they bring things together, forming and destroying worlds, bodies, souls, knowledge, etc. Turbulence isn't a system, because its constituents fluctuate, fluid and mobile. Rather, it is a sort of confluence, a form in which fluxes and fluctuation enter, dance, crisscross, making together the sum and the difference, the product and the bifurcation, traversing scales of dimensions.....one must concede that everything is not solid and fixed and that the hardest solids are only fluids that are slightly more viscous than others.[3]

With the Epicurean innovation that atoms may randomly swerve, solids—the discreet particles of the mechanists—become just variations of holistic fluids, unmeasureable and unquantifiable. David Abrams and other ecologists insist that our world of direct, sensate experience resembles this fluid, dynamic and essentially subjective world of flow much more closely than it does the mechanistic world of laboratories, legos or assembly lines.[4] And we move closer here to an Asian version of modernity in which "...all things are interrelated to form a network of interchange of processes."[5] In this variant of modernity the ideal is to transcend the notion of isolated selves and particles and recognize instead the basic unity of the universe.[6] But, once again we are getting ahead of our story. The contemporary insights of Serres and Abrams have deep roots in Western culture and to start our examination of those roots we turn to the Scots-American naturalist John Muir, explorer of Yosemite, founder of the Sierra Club and advocate for wilderness.

Sally Miller gives us a useful metaphor for John Muir, calling him the bridge between Jean-Jacques Rousseau and Aldo Leopold, the key figure in the modern environmental movement.[7] More than merely spanning a gulf, a bridge contains elements of both sides, adopting from the past and

186 | RE-THINKING CULTURE AND NATURE

anticipating the future. Muir, then, becomes our transitional figure between the Romantic critique of mechanism and the views of contemporary ecologists and environmentalists.

Arriving in the United States in 1849 from his home in Dunbar, Scotland, the 11 year-old John Muir was the product of a strict religious upbringing within an evangelical splinter group of the Church of Scotland. During his youth on the family farm in Wisconsin, Muir was an inventor, a student of science and a voracious reader. Influenced through his reading of natural theologists such as John Ray, the young Muir was determined to reconcile his interest in science and in nature with his religious faith. Suspicious of civilization, Muir went one better than Rousseau and chose deliberate and singular immersion in wilderness as the means of effecting this reconciliation. His journeys through the American South, Canada, California and Alaska convinced Muir that it was in wilderness that God displayed his goodness and wisdom in the most clear and distinct manner. But like Rousseau and others who aimed to reconcile God and nature, Muir often wavered in his more conventional religious beliefs, sometimes sounding more like a materialist than a creationist:

> Some of the days I have spent alone in the depths of the wilderness have shown me that immortal life beyond the grave is not essential to perfect happiness, for those diverse days were so complete there was no sense of time in them, they had no definite beginning or ending and formed a kind of terrestrial immortality.[8]

We can follow Muir's movement toward an early version of what we might today call ecocentrism or deep ecology by examining three epiphanies that mark his journey.

THE THREE EPIPHANIES

Muir sought in his books to become a kind of "everyman", someone who hoped that by telling the story of his personal encounters with nature and elaborating on the beneficial effects of these encounters on him he would be able to persuade others to follow his example in spirit if not in life. Thus all of his writing is intentionally autobiographical and often anecdotal. In his life-writing Muir focused on moments of crisis, through each of which he acquired the central ecological insights that shaped his life. They were "constructed" in the sense that he seemed to have discovered these moments or epiphanies after the fact by working through his journals and notes, organizing his life in order that it make sense to a public he was trying to influence.

The first epiphany occurs in 1864, while Muir was wandering through the forests, bogs and swamps north of Lake Huron in what is now Ontario. He had left Wisconsin that year in part at least to avoid being drafted into the army to fight in the American Civil War. A mechanic and inventor by trade and disposition, he worked at various jobs during this sojurn in Canada, but much of his time was spent exploring wilderness and "botanizing", an interest he had picked up while studying at the University of Wisconsin. He had been reading the journals of the Scottish explorer Mungo Park and was already making plans for a journey to South America in order to begin his own jungle explorations. Meanwhile, on an early Spring day in 1864, he found himself lost in a jungle of a different sort on a trail through an icy northern bog with no clear exit and no sign of human presence.

> ...I had been fording streams more and more difficult to cross and wading bogs and swamps that seemed more and more extensive and more difficult to force one's way through. Entering one of these great tamarac and arbor-vitae swamps one morning...struggling through tangled drooping branches and over and under heaps of fallen trees, I began to fear that I would not be able to reach dry ground before dark, and therefore would have to pass the night in the swamp...But when the sun was getting low...I found beautiful Calypso on the mossy bank of a stream, growing not in the ground but on a bed of yellow mosses in which its small white bulb had found a soft nest and from which its one leaf and one flower sprung. The flower was white and made the impression of the utmost simple purity like a snowflower....It seemed the most spiritual of all the flower people I had ever met. I sat down beside it and fairly cried for joy.[9]

For Muir the important insight, the epiphany, was that the orchids he had stumbled upon in that Canadian swamp were not related in any way to human beings. "Were it not for Muir's chance encounter, they would have lived, bloomed, and died unseen."[10] Nature, in the form of that small flower, was seen by Muir to transcend all specific beings, including himself. This was the start of Muir's revision of his decidedly anthropocentric religious background and his move toward pantheism.

Here, Michael Cohen argues, Muir was translating the Calvinism he had grown up with into a new kind of environmental ethic that preserved a foundation in religious faith. For the great American Calvinist Jonathan Edwards (1703–58), humility before God was central to the creed: "to lie low before God, as in the dust; that I might be nothing but God might be all." Just as important, for Edwards nature had not been complicit in the "Fall" and hence was not depraved as was man, but rather was "...pure and capable of reflecting moral

RE-THINKING CULTURE AND NATURE

truth." [11] Muir had picked up these ideas not just from Jonathan Edwards but also from the other exponents of "natural theology". Such humility before nature and the conflation of nature—particularly "wild" nature—with Beauty and with God was to become a central feature of Muir's work and the tradition of privileging "wilderness" that has come to characterize much of North American environmentalism. [12]

Later, following a serious industrial accident that left him temporarily blinded, he was off on another adventure—his Thousand Mile Walk. His trek through the U.S. South to Florida began in 1867 and was meant to end up following Alexander von Humboldt into the Amazon jungle. Muir's journal describing the walk was published posthumously in 1916 as the *Thousand Mile Walk to the Gulf*. It was on this walk that Muir had his second epiphany.

Approaching Savannah, Georgia in October 1867, Muir found himself without money, unable even to purchase food. While waiting for funds to be wired to him from Wisconsin, he spent five days living in a small shelter constructed amid the bramble bushes of the gothic Bonaventure Cemetery a few miles outside of the city. During those five days, in a state of reverie induced in part by hunger and exhaustion and in part by his surreal surroundings, Muir claims to have discovered some of the elemental truths that he would spend a lifetime pursuing. This process of dream-like discovery was capped a month later by an extended bout of malaria in Florida. What had Muir "discovered"?

Building on his epiphany in Canada, Muir moved even further away from an anthropocentric world view during his journey South. His journal for these weeks is a kind of polemic against his father's narrow view that the world was essentially made for humans. Instead, there begins to appear what Frederick Turner calls an "aboriginal tone" and Max Oelschlaeger a "Paleolithic awareness" based on a reverence for all forms of life and a firm sense of the human as simply one part of nature. [13] While recovering in Florida he confirmed this break from anthropocentrism:

> The world we are told was made for man. A presumption that is totally unsupported by facts. There is a very numerous class of men who are cast into painful fits of astonishment whenever they find anything, living or dead, in all God's universe, which they cannot eat or render in some way what they call useful to themselves....Nature's object in making animals and plants might possibly be first of all the happiness of each one of them, not the creation of all for the happiness of one. [14]

Muir did not shift from anthropocentrism to a materialist kind of ecocentrism, popular with many current environmentalists. Neither did he, as is claimed by deep ecologists who want Muir to be an early convert to a Zen

Buddhist perspective, merely "clothe" his pantheism in religious language. Rather, he sustained a focus on religion, on a God present in every particle of nature—what some have called a spiritual or mystical pantheism.[15] His experience living "off the land" in Bonaventure and his struggle with malaria confirmed for Muir he had no fear of death, that it was "...stingless indeed and beautiful as purest life."[16] He began to talk about the "web of life" and to focus his thoughts on the interrelationship of all Creation, and on the notion of life as "process" or "flow"—both of which would become especially important when he moved West to the mountains and glaciers of the High Sierra.

Muir did not make it to South America (though he did tour South America and Africa in 1911), but instead returned to New York and traveled to Panama and across to California, a locale almost as exotic. He arrived in San Francisco in March 1868, and spent his first year in California employed as a shepherd in the mountains, arriving in the Yosemite Valley in the Summer of 1869. Once there, his life took another dramatic turn, the epiphanies that followed over the next several years centered on three salient themes: a designed harmony, the uniqueness of all creation, and the ecstasy that stems from immersion in nature. It was there in the Yosemite Valley that all of Muir's ideas suddenly coalesced around these three ideas and his life's mission was set.

When Muir first looked down on Yosemite Valley he was "overpowered by its grandeur", experiencing a truly sublime moment which forced him to look beyond mere "explanation."[17] The dominance of harmony or mutuality in the midst of apparent variety and singularity was Muir's first insight, along with the conviction that he was himself in some way fused with everything around him. Here he responds to the sight (and sound) of Yosemite Falls:

> This was the most sublime waterfall flood I ever saw—clouds, winds, rocks, waters, throbbing together as one, And then to contemplate what was going on simultaneously with all this in the other mountain temples.... five hundred miles of flooded waterfalls chanting together. What a psalm was that![18]

It was the sound, the movement, "flow" of the water that spoke to Muir, even more than the spectacular sight of the Falls. And it was the idea of the myriad interconnections that were "out there" that so awed him, what we might today see as an eco-system conception of a watershed. For all these insights Muir retreated to the Romantic, spiritual realm of feelings:

> It is easier to feel than to realize, or in any way explain, Yosemite grandeur. The magnitudes of the rocks and trees and streams are so delicately harmonized they are mostly hidden....every attempt to appreciate any one feature is beaten down by the overwhelming influence of all the others.[19]

190 | RE-THINKING CULTURE AND NATURE

Stephen Fox calls this a psychic integration that Muir first experienced in Yosemite, the idea of an "other self" or an "unbidden presence" that guided his steps.[20] Nowhere is this clearer than in Muir's classic account of climbing Mt. Ritter in 1872:

> After gaining a point about half-way to the top, I was suddenly brought to a dead stop, with arms outspread, clinging close to the face of the rock, unable to move hand or foot up or down. My doom appeared fixed. I must fall. There would be a moment of bewilderment, and then a lifeless rumble down the one general precipice to the glacier below. When this final danger flashed upon me, I became nerve-shaken for the first time since setting foot on the mountains, and my mind seemed to fill with a stifling smoke. But this terrible eclipse lasted only a moment, when life blazed forth again with preternatural clearness. I seemed suddenly to become possessed of a new sense. The other self, bygone experiences, Instinct, or Guardian Angel—call it what you will—came forward and assumed control. Then my trembling muscles became firm again, every rift and flaw in the rock was seen as through a microscope, and my limbs moved with a positiveness and precision with which I seemed to have nothing at all to do. Had I been borne aloft upon wings, my deliverance could not have been more complete.[21]

One is immediately reminded here of Aldo Leopold's warning that one must "think like a mountain" in order to live in harmony with it. To accomplish this, to survive on the rock face of Mt. Ritter, Muir had to abandon the privilege of being a subject existing in a world of objects. These distinctions had to be blurred, dissolved as did the other dualities of knower and known, animate and inanimate, mind and body and an acceptance of mutuality put in their place. In moving through the Sierra Mountains, Muir came to feel that he was "...among friends, in company."[22]

If the discovery of Calypso borealis had shattered whatever remained of Muir's anthropocentrism, and if his days in the Bonaventure Cemetery had brought him a sense of the interconnection and equality of all elements of creation, then Yosemite provided the final seal of his transformation. In his Yosemite years (roughly 1870–1885) Muir was trying to live out Wordsworth's more abstract desire to have the hills "embrace and close him in". And, as many commentators on Muir have concluded, he seems to approach a kind of mystical consciousness in his prose explanations of this embrace.

But there may be no need to look only to the mystical or even spiritual realms to comprehend what Muir was doing with his Yosemite experience. One hears echoes of Smith, Hume and Rousseau in his striving for a sympathetic understanding, not just of fellow humans now, but of glaciers, mountains, trees and meadows. In doing so he was certainly following in the footsteps of Thoreau who, as Donald Worster points out, was searching for:

> ...sensuous contact, for a visceral sense of belonging to the earth and its circle of organisms. To smell or touch a tangible, palpable nature aroused in him a sense of 'vast alliances' and universal relatedness. He could then feel himself extended beyond the limits of his individual lump of matter, able to achieve access to the vital energy that is in nature.[23]

There is as much philosophy and science in this quest as there is mysticism and spirituality. Muir had been an avid student of chemistry at Wisconsin and from his study of atoms, elements and particles he gained a sense of the "...oneness of all the life of the world—the methods by which nature builds and pulls down in sculpting the globe; one form of beauty after another in endless variety."[24] There is a kind of Epicurean thought here, an extreme materialism coupled with a scientific and spiritual understanding of the essential unity of order, harmony and beauty. In the mountains, Muir sensed that the apparently inert masses of rock were in fact in a constant state of flux, continually re-creating themselves via glacial action and the movement within the earth.

It was during these Yosemite years, then, that via a series of epiphanies like the Mt. Ritter experience—each of which is described in his subsequent writing—Muir developed his unique blend of Christianity, Evolution and Aesthetics. Muir's synthesis included a number of highly controversial elements for his time as well as ours. He insisted, for instance, that plants and minerals shared sentience with humans and other animals, that every particle of creation had "...moods, cycles and behaviors that took no heed of human observation."[25] He insisted on the individuality of all beings, that forests were composed of individual trees, and rejected a requirement for an afterlife as the basis for happiness or fulfillment. He settled the old issue of Theodicy (i.e. the reconciliation of evil with an omnipotent God) by viewing all destruction in nature as simply another form of creation with neither beginning nor end—"a change from beauty to beauty."[26] Nature was a continuous flow and part of this flow involved strife and violence. Thus Muir found a kind of divine harmony within the violence of the natural world, even to the point of dashing out of his cabin at midnight to celebrate a tremor, crying out "A noble earthquake! A noble earthquake!"[27]

Tolerant of paradox, confusion and ambiguity, with Muir we have in one person a collection of approaches to nature and culture that will need serious "unpacking" by subsequent generations. He seems at once a conservationist, preservationist, Christian steward, misanthrope, mystic, pantheist,

192 | RE-THINKING CULTURE AND NATURE

evolutionist, creationist, materialist, poet and politician. As the first real "environmentalist", such a jumble of approaches may have been inevitable, but it was too contradictory to be sustainable. What remains most powerfully embedded in the Muir legacy (aside from his institutional offspring, the Sierra Club) is the poetic and often rapturous mysticism associated with his "feelings" for nature.

With Muir we have the oceanic experience writ large, a series of dramatic collapses of the distinction between self and other, whether mountain, tree or earthquake. "When we try to pick out anything by itself, we find it hitched to everything else in the universe."[28] Not uncommon in some Asian traditions in which human, rock, water and all creation share a common energy or spirit and in which nature and humankind are blended together harmoniously, these ideas posed a radical challenge to a tradition wedded to the mechanistic world view grounded in Bacon and Descartes. And Muir knew enough chemistry to realize that such ultimate kinship was in fact more real than imaginary.

OCEANIC FEELINGS AGAIN

But there is another aspect of Muir's story that bears closer examination and will lead us further into our subject. Muir repeatedly asserts an affiliation with wildness, the "Paleolithic awareness" that Oelschlager referred to. On Mt. Ritter he senses an "unbidden presence", an "other self" on the rock face with him. But this was not a voice from God or the intervention of an angel. This was a presence from within, akin to a genetic memory of a primitive human much more integrated with Nature—in this case a mountain.

This idea that we carry within us not only a DNA code but also a deep memory of our primitive past that manifests itself as dispositions or intuitions is, of course, quite controversial. For some it is a New Age myth in the same family as spoon-bending or loose talk about "Memes". But there are strong feelings on the other side as well. The psychologist Carl Jung argued for a "collective unconscious", an "inner self" buried from apprehension by the layers of culture that form our conscious mind. Jung's close friend the philosopher and anthropologist Laurens van der Post argued that at the centre of the human psyche was a "...natural intuitive and instinctive self at home in the natural world...our primitive or wilderness self."[29] The environmentalist and theologian Thomas Berry insists that we have a "genetic code" that manifests itself as a deep identity with nature and which we need to nurture and bring into balance with our parallel cultural codes.[30]

In the same tradition, Thoreau thought the aboriginals he met in the Maine woods might be "primal human beings" who could "...lead the *civilized* human back to realities which only lurk somewhere in the modern subconscious mind so subdued by the complex material concerns of industrial society."[31] Levi-Strauss, in his dramatic and determined search for the primitive which he "sensed in himself", found it in the form of tribes deep in the Brazilian forest but was unable to communicate with them, not just in the literal sense but also in the deeper sense of encountering a true "other". Paul Shepard refers to having a "luminous connection" with the wild based on a "biological heritage of the deep past" and the ecologist Stan Rowe posits that we surely have an "instinctive affection" for the natural world.[32] Even Prince Charles, the future Monarch, adds his always eclectic perception that modernity needs to balance scientific analysis with "instinctive wisdom", a "wisdom of the heart...a faint memory of a distant harmony."[33] More to the point perhaps, in earlier chapters we have noted the argument for predispositions such as a "a deep-rooted egalitarianism, a deep-rooted commitment to the norm of reciprocity, a deep-rooted desire for...the sense of community..." which, like the equally deep-rooted fear of snakes and attraction to sweetness may evoke some complex blending of genetic materialism with cultural memory.[34]

These are all aspects of that oceanic feeling identified by Rousseau, Wollstonecraft, Coleridge's Mariner, John Muir and Romain Rolland and that Freud could not find. The Zen poet Gary Snyder, himself having spent a Muir-like life immersed in the Rocky Mountains, captures this mood of the times:

> It seems evident that there are throughout the world certain social and religious forces that have worked through history toward an ecologically and culturally enlightened state of affairs. Let these be encouraged: Gnostics, hip Marxists, Teilhard de Chardin Catholics, Druids, Taoists. Biologists, Witches, Yogins, Bhikkus, Quakers, Sufis, Tibetans, Zens, Shamans, Bushmen, American Indians, Polynesians, Anarchists, Alchemists...primitive cultures, communal and ashram movements, cooperative ventures.[35]

Despite it's deeply subjective nature, this sense of oneness, of being intimately rather than just materially connected to the other may be a crucial element in our 21st century attempts to re-think our connection with the natural world. It is worth further exploration. But it is crucial we not be diverted by some Cartesean either/or dichotomy forcing us to choose between "culturalism" or "biologism". Culture in all its political, social and economic dimensions is a powerful force in determining or influencing human choices. Equally we have certain predispositions that are based in biology and which manifest themselves in specific behaviours and preferences. It is the combination that is important and the acknowledgement that both are present.[36]

There would seem to be at least three major paths into this idea of connection, each of which is quite complex and contains many branches. First there is the matter of evolution itself, the facts we now know about our biological connections with other animals and our own biological nature. Second, our long-standing anthropocentric focus is being challenged by a new interest in biocentrism, the idea that all life forms have value and even interests which must be acknowledged in some way. Finally, moving beyond this focus on "life" *per se*, many are thinking now in ecocentric terms, linking human, animal, organic and inorganic matter and aspects of "being" that exist in a non-hierarchical continuum.

THE BIOLOGY OF EMPATHY

The starting point for this exploration may not be *oceanic* at all, but merely our two kinds of memory, one personal and the other species-driven. We saw earlier that Wallace Stegner, when confronted with even an odor from childhood, found himself instantly transported back in time.[37] Here one type of holism is satisfied, the collapsing of time in a life of one person so that adult and child become one. And it seems somehow intuitively true that most of us do feel this drive, this compulsion to deny momentarily our cosmopolitan reality and go home again — if only for a sniff.

We can re-visit the claims made in Chapter One for qualities that through natural selection become innate to Homo sapiens, that manifest in behaviour across the millennia. We might think of these as oceanic phenomena of a social nature, predispositions that originate much deeper than our conscious minds or field of intentions. We know about the drive for self-preservation and that aggression may be hard-wired as a tool supportive of that drive. But Rousseau's insight that compassion may be an equally hard-wired drive leads us to altruism and a cooperative disposition as equally powerful tools that support that predisposition for compassion. Alison Jolly argues that: "If you are likely to meet the same companions day after day, and if you can be selectively nice to those likely to return the favor, altruism evolves so long as the cost to the giver is small compared to the benefit to the receiver."[38] So it is possible that in the millennia in which we existed in small hunter-gatherer groups, a prevalence of altruistic behaviour within a given group could have been a powerful advantage in terms of natural selection over groups "riven by selfishness".[39] But this need not imply a leap to some innate 'goodness'. Our desires and our equally innate selfishness are always contesting this potential altruism or sociality, but we remain able in calm and disinterested moments to "...know the difference between being human and being inhuman."[40]

How might this work? Darwin once again to the rescue. If being easy-going, tolerant of stress and having highly developed empathetic abilities are perceived as positive attributes for living in groups—which seems logical—then individuals with these qualities "...would enjoy enhanced reproductive success, and the genetic propensity would proliferate and become embedded in the species."[41] Our actions in the world demonstrate that we are a mixture of both altruism and aggression, singularity and cooperativeness, but according to evolutionists our propensities for social cooperation, sensitivity to social approbation and even ethical sensibilities are just as "biologically grounded" as our more salient aggressive and individualistic drives.[42]

Building on these assumptions, observers of animal behaviour like Frans de Waal and John Livingston argue convincingly that qualities such as gratitude, obligation, retribution, indignation and reciprocity precede both language and any established, albeit skeletal, systems of moral discourse.[43] These "innate mechanisms" are, then, deep in our animal natures and can be just as important as learning and reason in shaping the human possibility.[44] But in practice this sympathy or innate compassion only works for particular cases, requiring the development of human consciousness and its accompanying reasoning abilities to be transformed into a more generalized sense of compassion. As in the case of language which many believe individuals are "born into", it is this potential for an increasingly complex and sophisticated use of these innate predispositions (to feel and to communicate) within social groups that acts as the decisive bridge within humans nature and culture.[45] The innate predisposition of sympathy and the necessity of sociality, then, join with factors such as increasing knowledge, new technologies and more complex social processes to create a highly resilient adaptable species.[46]

So if we are to work toward a human culture that supports ecological sustainability (and not just human flourishing!) we need to acknowledge and nurture these predispositions as well as marshal cultural arguments to employ this resilience and adaptability in new ways.

This seems particularly important in these more ecologically conscious early years of the 21st century. The sensual and scientific evidence that human activity is producing potentially catastrophic climate shifts is leading to an increasing public demand that something be done. But merely fixing or patching up the current problems while failing to address the cultural and moral roots of the problem will only postpone a continuing and worsening conflict between humanity and nature. With the will for change and even for substantive sacrifice now seemingly more pervasive, it is crucial that the debate be carried beyond the immediate issue of climate, or species extinction, or pollution to the more complex issue of value change.

Biocentrism

The on-going debate about the relationship between humans and other life-forms centres on three key areas: the presence of feeling or sentience; the quality of reason or self-consciousness; and the ability to create culture or society.[47] In recent years advances in evolutionary thought and in genetics have drawn humans closer to other animal life in terms of our origins and our chemical composition. This has led to a blurring of boundaries, challenging the anthropocentrism that for so long has been pervasive in Western culture.

The combination of these discoveries with new sensibilities about animals in many parts of the world the conflict over the human/animal boundary has become intense. On the one hand animal rights advocates argue that humans are just another branch on an evolutionary tree, the differences between Homo sapiens and related mammals being merely quantitative, not qualitative—animals just engage in simpler versions of what we do.[48] Speciesism ["...a prejudice or attitude of bias toward the interests of members of one's own species and against those members of other species"[49]] and humanism are both rejected as bases for a valid moral theory regarding relations among living beings—"...an atavism that a sound ethics can no longer accept."[50]

The rhetoric on the other side is just as dramatic and, at times, apocalyptic. The idea that humans are "no more than a clever animal" is seen by the philosopher Martin Heidegger, for instance, as opening the door to a radically amoral culture in which there is only the present, no transcendent possibility, no human "calling" to create a civic culture built on our unique capacity for reason.[51] For Heidegger seeing humans as merely clever animals left us "absurd aliens in a dead, meaningless world", when in fact there was as "abyss" separating humans from animals and the rest of the natural world. This abyss was the inevitable and, perhaps in a rousseauean sense tragic, the result of consciousness and its insistence that we reflect on the world and in doing so step back from it, from nature.[52]

The debate surrounding the animal/human boundary is no doubt a current hot issue in philosophical thought and will re-surface when we turn to the issue of moral, legal and political rights. For now we need to turn to the specific components of the debate: sentience, consciousness and culture.

Sentience

Sentience or feeling has always been the easiest issue to resolve, despite the Cartesean insistence that expressions of pain in animals were mere "mechanical" reactions. Jeremy Bentham (1748–1832), the founding figure in utilitarian philosophy, set the mark for many when he insisted that the central question was not whether animals could reason or speak, but rather "Can they suffer?" In our time, Peter Singer's powerful arguments for animal rights hinge on sentience, the capacity for suffering and/or enjoyment or happiness. And he makes a key distinction between organic and inorganic phenomena that is particularly important for the issues being raised here:

> The capacity for suffering—or more strictly, for suffering and/or enjoyment or happiness—is not just another characteristic like the capacity for language, or for higher mathematics.....The capacity for suffering and enjoying things is a prerequisite for having interests at all, a condition that must be satisfied before we can speak of interests in any meaningful way. It would be nonsense to say that it was not in the interests of a stone to be kicked along the road by a schoolboy. A stone does not have interests because it cannot suffer. Nothing that we can do to it could possibly make any difference to its welfare. A mouse, on the other hand, does have an interest in not being tormented, because it will suffer if it is.[53]

This is a kind of minimalist case for the rights of certain life forms, an awareness that they can suffer and that they therefore have an "interest" in not suffering or in some way maximizing their well-being. One might imagine this as a kind of "nerve endings" issue which spawns all kinds of explorations in animal biology to determine evidence of pain. But it is also deeply subjective, often relying on our ability to imagine the feelings of the other before we extend our field of moral concern and thus remains somewhat mired in anthropomorphism. The recent film featuring the harsh life of Antarctic penguins (*The March of the Penguins*) apparently evoked a great deal of sympathy in the human audiences for these upright walking, tuxedo-dressed birds. As we saw earlier, in 55 BCE, a Roman crowd cursed Pompey who had staged a fight between humans and elephants after the beasts "...entreated the crowd, trying to win their compassion with indescribable gestures, bewailing their plight in a sort of lamentation."[54] More recently there are active debates about the ability of fish to feel pain when hooked by an angler.[55]

The sentience case is made more salient in contemporary times by the growth of factory farms, the mass processing of animals, and the effect of abstract packaging of animal parts for human consumption. PETA (People for the Ethical Treatment of Animals) has been very active in bringing these issues to public attention and animal rights activists like Peter Singer, Temple Grandin, and the prolific and passionate Jeffrey Masson are vocal in their demands for change in the way we treat farmed animals.[56] Noting that over 40 billion chickens are killed each year, Masson calls on us to reflect on their suffering and our obligations: "The suffering of almost all farm animals is unique, particular, mostly beyond language to describe or explain....If we give it no thought, and yet eat them for our meals, are we not morally blind, ethically dumb, and humanly remiss?"[57]

Perhaps the most popular case against causing pain to animals is solidly anthropocentric — that it may lead to a hardening of human morals and to a tolerance of cruelty toward other humans. Called redescriptivism — a re-describing of the act in order to apply it to humans — this argument has a long pedigree. Both Pythagoras (582–507 BCE) and Empedocles (492–432 BCE) made this case as did Erasmus, Thomas More, and Montaigne in the 16th century, Rousseau and Kant in the 18th and Shelley in the 19th century.

Consciousness

While sentience seems a quite visible and emotion-laden quality to consider in seeking to determine the boundaries between humans and the rest of nature, the issue of consciousness, reflection and reason is much more contentious. Responding to Masson's and Grandin's accusations of cruelty, some have addressed the problem by shifting to more humane ways of killing animal life and even considered extending a fundamental right to life to obvious relatives of Homo sapiens such as the Great Apes. Others have argued that sentience is not the right focus, being merely one "adaptive characteristic" of some organisms and that life itself is what we need to value.

Human anthropocentrism, however, values pre-eminently the existence of self-consciousness and rationality as keys to membership in the realm of moral discourse. Since empathy is one of the qualities that consciousness and rationality gives us, it may persuade us (eventually) to minimize the pain or even maximize the pleasure of other sentient but non-rational organisms, but they remain outside, the "other". The next level of debate, therefore, is the nature and quality of the mental life of these organisms. Erazim Kohak is adamant that animals are conscious beings and goes even further in extending a kind of knowing to all organic life:

> Of course animals know. Their knowing may be of a different order than human reflection, conceptualization, and articulation, but their purposeful

behavior in the context they treat as meaningful testifies both to a highly differentiated cognition and to an ability to store and transmit information. Not animals only: insects, trees, plants, though much more humbly, recognize the presence of a world of others and respond to the cycle of day and night and to the rhythm of the seasons. We have as little warrant for conceiving of them as mechanisms, as nature's artifacts, as we do of higher animals.[58]

Also weighing in on this point is the communitarian philosopher Alasdair MacIntyre, who argues that the practical reasoning of animals like dolphins, gorillas — the "knowing" that *this* is a good reason for doing *that* –is by analogy equivalent to human reasoning.[59] Martha Nussbaum, while stressing that there are significant differences between human and animal cognition, insists they still have a "conscious awareness....there is something the world is like to them."[60] She also insists that animals have intentionality, selective attention and the ability to appraise situations. The neurophysiologist Antonio Damasio calls this kind of awareness "core consciousness" and goes further than Nussbaum in arguing that some animals (apes, bonobos and his dogs) have what he calls "autobiographical consciousness", an "...extended consciousness [which] at its peak is uniquely human."[61]

The key phrase in Damasio's conclusions about consciousness is "at it's peak", the implication being clear that consciousness *per se*, like sentience, is an adaptive quality inherent in a great many life forms, not the sole possession of Homo sapiens. Contemporary ground-breaking work in neurophysiology demonstrates equally clearly that consciousness is the product of material processes, genetic adaptations shared by many species.[62]

Some animals, then, clearly have various forms or levels of consciousness and as a result exercise some form of reason and even self-reflection. The in-depth studies of chimps, bonobos and other animals by deWaal, Jane Goodall and others clearly shows reason at work. Maurice Maeterlinck saw all kind of consciousness in bees and John Muir found it in his dog Stickeen and in his oxen on the farm in Wisconsin.[63]

Animal Culture

The social dimension of animal life, the existence of forms of culture, language and other group phenomenon, seems an even more obvious basis for altering our anthropocentric fixation. Like humans, animals most often live in groups, mate and nurture offspring. But is this mere instinct or do they engage in behaviours we might call both social and moral? Do they reflect on the

200 | RE-THINKING CULTURE AND NATURE

well-being of others? Of the group? Joy Mench at the University of California argues that chickens, for instance, demonstrate sophisticated social behaviours, can recognize and remember more than 100 fellow chickens, and will alter behaviour in order to alleviate pain.[64] Joanne Lauck in her work on ants says that the fact that they "...display many so-called human emotions and engage in altruism has long been known. Ants display a range of emotions including anger, fear, depression, elation, and affection. They also demonstrate empathy when they help crippled and distressed sister ants."[65] Jeffrey Masson's work with dogs, elephants and other animals is replete with examples of social behaviours and deWaal who has studied animal behaviour for decades, says they display a range of emotional and cultural responses much like humans. "If animals can have enemies they can have friends; if they can cheat they can be honest, and if they can be spiteful they can also be kind and altruistic. Semantic distinctions between animal and human behavior often obscure fundamental similarities."[66]

While post-modern thought in most cases follows the humanist pattern of privileging the human, primarily on the basis of rationality and language, an important exception has been the work of Jacques Derrida. He has objected to the explicit ranking via an anthropocentric value hierarchy that places humans at the top of an imagined pyramid of life. For Derrida the treatment of animals is a moral issue akin to the Holocaust and as such has become "...one of the leading questions of our age...and uncircumventible for thought."[67]

Inevitably, one might imagine, the debates swirling about the human/animal question have come to focus on the issue of speciesism, flowing logically from earlier debates on racism and sexism. Cary Wolfe, an outspoken advocate for animal rights, justifies this focus by insisting that the human commitment to transcendence requires the "sacrifice of the animal" and, like Plutarch and Shelley in earlier versions of the argument, warns us that this can only lead, eventually, to the non-criminal putting to death of other *humans* as well by "marking *them* as animals."[68]

The increasing visibility of these ideas, the growth across many cultures of increased empathy for animals, and the rising awareness of human health issues associated with factory farming has once again brought to popular attention the issue of vegetarianism. Echoing Derrida, in Michael Coetzee's *The Lives of Animals* his main character, an outspoken vegetarian, raises the comparison of animal killing and the Holocaust: "...we are surrounded by an enterprise of degradation, cruelty, and killing which rivals anything that the Third Reich was capable of. Indeed dwarfs it, in that ours is an enterprise without end, self-regenerating, bringing rabbits, rats, poultry, livestock ceaselessly into the world for the purpose of killing them."[69] Amy Guttman, in the Introduction to Coetzee's book, reminds us that "Unlike some animals, human beings do not need to eat meat. We could—if only we tried—treat animals with due sympathy for their *sensation of being*."[70] The media reminds us constantly that this might be a wise move, pointing out that the consumption of animals is both inefficient and unhealthy:

> From a health perspective, eating meat is associated with health problems that are widespread in Canada: heart disease, stroke, obesity, cancer and diabetes.....In Ontario and Quebec alone, livestock produce a volume of manure equal to the sewage from 100 million people...It takes about 40,000 litres of water to produce a kilogram of beef, 6,000 produces a kilogram of pork, and 3,500 to produce a kilogram of chicken. Far less water is required to grow grains, legumes, fruits and vegetables.....We can't simply replace red meat with fish either. The human quest for animal protein is devastating the Earth's oceans....A healthier diet for both people and the planet involves dining lower on the food chain and only eating neat that is raised responsibly.—that is, organic or free range.[71]

The same points were made in the 18th century and in the 19th century by Shelley.[72]

It seems clear that the view that there really is no "essential" difference between humans and other animals (if not organisms) is gaining credence. There are differences of degree, but not of kind. This awareness has persuaded the sociobiologist and entomologist Edward Wilson to advance the "biophilia hypothesis" which he defines as "...the innate tendency [of humans] to focus upon life and lifelike forms, and in some instances to affiliate with them emotionally."[73] In terms of helping us to re-examine and even re-vision our actual relationship with nature, this kind of acceptance of being in a continuum with other life forms edges us closer to the ideal of an I-Thou relationship with the "other" as envisaged by Martin Buber.

ECOCENTRISM

Buber's *Thou* remains a life form, a creature partaking of a "primordial divine unity" characterized by its organic quality, it's participation in life.[74] This biocentric approach in which all living things hold moral value is an improvement, one might argue, over an anthropocentrism in which only humans can have moral claims. The inanimate realm, however, is still seen as mere means or resources that are instrumental to human purpose or to the purposes of other living things that depend on it. To move toward an ecocentric view that expands the field of care, respect and rights beyond creatures and beyond the

202 | RE-THINKING CULTURE AND NATURE

organic itself we need to explore yet another set of ideas.

The Epicureans were perhaps the first to view the world in this non-anthropocentric way, but Baruch Spinoza (1632–1677), known as the first modern pantheist, comes closest to a modern ecocentrism. Spinoza held that every being and every object—organic or inorganic—was a temporary manifestation of a common God-created substance. There was no hierarchy in these various formations, "...a tree or a rock had as much value and right to exist as a person."[75] Again, this is very close to the Confucian view that life is a "...continuity of being...a continual process and transformation, linking inorganic, organic, and human life-forms."[76]

Does this mean that rocks have rights? Do rocks or rivers have what Spinoza called *conatus*, a tending or striving manifested as inertia, resistance, force appetite or desire? An obvious common sense 'no' for most people who are still wrestling with the issue of whether all, some or no animals should have rights. These boundaries—blurred for biocentrists—completely disappear within the ecocentric perspective in which scientific as well as poetic and aesthetic sensibilities are employed in making the case for a seamless moral world. Erazim Kohak, in assessing the relative moral worth of the rocks he encounters in digging a well, recalls that his geologist friends do speak of the "life of rocks", recounting their history from their composition. He admits to fleeting moments of kinship with these very solid and stolid chunks of Being:

> Digging my well, I had a distinct sense that the rocks that for centuries had lain beneath the earth—deep, at least, as a pick and shovel depth in rocky soil—welcome being brought to the surface to which so many of them press their way even in climates where there is no frost to push them. It seemed to me that they are pleased, and most at ease in the shallow stream, washed by the river and warmed by the sun. Those, to be sure, were no more than fleeting moments, to which I attribute no cognitive significance.[77]

Kohak's experience gives us an entry point into an important dimension of ecocentrism; its strong connection with the direct experience of nature. Of course people often feel kinship with animals, especially pets, but this is often based on compassion for a fellow living being. Generating the same kind of feeling for a rock or for nature *per se* requires a more imaginative leap, often flowing from an intense immersion in nature. Wordsworth signals this early in his autobiographical poem *The Prelude* when he gives the loose stones on a path a moral life and a sense of feeling.[78] We saw this with John Muir and had a sense of it as well from Mary Wollstonecraft. The key issue for John Muir and others whom we might consider ecocentrists is that no aspect of creation should be privileged over any other. Species and eco-systems have a specific kind of being and therefore worth, and that worth could rank them higher than individuals of higher species, including humans.[79] Henry David Thoreau, famous for his residency at Walden Pond, insisted that we needed to "...re-ally ourselves with Nature every day. We must make root, send out some little fibre at least."[80]

Neil Evernden in his book *The Natural Alien*, develops the link between this kind of engagement with the world "as it actually is" and phenomenology as developed in the work of philosophers like Heidegger and Merleau-Ponty. He cites the philosopher and literary critic George Steiner who in his book on Heidegger recalls Heidegger's account of such a Thoreau-like encounter with the world in which the "hidden being that gives the rock its dense thereness" becomes clear, a bringing forth of Being from concealment in an "intensity of presentness."[81] In a Heideggerian sense, this seeing into the object, accomplished by immersion and observation, reveals a kind of universal inwardness that serves to erase the boundaries between categories.

This focus on the concrete results in this fleeting or even continuous sense of oneness with Nature. In Asian culture, particularly in Buddhist, Confucian and Taoist thought, it is the norm that nature is all-inclusive, a continuum linking inorganic, organic and human life-forms, a chain of being that is never broken. In Western culture, ecocentrism borrows heavily from the spirit if not the substance of these Asian traditions. The Zen poet Gary Snyder is one of the better known travelers between these two worlds:

> This living flowing land
> is all there is, forever
> We are it
> it sings through us —
>
> We could live on this Earth
> without clothes or tools![82]

Poetry and science and philosophy all merge in this ecocentric world view. Heidegger accuses Western culture of abandoning a concern with *Being* and obsessing instead on *beings*, which are mere particles of an underlying whole. While it seems that we must, given the evolution of our brain, think in categories, there is a world beneath the categories to which we must also attend. The ecologist Stan Rowe expands on this critique of our focus on the abstract at the expense of the concrete:

> For thousands of years we people have been viewers immersed in the Ecosphere, deep-air animals living at the phase boundaries where air and water meet land, mistakenly identifying all manner of things as organic'and

inorganic, biotic and abiotic, animate and inanimate, living and dead. Dictionaries full of nouns show the efficiency with which we have thought the world to pieces. Around our ignorant taxonomy we have constructed religions, philosophies and sciences that fragment and compartmentalize a global ecosystem whose *aliveness* is as much expressed in its improbable atmosphere, crystal rocks, seas, soils and sediments as in organisms. When did life begin? When the Ecosphere itself was born, if not even earlier.[83]

Here, in a complete break with the Baconian system, *Life* is the ecosystem itself, the totality of all *Being*. While Rowe the scientist may use the techniques of modern science in exploring the particulars of prairie grass, he refuses to equate those particulars with the reality of a holistic world. The philosopher and literary naturalist Loren Eisley takes us further into this blending of a poetic sensibility and scientific understanding with this meditation on birds, atoms and life:

As he watched the birds fly unerringly through the fading light, he realized that all around him in the rocks of the desert the chemicals that made up living bodies lay inert. Beneath his feet were iron, calcium, carbon, etc., in passive, lifeless form. But in the air above him were those same chemicals in the form of birds. Through some mysterious synthesis, those chemicals were now warm, intelligent, sensitive creatures, endowed with a power of flight as beautiful as it was remarkable. If there was a reason why these lifeless elements had so mysteriously taken wing, Eisley does not think human intellect can discover it. If he caught one of the birds and dissected it, he would ultimately find only the dead chemicals at his feet.....he would never be able to find out what made those elements perform the complex, unlikely transformation that made them into life.[84]

We arrive here very close to a Buddhist/Confucian/Taoist treatment of nature, with a seamless connection between the divine, the human and the natural realms. Like Rowe, Eisley and Heidegger, there is no need here for the radical transcendence so central to the post-Platonic and Christian West.[85]

And so we end our review of these three approaches with a focus on a reverence for nature as a whole (humans included!) as *Being*, and a respect for its processes "...over which we humans will never have control."[86] And instead of radical transcendence, we end with an echo from Shelley, Coleridge and Rousseau, a call for a new radical sensibility based on sensing and feeling coupled with wonder and reverence rather than just knowing.[87]

SOME PRAXIS IMPLICATIONS

And where do these biocentric and ecocentric perspectives, with their undermining of established paradigms and flirtations with alien traditions, lead us? We have already dismissed one possibility, the idea that we should somehow return to the primitive, to a state of nature. We may have lessons to learn from aboriginal peoples who remain in our midst or whose past ways we can read, but despite the grim images from post-apocalypse films and novels it seems unlikely we will attempt a retreat to primitivism.[88]

Reconsidering A Culture of Consumption

There may, though, be intermediate steps that could prove palatable even to people embedded in heretofore consumption-driven cultures. The homo economicus/human-as-consumer model which has been so effective in terms of the scale and variety of production has some obvious shortcomings, particularly in the area of environmental impact and in its reliance on ever-increasing focus on employment and labour both to produce and consume. As a result in the more advanced industrial cultures there are increasing indications of interest in scaling back, in focusing more on the stewardship of nature and the cultivation of simpler lifestyles.[89] Accompanying these initiatives are the twin correlates of a new attraction to the local, to place, and to a consideration at least of stability in place of growth. These ideas are of special interest here because they have obvious roots in the philosophical and political traditions I have identified as part of a shadow or alternative modernity.

Crafted in response to the Romantic critique's challenge to both anthropocentrism and to progress, the suggestion that we might choose to radically simplify the way we live centres on shifting to a steady-state human economy with a renewed focus on place and on community, voluntary limits on consumption, and a slowing down of the pace and nature of change. In assessing the viability and strength of such a dramatic shift in cultural norms and individual priorities, a key issue will be identifying historically derived cultural tools that might be marshaled on its behalf. And, just as important, to assess whether this strand spun off directly from the Romantic critique of modernity has "legs"; does it have the necessary persuasive power to engage individuals heretofore immersed in a culture built on ever-expanding consumption to accept a dramatic paradigm shift toward ecological sustainability?

This may be the central question that lies behind my argument for the importance of a dramatic shift in theorizing our relationship with nature. Will our response to the current fears and anxieties about climate change

206 | RE-THINKING CULTURE AND NATURE

be merely tactical or will we consider undertaking more substantive changes? Will we, in other words, look to the deeper cultural causes of phenomena like climate change, or be content with whatever remedial steps allow us to continue on our current path, albeit more prudently? These suggestions of a more steady-state economy and an enhanced focus on place, on community and on stewardship are a test of our commitment to this kind of substantive change.

This issue was posed forcefully in a 2002 Report of the *Global Scenario Group* based in Stockholm. In their report, *The Great Transformation*, Paul Raskin and his colleagues concluded that a sustainable political and economic culture was attainable without either a social revolution or technological miracle, but the political will within existing government bodies was "no where in sight". Hence, they concluded, there must be "fundamental changes in popular values, lifestyles and political priorities...[and we must] revise the concept of progress toward non-material dimensions of fulfillment."[90]

Stewardship

We can start with the idea of stewardship, the still essentially anthropocentric idea that it is our duty as more advanced, conscious and rational beings to ensure the future well-being of the planet. The steward's (or ward of the sty!) role is to ensure that the human footprint is as gentle as possible given the continued necessity for human exploitation of nature. But the Biblical dominion is here replaced by a more benign idea of trusteeship with an accompanying focus on the long-term viability of the land. For John Muir, the steward's role was still grounded in religious faith, but for contemporary advocates of stewardship like Wendell Berry, who writes in a more secular and Jeffersonian tradition, the steward's role is grounded in more ecological considerations and in a concern for conservation and preservation. Thus for Berry: "...every person exercising the right to hold private property has an obligation to secure to the rest of us the right to live from that property...to use it in such a way as to not impair or diminish our rightful interest in it."[91] In a way it is our very power to destroy that makes such a persuasive case for stewardship. Thanks to science, we have the means to destroy but also possibly to render our presence more benign, to preserve and protect and to use in a more sustainable way.[92] The idea of stewardship may offer us one way to draw upon an aspect of our Judeo-Christian heritage in order to blunt the impact of the technological prowess that other aspects of that heritage have given us.

Conservation

Stewardship is closely linked to the idea of conservation, an important part of efforts to establish limits on human exploitation of nature. Conservation is one of three general approaches to contemporary environmentalism:

1. Preservation: a defence of wilderness
2. Pastoralism: promotion of rustic lifestyles and return to simpler pleasures
3. Conservation: wise use of resources in perpetuity[93]

Preservation taken to extremes would involve a substantial reduction in human population and a return to rural or pastoral economies, an option that we have already set aside as likely to be both impractical and unpopular. But its more common expression involves setting aside from human use large tracts of land or water.[94] Pastoralism may be more complex and we will return to it shortly. Conservation, the most popular option in terms of public awareness and participation has its basis in the ideal of stewardship.

John Muir, who resonates with all three options but was at heart most likely a preservationist, succumbed toward the end of his life to public and political pressure and linked his reputation to the conservation movement. In many ways the great tragedy of his career, he was forced to set aside his sense of the fundamental interconnections between various forms of wilderness in the mountains and valleys of California and focus his attention instead on the picturesque, on landscapes that appealed to a public aesthetic still tied to human use and/or enjoyment.[95] As a result of his efforts we do still have the Yosemite Valley, but not the surrounding watershed areas that comprise its ecosystem. We have instead a "park".

The conservation movement of the late 19th and early 20th centuries was largely an American phenomenon, triggered in part by the interest raised by the widely read *Man and Nature* by George Perkins Marsh (1864), the many books and other publications by Muir, the work of Gifford Pinchot in the government and by community based groups like the Sierra Club. "Wise use" was the watchword and human needs for both prosperity and pleasure the first priority. The tradition of stewardship at the base of conservation called for practical steps which "...bind men to nature in the capacity of a loving caretaker and cultivator, not in a symbiotic union that simply denies the reality of man's separation from nature."[96] Given the earlier discussion, this characterization of conservation as an option that avoids any hint of "symbiotic union" should make more clear the important boundary between environmental action grounded in anthropocentric thinking and that based on biocentric and especially ecocentric approaches.

Conservation remains the most popular and, in the short-term at least, the most effective environmental activity and as such plays an important role in the long term struggle to achieve an ecological sustainable human culture. While conservation activities are often perceived as being local in origin, it is government that plays the key role in their success. In the last decades of the 20th century the idea of conservation (which had been linked until then to parks and resource management) became inextricably connected to the call for sustainable development. It became commonplace then (and more so now) to acknowledge that the capitalist economic system had a basic weakness in terms of the environment, namely its failure to take into account externalities such as pollution and waste in setting market prices.[97]

Sustainable development is the anthropocentric response of governments and of much of public opinion to the ecological problems now widely seen as being caused by human economic activity. It is based on the creation of more enlightened policies that will "...allow for human use of ecological services in perpetuity without their diminution, avoiding those unsustainable ones that are incompatible with that aim."[98] Critics—and there are many—call this objective impossible, noting the oxymoronic nature of the term itself and insist that "...there is no way of ever *getting it right*, no stable or sustainable situation, no way of bringing the social into long-term harmony with the natural."[99] Governments, recognizing the need to avoid the social implications of continued non-sustainable development, have turned to conservation, ecological management and the promotion of environmental citizenship as a way to preserve existing systems economic and social systems.

These new initiatives do not, of course, address directly the larger issues that stem from a nature seen as holistic, inter-connected, sensible or at least sensitive and inevitably reciprocal—a nature in which one cannot stop what is going around from "coming around". It is argued by many ecological thinkers that the effectiveness of conservation and policies tied to sustainable development must at best be ephemeral or tied to only very specific problems or issues. Indeed, political responses in general may be insufficient given the nature of the ecological problems we face now and in the near future given the short-term thinking so prevalent in political life.

The Importance of Place

The tendency among conservationists to focus their efforts on specific issues and the specific sites of environmental crises has had the interesting effect of re-focusing people on the importance of *place* and of *community*, two concepts that will occupy us in the remainder of this chapter. To explore these two topics we can turn to the work of Aldo Leopold who, if John Muir was a bridge between Rousseau and Leopold, becomes the bridge between Muir and Greenpeace.

A botanist, forester, conservationist and naturalist, Leopold's *A Sand County Almanac* (1949) was one of the foundational texts for the modern environmental movement, the other being Rachel Carson's *Silent Spring* (1962). Over a lifetime of fieldwork in forestry and conservation Leopold developed his idea of the Land Ethic, a systems approach to the issue of a sustainable human relationship with nature. Instead of focusing on the particular, whether species, individual or landform, Leopold built on Muir's insistence that "everything was connected to everything else" in order to shift our attention to ecosystems, habitats, wilderness areas and natural communities:

> The extension of ethics to this third element in human environment is...an evolutionary possibility and an ecological necessity.....The land ethic simply enlarges the boundaries of the community to include soils, waters, plants, and animals, or collectively: the land....In short, a land ethic changes the role of Homo sapiens from conqueror of the land community to plain member and citizen of it. It implies respect for his fellow members, and also respect for the community as such.[100]

Put in terms of human ethical action, Leopold's extension of ethics was based on the idea that "...a thing is right when it tends to preserve the integrity, stability, and beauty of the biotic community. It is wrong when it tends otherwise."[101] Of particular significance for many followers of Leopold's vision was the inclusion of beauty, a nod in the direction of his naturalist's sensibilities and a lifetime spent more in the field than in lab or classroom.

The change of focus from particulars to systems implied in Leopold's work meant a shift away from a concern with the welfare of individual animals or even in some cases species, and a shift away from picturesque sites like Yosemite and toward the harsh but necessary realities of natural balances and trade-offs, including a respect for predation, fire, erosion and other *natural* phenomena. But to make his point accessible to a public audience Leopold illustrated the workings of the ecosystem by looking at particulars. Early on in *A Sand County Almanac* he reflects on how the annual arrival of snow provides essential cover for the meadow mice who built tunnels in the snow linking various caches of food. In the Spring, as the snow melts, the rough-legged hawk migrates from the north, knowing that the melting snow makes visible the previously hidden mice. "To the mouse, snow means freedom from want and fear....[for the hawk] a thaw means freedom from want and fear."[102] Gary Snyder, writing in the same tradition, puts it this way:

210 | RE-THINKING CULTURE AND NATURE

Baby jackrabbit on the ground
Thick furry brindled coat
Little black tailtip
Back of neck ate out
Life for an owl.[103]

Leopold drove his lesson home with his challenging metaphor that we need to "think like a mountain" which came directly from his experience in the American Southwest in observing the impact of the campaign by farmers and ranchers to eradicate the wolf population. In doing so, they created an explosion of the deer population. "I now suspect that just as a deer herd lives in mortal fear of its wolves, so does a mountain live in mortal fear of its deer. The cowman who cleans his range of wolves ....has not learned to think like a mountain."[104] Likewise, John Livingston in a line anticipating the current malaise of the pine beetle carnage in Western Canada, warned that "... it is the spruce budworm's job to prevent the overriding dominance by the spruce trees over the rest of the living community of the forest. Dominance in too great a degree makes for greater vulnerability of both the dominated *and* the dominator."[105]

Over the past fifty years these lessons, radical in their time, have slowly become mainstream. We now sense that Muir was right in seeing fire as a creative force, "...the maker and preserver of prairies..", and that Leopold was right when he cautioned against even well-intentioned wilderness conservation measures because "...all conservation of wildness is self-defeating, for to cherish we must see and fondle, and when enough have seen and fondled, there is no wilderness left to cherish."[106]

As one reads Leopold's *Sand County Almanac* it is as if we are walking together on his farm, past the wood lot, noticing as he does a tree that needs cutting, a bird nest despoiled by predators and a pond alive with insects and fish. We share with him across time and space an intimate acquaintance with place. Much the same occurs in reading Gilbert White's 18th century rambles around Selborne and in countless other writings by naturalists.[107] Closely linked, then, to the conservation movement is a renewed interest in the importance of place and, as well, in community in the broad sense inclusive of humans, animals and land. Important conservationists like Wendell Berry are convinced that those who live and depend on a place will not harm it.

Rousseau on his island, Wollstonecraft in Norway, Muir in Yosemite and Leopold in Sand County have served to remind us of this importance of place in human affairs. And it is the very absence of that sense of place in the works of Bacon, Descartes and Kant that disconnects in important ways their ideas from lived reality and the sensuous. They might argue that such a removal from the deep subjectivity of a specific place or from the body itself is a necessary part of achieving clarity of thought, but this abstraction has in fact contributed to the discomfort so many feel with the instrumental reason bequeathed us by that separation.

This discomfort — manifested in the rise of a very cosmopolitan and culturally relativist post-modernity, in a revival of fundamentalist movements across all cultures, in widespread feelings of alienation in the so-called developed world, and in a deeply troubled and ecologically dangerous culture of consumption — I argue is inextricably linked with the erosion in modernity of an authentic connection to place and to community. Val Plumwood in her most recent work puts considerable emphasis on this issue of "spatial remoteness" in modern culture, noting that: "The most obvious way of avoiding the ecological consequences of your decisions is living somewhere remote from the places and people they affect."[108]

Early in the 19th century the poet Shelley observed this dislocation, praising the culture of the ancient Greeks who "...lived in a perpetual commerce with external nature and nourished themselves upon the spirit of its forms."[109] Shelley was touring the ruins of Pompei, thrilled with the openness of the architecture preserved in the lava, an architecture that allowed the light, wind, odors and rhythms of the natural world to become part of human domestic life. The 2500 years that followed saw the increasing enclosure of human spaces, ending with these enclosed spaces becoming portable as we developed a culture that denied the importance of any link with "home places". To be modern means "...to be on the move", observes the sociologist Zygmunt Bauman, to be nomads "who wander in order to settle."[110] This movement, especially in Western culture, is grounded in the key institutions of modernity — metropolis, economy, media, ideology — which in effect are institutions of dislocation, institutions that promote the fluidity of existence and the irrelevance of place.[111] The natural world becomes less a place the individual inhabits or belongs to than a locale to visit, a tourist destination, a last resort to try out.[112]

Accompanying this material reality of a rootless mass of labourers shifting ground in response to abstract market and geopolitical forces is an intellectual frame that posits an Heideggerian abyss separating humans from nature which is seen as an absolute "other". Faith in technologies, in consumption, and in a lingering spirituality that segmented mind/soul from body all worked to cement this discontinuity between humans and the environment of which they are a part. Meanwhile, feedback loops are disrupted and as we become more physically and philosophically remote from nature

212 | RE-THINKING CULTURE AND NATURE

the observed, the lived link between actions and consequences is weakened.

But, of course ,there is at the same time that alternative modernity that we have been tracing, the Romantic or holistic version that focused on the fundamental continuity between humans and nature. Stan Rowe's *Home Place*, Wallace Stegner's *Wolf Willow* prairie home, Wordsworth's Lake District, White's Selborne, Thoreau's Walden Pond, Leopold's Sand County...these are all "places" that demonstrate that "...our natural tendency to love and revere [nature] is a wholly appropriate part of our emotional constitution. We are not alien invaders of this world. We are home in it."[113]

In practice this "home" is not with nature *per se*, but with a specific place or territory. People often express the sentiment of feeling "at home" in particular landscape, ecosystem or bioregion based on aesthetics, landforms, or climate. For many environmentalists, reacting to the increasing universalist claims of the modern cityscape, ties to the land itself in the tradition of Leopold or Wordsworth becomes central to their hopes for ecological sustainability and human flourishing. In one of its strongest formulations based on the Jeffersonian tradition, this focus on place becomes in turn the key to democracy and to community which in turn are linked to ecological sustainability. No one expresses this meta-linkage better than the modern day Jeffersonian Wendell Berry.

For Berry, true citizenship means "...being a citizen of a community and a place; it begins at home."[114] This political tradition, part of the path-not-taken in modernity, is based on the idea that:

> ...as many as possible should share in the ownership of the land and thus be bound to it by economic interest, by the investment of love and work, by family loyalty, by memory and tradition.....It proposes an economy of necessities rather than an economy based upon anxiety, fantasy, luxury, and idle wishing. It proposes the independent, free-standing citizenry that Jefferson thought to be the surest safeguard of democratic liberty.[115]

Berry sees this kind of human culture, one based on deep knowledge of the land and the ecosystem and a commitment to sustaining both beyond one human lifetime, as in fact the only way humans can create an ecologically sustainable culture.

Just as Aldo Leopold laboured over a lifetime to learn the rhythms, needs and foibles of his Sand County farm, so Berry takes the long view that we need to develop a sense and knowledge of place. He argues for "kindly use", a stance toward the land which he argues can dissolve the sense of otherness that people often feel toward nature.

The land is too various in its kinds, climates, conditions, declivities, aspects and histories to conform to any generalized understanding or to prosper under generalized treatment. The use of land cannot be both general and kindly—...To treat every field, or every part of every field, with the same consideration is not farming but industry. Kindly use depends upon intimate knowledge, the most sensitive responsiveness and responsibility.[116]

It is incumbent upon us, Berry argues, to hear and sense the rules of nature, not to domesticate it, but to becomes domestic to it.[117] David Abrams builds on this idea in stressing the sensuous connection between humans and nature and our need to listen to the land in order to comprehend its needs and our need for it.[118]

This assertion of the importance of place is linked, via Jeffersonian ideas later supported by deTocqueville, to the vibrancy and authenticity of the human communities that occupy these places. These grounded communities, in turn, are seen as the best means through which humans will be able to devise and employ ecologically sustainable practices. At the local, community level abstractions such as moral or ethical rules begin to fade in authority in comparison with the observed and felt practical realities. Global warming is no abstraction to people living near the Arctic or on a Pacific island. Kirkpatrick Sale, Mike Carr and others argue that the proper scale for human communities is the bioregion, "...not so small as to be powerless and impoverished, not so large as to be ponderous and impervious, a scale at which at last human potential can match ecological reality."[119]

Community

So we return to the issue of size as a key variable in the assessment of quality of life and of culture, the smaller, more coherent bioregion as "...not only more ecologically sustainable, but also more intrinsically satisfying to human beings."[120] This is not uncontested ground and we will return to review some of the more critical assessments of this "small (and local) is beautiful" position. But first we need to explore in more detail the substance of the argument.

From the insight that size of community, attachment to place, ecological health and human happiness are all of a piece, two movements or cultural impulses arise: first the urge to simplicity and a return to a rural/village lifestyle; and secondly the desire to establish limits to growth and to move toward a steady-state economy. The fact that each of these options may be in remission in most parts of the more highly developed world need not negate their importance as long-standing components of an alternative modernity.

214 | RE-THINKING CULTURE AND NATURE

The desire to decentralize, the shift from the supposedly alienating and environmentally destructive industrial and bureaucratized urban culture to a bioregional culture based on agriculture, craft production and self-sufficiency has a long tradition. Some link it to an idealized feudalism, an approach Herman Daly and John Cobb take quite seriously in their book on sustainability, calling the Medieval era in European history "...far more successful and affluent, as well as more just and humane..." than we are willing to admit.[121] Their support for a feudal-type model for modern society is based on the idea of subsidiarity, the principle of political organization that shifts power to the lowest level of competence: "The higher level must not absorb the functions of the lower one, on the assumption that, being higher, it will automatically be wiser and fulfill them more efficiently."[122] For critics of the centralizing and urbanizing trends often described as inherent in modernity, a form of social organization informed by subsidiarity would make possible a "...decentralized Jeffersonian polity of relatively small, intimate, locally autonomous, and self-governing communities rooted in the land."[123]

At the base of this decentralist approach to "fixing" modernity is the ideal of the village, an ideal which is perhaps utopian, filled with nostalgia and like Rousseau's State of Nature, may never have existed. But none of this lessens its on-going appeal. In its 19th century version, fueled by books like Gilbert White's on his village of Selborne and Mary Russell Mitford's *Our Village*, the ideal village was "...almost completely self-sufficient, producing its own goods, generating its own power, growing its own food, and boasting a pond for recreation and water power. The village ideal is a self-sustaining and repairable, regenerating system, the whole community operating as smoothly as the mill."[124] For the English writer John Ruskin it was a "...contented, self-sufficient rural community where the good and the beautiful are one..." and for the American socialists Helen and Scott Nearing it was a place for "...individuals, householders, villagers and townsmen living together and cooperating day in, day out....a simple, satisfying life on the land, to be devoted to mutual aid and harmlessness, with an ample margin of leisure in which to do personally constructive and creative work."[125] In all these examples and in the work of Thoreau and the modern neo-Thoreauvian Bill McKibben the central objectives are simplicity, close and sensual contact with nature and increasing leisure time. The poet Robert Southey in planning the stillborn Pantisocracy community with Coleridge and Wordsworth in 1796, captured the spirit of the ideal when he asked:

> ...would that state of society be happy where every man laboured two hours
> a day at some useful employment, where all were equally educated, where

the common ground was cultivated by common toil, and its produce laid in common granaries, where none were rich because none should be poor, where every motive for vice should be annihilated and every motive for virtue strengthened?"[126]

This imagined world of rural, decentralized, largely self-sufficient communities would involved massive de-industrialization, thereby eliminating several major environmental threats. It is supposed that animal cruelty would be eased or eliminated as people raised food locally and labour would assume once again the dignity it was supposed to have had in pre-modern times.[127] Craft work would replace industry as the "cornerstone of social life" William Morris claimed, and wants and pleasures would be modified to assure minimal reliance on mechanical means of production.[128] In taking these steps we would, Wendell Berry assures us, "...achieve the character and acquire the skills to live much poorer than we do."[129] Kirkpatrick Sale refers to this as part of the "decentralist tradition" in Western culture, that part of shadow modernity with "...harmonious self-regarding communities run by participatory democracy and fueled by self-sufficient regional economies deeply conscious of the limits of natural resources and the need for ecological harmony."[130]

Part of the appeal of this village ideal goes back to the issue of leisure and the cultural memory—real or imagined—that in more rural and more primitive times humans laboured less to meet their basic needs, a claim substantiated, as we have seen, by current research on hunter-gather groups. This in turn reveals something about the nature of human flourishing and happiness. Recall that Rousseau tells us that on the Isle St.Pierre "...precious *far niente* was my first and greatest pleasure."[131] This celebration of an innate kind of laziness goes back at least to the ancient Stoic philosophers, was bemoaned by Captain Cook when he was frustrated by a "happy mediocrity" in Tahiti, and was the subject of Bertrand Russell's *In Praise of Idleness* written in 1935. A human preference for working the minimal amount in order to provide the basic needs of life has been a continuous problem for a growth-oriented culture. Max Weber noted the difficulty in persuading workers to increase production with incentives that were superfluous to their needs, observing that "...a man does not by nature wish to earn more and more money, but simply to live as he is accustomed to live and to earn as much as is necessary for that purpose."[132] In his massive survey of the history of civilization, Felipe Fernandez-Armesto concludes that "Idleness and willfulness are more generalized human characteristics than enlightened self-interest and people rarely opt, in the search for long-term gains, for solutions which involve an immediate sacrifice of time or liberty."[133]

But what about the complaint that this is only nostalgia? Or that these images of rural life might be hopelessly utopian? And is the decentralized model really the most effective way to address the ecological problems we face, especially if these small units have to cope with the power of large corporations or foreign governments? Plumwood, for one, remains skeptical since the current range of ecological problems are not likely to be confined to coherent political communities nor can we be assured that local groups will be aware of ecological relationships and dependencies.[134] Even more telling for many people when considering this approach is whether we really do want to return to what Marx in the *Communist Manifesto* so disparagingly called "the idiocy of village life"? And could we do it? It does seem dangerously utopian to think that we could—despite the recent travails of General Motors—dismantle our industrial and post-industrial economy.[135] And do we have convincing evidence that these rural communities would in fact generate the values necessary for ecological sustainability or would they mimic some existing local communities and become "...avid despoilers of wilderness as they eagerly vie for new dams, roads, and clear-cuts on public lands.." or be "...more concerned with extractive-industry jobs or four-wheeling opportunities than with endangered species and unsullied landscapes."[136]

Proponents like Daly and Cobb insist that despite these concerns, their ideal of pluralistic, democratic communities could overcome the more parochial and seemingly impractical qualities of a decentralized culture. Still, despite its emotional and nostalgic appeal, and despite the common sense notion that being in closer contact with nature would likely make us more sensitive to ecological concerns, it does not seem likely that this will emerge as a practical response to the ecological crises we face in the 21st century. Indeed, for some critics such a retreat to the soil and the village is akin to returning to an "...age of tribes and tribalism."[137]

The Steady State

What about the proposals for a steady-state economy, one that stops or dramatically slows down growth rather than actually retreating? The healthiest condition for an eco-system is a state of dynamic equilibrium, one in which birth and death, growth and decay are more or less equal. And as we saw in the first two chapters, for most of humanity's time on the planet this kind of equilibrium or steady state was the order of the day—an on-going and evolving equilibrium between social appetite, technical capacity and accessible resources.[138] Indeed one could argue that this equilibrium persisted until the early 1800's when new technical knowledge, attitudes, institutions and processes led to an unprecedented era of quantitative and exponential growth.[139] We live today with the social and ecological implications of this very unusual growth pattern and while it may seem utopian to question "growth" or contemplate "stagnation", in the broad sweep of human history this recent period of dramatic quantitative expansion is clearly an anomaly.

We have looked at one response to this era of growth or progress, the desire to re-invent the lifestyle and economic patterns of a previous, more agrarian era. Similar in many ways, the idea of a steady-state economic system with a corresponding lowering of consumption expectations is seen by many to be a more practical response. Advocates for the steady state approach became more prominent in the 1970s, partly in response to the Club of Rome's *Limits to Growth* report (1972), the 1973 Energy Crisis, pollution disasters like the 1978 Love Canal in New York, and the resulting public engagement with environmental issues. These were followed by the Bhopal disaster in 1984 and Chernobyl in 1986. Shocked at the prospect of runaway pollution, resource depletion and economic collapse, people were more open to considering more unconventional options, including slowing or even stopping growth.

One of the first unconventional options to garner widespread interest came from England in the form of E.F. Schumacher's *Small Is Beautiful* published in 1973. Building on the parallel interest in the agrarian ideal, Schumacher called for massive decentralization and for "...institutions built to the human scale within harmonious self-regarding communities run by participatory democracy and fueled by self-sufficient regional economies deeply conscious of the limits of natural resources and the need for ecological harmony."[140] This was a full frontal attack on the industrial model of modernity, attempting to bring under one roof the wide variety of dissident voices. Schumacher called it Buddhist economics, potentially widening the net even further, identifying the problem as not wealth itself but the attachment to wealth, not our attachment to pleasurable things, but our craving for them:.[141]

> ...since consumption is merely a means to human well-being, the aim should be to obtain the maximum of well-being with the minimum of consumption .....The ownership and the consumption of goods is a means to an end, and Buddhist economics is the systematic study of how to obtain given ends with the minimum means.[142]

The lack of a "limiting principle" in either production or consumption would, according to Schumacher, lead to a crisis in a world of finite resources. Since the forces of production are not likely to limit themselves and would resist imposed limits, it was necessary to focus on consumption which over the past century had become an end in itself rather than a means to an end.

This idea has deep roots in Western culture. There are obvious links to the Epicurean and Stoic philosophies discussed at several points in this book, the preference for peace of mind and social stability over constant change and endless desire. Rousseau argued for slowing down the pace of change, permitting progress only by "..slow and sure steps."[143] In his critique of technological culture, Heidegger urged a general policy of "letting things be", allowing each component of nature to pursue its own destiny. [144] Deep ecologists insist that there are "joyful alternatives to consumerism" and that a pursuit of simplicity would not mean just giving things up, but rather finding new sources of happiness.[145] Fritjof Capra reminds us that in Buddhist thought it is "... better to have too little than to have too much, and better to leave things undone than to overdo them, because although one may not get very far this way one is certain to go in the right direction."[146] Finally, in a recent book on bioregionalism, Mike Carr raises the spectre that the culture of consumerism that we (and increasingly the rest of the world) revel in is becoming in fact a "culture of extinction."[147]

Schumacher's book had a major impact world wide in the sectors of society that were concerned about an apparent environmental crisis. Support began to spread for community economic development (CED) and Alternative Technology (AT) initiatives, especially in marginal areas of the developed nations and within developing countries. Alternative energy sources, some locally based, began to be discussed and tried out. All these initiatives were given additional boost by the 1987 report from the World Commission on Environment and Development (The Brundtland Report) with its famous call for "sustainable development".

In 1976, the Canadian philosopher Charles Taylor wrote an important essay on the *Politics of the Steady State*, outlining what would have to be done to convince people, consumers and producers, to opt for the kind of approach Schumacher was proposing, a no-growth or steady-state economy. Assuming the issue of limits to growth being a given, Taylor was concerned with avoiding an authoritarian response, one that either imposed limits from above or imposed massive inequalities as a response to limits on growth. Summoning up memories of the popular responses to crises like Dunkirk, Taylor posited a "...sudden rediscovery of social solidarity and common purpose..." with state guaranteed standards for a materially reduced but "decent life."[148]

Taylor's steady state was not to be a stagnant state. Rather, he envisaged a focus on qualitative change and limited quantitative growth using new technologies, designs and renewable resources. To make this possible he suggested that a new set of ideals or collective values would be necessary:

FINDING ECOLOGICAL SENSIBILITY IN A MECHANISTIC CULTURE | 219

It is only of we come to see it as the occasion to realize a new kind of life in which the balance of nature is not a regrettable limit but an opportunity to express and realize new human possibilities that we shall be able to sustain the kind of public morale we will need to maintain the discipline of equality and hence free institutions.[149]

Hence the need for leadership, for a recognition of the serious nature of the crisis, and a renewed sense of solidarity among the people who will be engaged with the change. A tall order, perhaps viewed as more likely in the post-1960s era Taylor was writing in than in recent more conservative times.

CONCLUSION

By the 1980s, despite ever-increasing empirical evidence that the limits to growth argument was valid, popular opinion and government leaders in most developed nations shifted away from these kinds of solutions. While ecosystem rhetoric and the science that supports it was pervasive in schools, universities and environmental groups, the political culture moved into a renewed celebration of individualism with an even greater focus on consumption. This was, after all, the era of Reagan and Thatcher, of a renewed conservatism, a renewed religiosity and a still distracting Cold War.

By the 1990s, the resource crisis predicted in the 1970s had not happened, global warming was still not clearly evident, and ozone depletion seemed abstract with sunblock offering an easy solution to the now cancer-causing rays of the sun. In North America, Green politics, CED and environmental groups remained marginal and while many might still think of small as beautiful, bigger seemed sublime as consumers shifted to an SUV culture. In Europe, Green politics was more prominent, community-based cooperatives had deeper roots and Smart cars appeared, but consumption levels remained oblivious to the dire warnings of the previous decades. The Asian "Tigers" and China seemed positively Victorian in their commitment to growth. What had happened?

For the past 25 years the simplicity/limits to growth option has been under full attack by politicians, academics and a host of media pundits. The prominent University of Maryland conservative economist Julian Simons published his *Resourceful Earth* in 1981, arguing quite powerfully that the best course for humanity was to encourage both population growth and economic expansion—a direct assault on the case for over-population and the need for limits to growth. Even the Brundtland Commission Report in 1987, though it certainly called for more moderate growth, furthered the conservative cause by implicitly agreeing that "development" was potentially "sustainable". The case for aggressive social policy as the best instrument for addressing social and ecological problems slowly gave way to a 20th century version of Smith's invisible hand. Charles Schultze, economic advisor to Ronald Reagan, expressed bluntly the new way of thinking: "Market-like arrangements...reduce the need for compassion, patriotism, brotherly love, and cultural solidarity as motivating forces behind social improvement....Harnessing the *base* motive of material self-interest to promote the common good is perhaps the most important social invention mankind has achieved."[150]

Perhaps the most telling indication of the failure of these visions in reshaping modernity in a more ecologically sustainable and human-scale manner was the arrival on the scene of Ted Kaczynski, the Unabomber. A self-styled small-is-beautiful primitivist who dropped out of society, Kaczynski wreaked the revenge of the terrorist on those whom he saw as supporting the "system". For him, wild nature was in a life and death struggle with a culture of technology, a culture which needed to be eliminated completely in order for small groups of autonomous individuals to co-exist in ways that would not threaten nature.[151] And so he sent letter bombs to his enemies and lived the life of a technophobic primitive.

But with a new millennium bringing with it a clearer sense of the dire consequences we face, there remain other options than primitivism, benign or violent. The final decades of the 20th century opened up the debate about the nature of the modernity-to-be and in the 21st century we need to heed the advice of the poet who sensed an equivalent crisis and opportunity in the early years of the 19th century.

> ...We might be otherwise, we might be all
> we dream of: happy, high, majestical,
> Where is the love, beauty, and truth we seek
> But in our mind? And if we were not weak
> Should we be less in deed than in desire?
> ... and those who try may find
> How strong the chains are which our spirit bind:
> Brittle perchance as straw... We are assured
> Much may be conquered, much may be endured
> Of what degrades and crushes us. We know
> That we have power over ourselves to do
> And suffer — what, we know not till we try;
> But something nobler than to live and die.[152]

NOTES

1 Adam Smith, *The Wealth of Nations* (London: Penguin, 1986), p 110.

2 Anthony Quinton, "The Right Stuff", *The New York Review of Books*, 32 (December 5, 1985), p 52. Cited in J. Baird Callicott, "The Metaphysical Implications of Ecology" in Callicott and Ames, p 55.

3 Latour, p 107.

4 Abrams.

5 From Chung-ying Cheng, "Model of Causality in Chinese Philosophy", *Philosophy East and West*, v. 26, 1976, cited in Callicott and Ames, p 11.

6 Fritjof Capra, *The Tao of Physics* (Boston: Shambhala, 1991), p 24.

7 Sally Miller, *John Muir: Life and Work* (Albuquerque: University of New Mexico Press, 1993), p 9.

8 Thurman Wilkins, *John Muir: Apostle of Nature*, (Norman: University of Oklahoma Press 1995), p 272.

9 Millie Stanley, *The Heart of John Muir's World* (Madison, Wisconsin: Prairie Oak Press, 1995), p 118.

10 Nash (1989), p 39.

11 Dennis Williams, *God's Wilds: John Muir's Vision of Nature* (College Station, Texas: Texas A&M Press, 2002), p 18.

12 Michael Cohen, p 53.

13 Frederick Turner, *Rediscovering America: John Muir and His Times* (San Francisco: Sierra Club, 1985), p 142; Oelschlaeger (1991), p 198.

14 Fox, p 53.

15 Wilkins, p 61.

16 Williams (2002), p 28

17 Michael Cohen, p 37.

18 Michael Cohen, p 263.

19 John Muir, (1978), p 132.

20 Fox, p 13.

21 John Muir, *Mountains of California* (New York: Anchor Books, 1961), p 51-2.

22 Turner, p 191.

23 Worster, p 78-9.

24 Steven Holmes, *The Young John Muir: An Environmental Biography* (Madison: University of Wisconsin Press,1999), p 105.

25 Turner, p 153. Here again we find tantalizing East/West connections. In the 9th and 10th centuries there had been a vigorous debate in China about whether all things, plants, included, could have a potential Buddha nature. The Buddhist scholar Ryogen (912–985) settled the issue by arguing successfully that "...the life cycle of plants moved thru 4 stages which corresponded with the process, of human enlightenment: sprouting out, residing/ growing,, changing/reproducing, and dying. Thus plants aspire for the goal, undergo disciplines and reach enlightenment and enter into extinction. They are, therefore, sentient beings." William Lafleur, "Buddhist Value of Nature" (pp 183-209) in Callicott and Ames, p 188.

26 John Muir, (1978), p 229.

27 John Muir, *The Yosemite* (San Francisco: Sierra Club Books, 1988), p 58.

28 John Muir, (1978), p 157.

222 | RE-THINKING CULTURE AND NATURE

29 Lauck, p 25. For a sympathetic critique describing this as a romantic or nostalgic sentiment, see Tu Weiming, "Beyond the Enlightenment Mentality" (pp 3-21) in Mary Tucker and John Berthrong, eds., *Confucianism and Ecology* (Cambridge: Harvard University Press, 1998).

30 Berry, (1988), p 199.

31 Fleck, p 2.

32 Shepard, (1982), p xiv; Rowe, p 1.

33 HRH Price Charles, "Millennium Reith Lecture", *Toronto Globe and Mail*, 31 May 2000, p A13.

34 Lee, (1992) in Gailey, p 90.

35 In Trevor Carolan, "The Wild Mind of Gary Snyder", *Shambhala Sun*, May 1996 (18-26), p 23.

36 Stephen Quilley even suggests that these cultural accretions may become part of an evolutionary process: "Across the field of evolutionary biology there is ...a growing recognition that the neo-Darwinian paradigm has to be opened up to incorporate a growing body of evidence as to the evolutionary significance of cooperation, association and symbiosis, group and multilevel selection, and—the ultimate Lamarckian heresy—the (albeit limited) possibility of the inheritance of acquired characteristics." Stephen Quilley, "Ecology, 'human nature' and civilizing processes: biology and sociology in the work of Norbert Elias" (pp420-58) in Quilley and Loyal, p 43.

37 Stegner, p 19.

38 Jolly, p 106. Randolph Nesse says sexual selection may have a role in qualities like commitment and altruism: "When potential mates are assessing each other, one of their top priorities is to find someone who is kind and honest....Otherwise, when bad times come, all the good genes and resources in the world will not be of much use......If men and women each have a genetic tendency to seek mates who are kind and honest, this will select strongly for acting kind and honest, and the best way to accomplish this is to actually be kind and honest....[thus] in addition to all the other selective advantages of a capacity for commitment, it may have been subject to sexual selection." Randolph Nesse, "Natural Selection and the Capacity for Subjective Commitment" in Nesse, p 33.

39 Jolly, p 108.

40 James Wilson (1993) p 11.

41 Thomas Schelling, "Commitment: Deliberate Versus Involuntary" in Nesse, p 54.

42 Peter Corning, "The Sociobiology of Democracy: Is Authoritarianism in Our Genes:", *Politics and the Life Sciences* March 2000, (pp 103-108), p 105.

43 Frans de Waal, *Good Natured: The Origins of Right and Wrong in Humans and Other Animals* (Cambridge: Harvard University Press, 1997); John Livingston in CBC *Ideas* broadcast, 29 May 1986 (0 Transcripts, Montreal, 1986).

44 Megarry, p 10. See also Pierce and Bekoff.

45 Quilley, p 50

46 Van der Leeuw and B. deVries, "Empire: the Romans in the Mediterranean", in Goudsblom and deVries, p 211.

47 The issue of the presence or absence of a soul is less a factor today than in the great debates of the 17th century over these distinctions.

48 Elizabeth Skakoon, "Nature and Human Identity", *Environmental Ethics*, v. 30, Spring 2008 (pp 37-49), p 39. See also Carey Wolfe: "...the traditionally distinctive marks of the human (first it was possession of a soul, then *reason*, the tool use, then tool *making*, then altruism, then language, then the production of linguistic *novelty*, and so on) flourish quite reliably beyond the species barrier." In *Animal Rites* (Chicago: University of Chicago Press, 2003), p 2.

FINDING ECOLOGICAL SENSIBILITY IN A MECHANISTIC CULTURE | 223

49 From Richard Ryder, *Speciesism: The Ethics of Vivisection* (Edinburgh: Scottish Society for the Prevention of Vivisection, 1974) cited in Robyn Eckersley, *Environmentalism and Political Theory* (Albany: SUNY Press, 1992), p 43.

50 Paola Cavalieri, *The Death of the Animal* (New York: Columbia University Press, 2009), p 33.

51 Michael Zimmerman, "Martin Heidegger: Antinaturalistic Critic of Technological Modernity" (pp 59–81) in Macauley, p 74. Zimmerman points out that Heidegger is equally critical of those who deny a fundamental connection of humans with nature.

52 Soper, (1995), p 47–49. Only the poet or artist, Heidegger felt, could regain an unthinking, sensory or aesthetic relationship with nature.

53 Peter Singer, from *Practical Ethics* (2nd ed. Cambridge University Press, 1993, pp 49–51), in Gruen and Jamieson, p 54.

54 Nussbaum (2001) p 89. (citing Cicero).

55 Keith Thomas notes that since fish were virtually bloodless and did not cry out or change expression when caught, angling was able to retain its reputation as a philosophical, contemplative and innocent pastime." p 177.

56 Every year, worldwide, tens of *billions* of animals suffer—and, one could add, are unable to exercise their most basic capabilities—through being crowded indoors, unable to form the social groups natural to them, in many cases unable even to stretch their limbs, some of them so tightly caged that they are unable even to turn around or walk a single step Undoubtedly, in terms of the sheer numbers involved and the vast amount of suffering that results, ending factory farming should be the priority issue for all concerned with either the welfare, the preference satisfaction, or the capabilities, of nonhuman animals. Peter Singer, "A Response to Martha Nussbaum", http://www.petersingerlinks.com/nussbaum.htm, p 1.

57 Jeffrey Masson, The Pig Who Sang to the Moon: The Emotional World of Farm Animals, Reviewed by Elizabeth Abbott, *Globe and Mail*, 10 Jan. 2004, p D7

58 Kohak, p 185

59 MacIntyre, *Dependent Rational Animals* (Chicago: Open Court, 1999), p 61.

60 Nussbaum (2001), p 126.

61 Antonio Damasio, *The Feeling of What Happens* (New York: Harcourt, Brace & Co., 1999), p 198.

62 Oerlemans, p 65.

63 Maurice Maeterlinck in *La vie des abeilles* (1943) cited in Peter Scheers. "Human Interpretation and Animal Excellence" (pp 56–64) in Willem Drees, ed., *Is Nature Ever Evil?* (New York: Routledge, 2003), p 61. As a boy on the farm in Wisconsin, Muir developed a special relationship with the oxen, learning early on that "...each ox and cow and calf had individual character....The humanity we found in them came partly through the expression of their eyes when tired, their tones of voice when hungry and calling for food, their patient plodding and pulling in hot weather, their long-drawn-out sighing breath when exhausted and suffering like ourselves, and their enjoyment of rest with the same grateful looks as ours. We recognized their kinship also by their yawning like ourselves when sleepy and evidently enjoying the same peculiar pleasure at the roots of their jaws; by the way they stretched themselves in the morning after a good rest; by learning languages—Scotch, English, Irish, French, Dutch—a smattering of each as required in the faithful service they so willingly, wisely rendered; by their intelligent, alert curiosity, manifested in listening to strange sounds; their love of play; the attachments they made; and their mourning, long continued, when a companion was killed." John Muir, *The Story of My Boyhood and Youth* (Edinburgh: Canongate, 1987), p 47.

224 | RE-THINKING CULTURE AND NATURE

64   In Michael Specter, "The Extremist", *The New Yorker* 14 April 2003 (pp 52–67), p 64.

65   Lauck, p 104.

66   Frans de Waal, 1997, p 19.

67   Matthew Calarco, "Thinking through Animals: Reflections on the Ethical and Political Stakes of the Question of the Animal in Derrida", *Oxford Literary Review*, v. 29, 2007 (pp 1–15), p 10.

68   Wolfe, p 6. Italics in original. This seems an obvious reference to the Holocaust.

69   Coetzee, p 21.

70   Coetzee, p 4.

71   David Boyd, *Toronto Globe and Mail*, 8 April 2004. Bee Wilson in the *New Yorker* points out that it takes an average of four pounds of grain to make a single pound of meat and "...as the world becomes richer, people eat too much, and too much of the wrong things—above all meat." "The Last Bite", *New Yorker*, 19 May 2008, p 77.

72   From a note to Queen Mab: "The change which would be produced by simpler habits on political economy is sufficiently remarkable. The monopolizing eater of animal flesh would no longer destroy his constitution by devouring an acre at a meal....The quantity of nutritious vegetable matter, consumed by fattening the carcass of any ox, would afford ten times the sustenance, undepraving indeed, and incapable of generating disease, if gathered immediately from the bosom of the earth." Cited by Christine Kenyon-Jones, *Kindred Brutes*, Aldershot: Ashgate Press (2001), p 133.

73   Edward Wilson (2002), p 134. Wilson clearly remains biocentric here, agreeing with Peter Singer that "Human beings sharply distinguish the living from the inanimate" p 134.

74   Laurence Silberstein, *Martin Buber's Social and Religious Thought* (New York: New York University Press, 1989), p 49.

75   Nash, (1989), p 20.

76   Tucker and Berthrong, , p xxxvi.

77   Kohak, p 76.

78   William Wordsworth, *The Prelude*, Book III (London: Penguin, 1995). p 110.

79   Collier, (1999), p 78.

80   Worster, p 79

81   Evernden, p 70.

82   From *Frazier Creek* lines 14-19 cited in Susan Kalter, "The Path to *Endless*: Gary Snyder in the Mid-1990's", *Texas Studies in Literature and Language* v. 41:1, 1999 (pp 16–46) p 23.

83   Stan Rowe, "The Changing World: An Ecological Perspective", un pub mss, "Nature and Human Community" series, Simon Fraser University, December 1992, p 2.

84   In David Gamble, "Loren Eisley: Wilderness and Moral Transcendence", *The Midwest Quarterly* v. 33:1, 1991 (pp 108–123), p 112.

85   Tucker and Berthrong. xxvii.

86   Caldwell, p 276.

87   Hwa Yol Jung, "The Ecological Crisis: A Philosophic Perspective, East and West", *Bucknell Review*, v. 20:3, 1972 (25–44), p 41.

88   There are exceptions, of course. Edward Abbey, speaking for hard-line preservationists, says "Cut the bloody cord, that's what we feel, the delirious exhilaration of independence, a rebirth backward in time into primeval liberty, into freedom in the most simple, literal, primitive meaning of the word...". Cited in Kimberly Smith (2003), p 44.

89   With the recent global economic crisis the interest in scaling back is even more widespread.

90   Paul Raskin, et al, *The Great Transformation*, (Stockholm: The Global Scenario Report, 2002),

FINDING ECOLOGICAL SENSIBILITY IN A MECHANISTIC CULTURE | 225

p 41.

91 Smith, (2003), p 63.

92 Kohak, p 101.

93 Evernden, p 4.

94 Roderick Nash, "Island Civilization: A Vision for Human Occupancy of Earth in the Fourth Millennium", *Confluence*, v. 14:2, 2009 (pp 42–55).

95 Michael Cohen, (1984), p 273.

96 Christopher Lasch, *The Minimal Self: Psychic Survival in Troubled Times* (New York: Norton, 1984), p 256.

97 Speth, p 53; "An externality occurs when production or consumption by one firm or consumer directly affects the welfare of another firm or consumer, where directly means that the effect is not mediated through any market and is consequently unpriced". Daly and Cobb, p 53.

98 Steve Vanderheiden, "Two Conceptions of Sustainability", *Political Studies* v. 56, 2008 (pp 435–455), p 435.

99 DeVries, et al, p 299.

100 Leopold, p 203.

101 Leopold, p 224.

102 Leopold, p 4.

103 Gary Snyder, "Baby Jackrabbit" in *Danger on Peaks* (Washington, D.C. :Shoemaker Hoard, 2004), p 27. John Livingston agrees, observing that "When a predator takes one of 'my' grouse in the bush, it is not a crime against me or against God. It is co-operation on the part of the grouse." Livingston (2007), p 91.

104 Leopold, p 132.

105 Livingston (2007), p xxxi.

106 Holmes, p 113; Leopold, p 101.

107 Gilbert White, *The Natural History of Selborne* (London: Penguin, 1977)

108 Plumwood, (2002), p 72.

109 *Letters of Percy Bysshe Shelley*, vol. 2 (London: Pitman and Sons, 1909), p 666.

110 Zygmunt Bauman, "Parvenu and pariah: heroes and victims of modernity" (pp 23–35) in James Good, ed., *The Politics of Postmodernity* (Cambridge: Cambridge University Press, 1998), p 23.

111 Harrison, p 198.

112 As mentioned earlier, these past times are "images" not realities and we must be wary of this calling on the past to give us different images of future possibilities devolving into mere nostalgia, a literal 'going back'. Jonathan Bate suggests that to avoid this we restrict our sense of the past to allegory. Bate (2000), p 36.

113 Mary Midgley, "Criticizing the Cosmos" (pp 11–26) in Drees, p 21.

114 Smith, p 193.

115 Wendell Berry,(1977), p 13

116 Wendell Berry, p 31.

117 Smith, p 144.

118 Abrams, p 68

119 Kirkpatrick Sale, (1991), p 55.

120 Donald Alexander, "Bioregionalism: Science or Sensibility?", *Environmental Ethics* v. 12:2, 1990 (pp 161–173),p 169.

121 Daly and Cobb, p 15.

226 | RE-THINKING CULTURE AND NATURE

122 E.F. Schumacher, *Small Is Beautiful* (New York: Harper, 1989/1973), p 260.

123 David Shi, *The Simple Life* (New York: Oxford University Press, 1985), p 266. There are endless odes to this bucolic and communitarian ideal, one of my favourites being from the Irish nationalist Eamon DeValera in 1943: "That Ireland which we dreamed of would be the home of a people who valued material wealth only as the basis of right living, of a people who were satisfied with frugal comfort and devoted their leisure to things of the spirit—a land whose countryside would be bright with cosy homesteads, whose fields and villages would be joyous with the sounds of industry, with the romping of sturdy children, the contests of athletic youths and the laughter of happy maidens, whose firesides would be forums for the wisdom of serene old age. It would be, in a word, the home of a people living the life that God desires that man should live." radio broadcast, St.Patrick's Day, 1943. cited in Richard Douthwaite, *The Growth Illusion* (Bideford, Devon: Resurgence, 1992), p 249. Mary Russell Mitford, *Our Village: Sketches of Rural Character and Scenery* (London: G. Bell & Sons, 1877).

124 Daniel Botkin, (1990), p 102.

125 John Batchelor, *John Ruskin: No Wealth but Life*, (London: Chatto & Windus, 2000), p 178; Kimberley Smith, p 31.

126 Nicholas Roe, *The Politics of Nature: William Wordsworth and Some Contemporaries*, (London: Palgrave, 2002), p 57. In a more cynical contemporary vein, the French novelist Michel Houellebecq offers a more jaundiced view: "A detailed description of this pastoral 'idyll' is of limited interest, but to be comprehensive I will outline it broadly. You are at one with nature, have plenty of fresh air and a couple of fields to plow (the number and size of which are strictly fixed by hereditary principle). Now and then you kill a boar; you fuck right and left, mostly your wife, whose role is to give birth to children; said children grow up to take their place in the same ecosystem. Eventually, you catch something serious and you're history." In *The Elementary Particles* (New York: Vintage International, 2000), p 19.

127 Keekok Lee, "To De-Industrialize—Is it so irrational?" (pp 105–117), in Andrew Dobson and Paul Lucardie, *The Politics of Nature*, (London: Routledge, 1993), p 115.

128 Leiss (1990), p 46.

129 Wendell Berry, "Word and Flesh", commencement address at College of the Atlantic, Bar Harbor, Maine, reprinted in *Harper's*, Sept., 1989, p16–22. For a critique of this position, see Richard White, "Work and Nature: Are you an environmentalist or do you work for a living?" (pp 171–185) in Cronon, and Evan Eisenberg, "The Call of the Wild", *The New Republic* 30 April 1990 (pp 30–38).

130 Kirkpatrick Sale, "Preface" to E.F. Schumacher, *Small Is Beautiful* p xx.

131 Rousseau, *Reveries*, p 83. e.g. "doing nothing".

132 Elizabeth Kolbert, "Why Work: A Hundred Years of The Protestant Ethic", *New Yorker*, 29 Nov 2004 (pp 155–160), p 156. In an effort to increase efficiency at harvest time, farmers tried paying their workers more in order to increase production: "Raising the piece-rates has often had the result that not more but less has been accomplished in the same time, because the worker reacted to the increase not by increasing, but by decreasing the amount of his work. A man, for instance, who at the rate of 1 mark per acre mowed $2\frac{1}{2}$ acres per day and earned $2\frac{1}{2}$ marks, when the rate was raised to 1.25 marks per acre mowed, not three acres, as he might easily have done, thus earning 3.75 marks, but only 2 acres, so that he could still earn the $2\frac{1}{2}$ marks to which he was accustomed."

133 Fernandez-Armesto, p 205.

134 Plumwood (2002), p 77.

135 Lewis and Sandra Hinchman, "Should Environmentalists Reject the Enlightenment?",

FINDING ECOLOGICAL SENSIBILITY IN A MECHANISTIC CULTURE | 227

Review of Politics, v. 63:4, 2001, The Hinchman's say we need the heritage of the Enlightenment to protect is from large economic interests: "Like it or not, we cannot defend the environment without deploying the modern state as a counterweight to industry and — unfortunately — the anti-environmentalism of people more concerned with extractive-industry jobs or four-wheeling opportunities than with endangered species and unsullied landscapes." p 674.

136 Hinchman, p 674.

137 Bauman (1998) p 31.

138 Boyden, p 93.

139 Maurice Lamontagne, "The Loss of the Steady State", in Abraham Rotstein, ed. *Beyond Industrial Growth* (Toronto: University of Toronto Press, 1976) pp 1–21), p6.

140 Sale (1989) p xx.

141 Schumacher, p 60.

142 Schumacher, p 61.

143 Rousseau, *Emile*, p 232.

144 Zimmerman, p 60.

145 Andrew McLaughlin, *Regarding Nature* (Albany: SUNY Press, 1993), p 201.

146 Capra, p 106

147 Mike Carr, *Bioregionalism and Civil Society* (Vancouver: UBC Press, 2004), p 21.

148 Charles Taylor, "The Politics of the Steady State" in Rotstein, p 55.

149 Taylor (1976), p 64.

150 Cited in Daly and Cobb, p 139.

151 Tim Luke, "Re-Reading the Unabomber Manifesto", *Telos*, #107, Spring 1996 (81–95), p 89.

152 Percy Shelley, "Julian and Maddalo", in Hutchinson (1970), p 329.

CHAPTER SIX

NEW INSIGHTS ABOUT CULTURE AND NATURE

"The man of culture makes a friend of nature"

FRIEDRICH SCHILLER

In the midst of the most violent and bloody years of the French Revolution the young German philosopher, playwright and poet Friedrich Schiller (1759–1805) posed a set of questions and suggested some answers that ring as true today as they did in 1795. Schiller was 36 when he composed his *Letters on the Aesthetic Education of Man*, sharing a generation with Mary Wollstonecraft who was 36 when she left Paris for Norway. Both were avid readers of Rousseau and Goethe and participants in the wave of rational thinking associated with the Enlightenment, part of a generation at first enthralled by the utopian possibilities opened up by the French Revolution and then traumatized by the violence that followed. Rejecting the subsequent shift to conservatism that was adopted by many of their peers, the lesson they learned from the failure of attempts to impose democracy and peace via reason and force was that changing ideas and values had to precede or at least parallel attempts at political change.[1] There seems an obvious connection here to our current impasse in responding to an increasingly pressing and perhaps violent ecological crisis.

In the 8th of his *Letters* Schiller suggested a path through what was already an obvious schism between reason and feeling that was to so imbalance Western culture for the next 200 years. *"Our age is enlightened"* Schiller proudly announced. *"Reason has accomplished all that she can... The spirit of free inquiry has dispatched those false conceptions that for so long barred the approach to truth, and undermined the foundations upon which fanaticism and deception had raised their throne..."* This is Schiller's proud recognition of the contribution toward

230 | RE-THINKING CULTURE AND NATURE

the "triumph of reason" made by his mentor Immanuel Kant, the impact of the French Revolution, and the new secular politics of the era—the real start of the modernity we know today. But, he continued, *"How is it, then, that we still remain barbarians"?*[2]

Schiller had an answer, and it comes very close to my own response to the question why we remain in so many ways ecological barbarians in the 21st century despite our enlightened awareness of the catastrophe we court. It is the *"disposition of men"*, Schiller argued, that stands in the way; their reluctance to face up to the truths that they now can see, to *"dare to be brave"*. And this faulty disposition in turn stemmed from the epistemological error of a dualism that had become dominant in Western culture, a dualism that created a competitive rather than cooperative or symbiotic relationship between reason—what Schiller called the formal drive—and feeling or the sensuous drive. I have argued in earlier chapters that this sensuous drive had been dominant in Homo sapiens during the long Paleolithic era, followed by a brief era of unity—or so the Romantics wanted to believe—in classical times and then the gradual repression of the sensual and the increasing domination of the more abstract instrumental or formal drive.

Schiller acknowledged that with sensual bodies and rational, self-conscious minds we are, thanks to evolution, inherently dualistic and always have the potential to be at odds with ourselves, *"...either as savage, when feeling predominates over principle; or as barbarian, when principle destroys feeling. The savage despises civilization, and acknowledges nature as his sovereign mistress. The barbarian derides and dishonours nature...The man of culture makes a friend of nature, and honours her freedom while curbing only her caprice."*[3]

And what had happened to this idealized "man of culture?" Again, we see Schiller's Rousseauean inspiration: *"It was civilization itself that inflicted this wound upon modern man. Once the increase of empirical knowledge, and more exact modes of thought, made sharper distinctions between the sciences inevitable, and once the increasingly complex machinery of state necessitated a more rigorous separation of ranks and occupations, then the inner unity of human nature was severed too, and a disastrous conflict set its harmonious powers at variance. The intuitive and the speculative understanding now withdrew in hostility to take up positions in their respective fields, whose frontiers they now began to guard with jealous mistrust... in the one a riotous imagination ravages the hard-won fruits of the intellect, in another the spirit of abstraction stifles the fire at which the heart should have warmed itself and the imagination been kindled."*[4]

Just as Rousseau had insisted on the moral sterility of reason alone and strove to re-connect with the sensual via reverie, Schiller insisted that *"the way to the head must be opened through the heart. The development of man's capacity for feeling is, therefore, the more urgent need of our age, not merely because it can be a means of making better insights effective for living, but precisely because it provides the impulse for bettering our insights.*"[5] Like Shelley a few years later and critical theorists like Adorno, Horkheimer and Marcuse in the 20th century, Schiller was no naive "romantic" decrying the reign of reason.[6] He was calling instead for a unity between the rational and the sensual, the scientific and the humanistic, a renewal of the more holistic and harmonious culture that was believed to have existed in ancient Greece — and, we now believe, in centres of culture as diverse as China and India.

The process of enlightenment that Schiller both celebrated and critiqued in 1795 has since spun off in a new, even more powerful instrumental direction. Appropriated by science and its various prosthetic technologies, it is now committed to a culture built on constant change and material progress. His hope for a rejuvenation of the sensuous via art and culture remains muted but his objective of a unified, healthy culture built on the twin pillars of reason and feeling remains a 21st century imperative. The contributions of Schiller, like those of Rousseau, the Shelleys and other figures referred to here are not "merely" historical, but are part and parcel of the on-going critique of and prescription for a modernity that has gone astray. Having been "present at the creation" they may have a stronger claim to relevance than those of us who know modernity only by habit and inheritance.

"The way to the head must be opened through the heart". What does this 18th century lesson have to teach us in the 21st? The dualism of modernity has been a central theme throughout this exploration of the nature and culture issue. Indeed the entrenched notions of *nature* and *culture* are at its heart. But as thinkers from Plato to Schiller have noted, while such a conceptual duality is inevitably central to a self-conscious Homo sapiens, it need not inevitably result in competition or conflict.[7] But the evidence of continuing alienation, exploitation, violence and discrimination in human relations with humans, coupled with a looming ecological catastrophe caused in part at least by our persistent refusal to see humans as an integral part of nature, must give us pause to reconsider the way we have approached this duality. We, like Schiller's peers, seem to remain barbarians, our accomplishments leaving "...a trail of devastation across the face of the earth."[8]

What to do? I have argued here that during the millennia of the Paleolithic and well into the Neolithic there was among humans a symbiotic approach to nature and the other that rested on sensual appreciation, spirituality and respect. We have also seen that once lost, there is no going back to *that* unity, nor would it be desirable to abandon the many benefits that our enhanced consciousness has given us in a vain quest for a lost holism. We return, then,

232 | RE-THINKING CULTURE AND NATURE

to Schiller's question: *"How...are we to restore the unity of human nature that seems to be utterly destroyed by this primary and radical opposition?"*[9] Being more poet than politician or philosopher, Schiller's response is an aesthetic one. It will be "beauty" or "culture" manifested through a "play drive" that will bring reciprocity to the rational and the sensuous. In a recent book on bioregionalism, Mike Carr's focus on dance, music and other "carnivalesque" features of contemporary environmentalism has elements of Schiller's Play Drive.[10] But while this seems a less promising solution in the 21st century than it may have in the 18th, Schiller has an important point for us to consider: the issue of reciprocity. Neither the rational nor the sensuous should dominate; we should neither abandon science and retreat to the primitive nor should we give in to the siren call of a nature-free technological future. Reciprocity between the rational and the sensuous dimensions of the human would ensure that *"...the activity of the one gives rise to, and sets limits to, the activity of the other, and ...each in itself achieves its highest manifestation precisely by reason of the other being active."*[11]

In the last chapter I referred to the enhanced understanding we can reach about our relationship to nature when we are guided by a combination of the sensibilities of the naturalist, the passionate concerns of the environmentalist, and the scientific knowledge of the ecologist. The likes of Leopold, White, Berry, Thoreau, Clare and Rousseau heighten our appreciation of the natural world by allowing us to share the intimate and personal dimension of their feelings for nature. Environmentalists like Evernden, Rowe, Livingston and Bookchin model for us the political dimension of a deep concern for the well-being of the natural world and human societies. And scientists like Darwin, Carson and Wilson provide the essential data and understandings through which passion and politics must be filtered.[12]

But this is a complex issue. Contemporary environmental studies is, by necessity, what is often referred to as "interdisciplinary", that is, combining the expertise and perspectives of a variety of academic disciplines. Mobilized in this way in both academic and corporate bodies, this new interdisciplinary endeavour quickly becomes very applied, responding as it must to pressing social and economic issues. Given the nature of the issues, this academic expertise often becomes closely linked to politics and bureaucratic bodies, economic interests and/or popular organizations such as the Green Party or Greenpeace. Science itself, therefore, can easily stray from its vaunted objectivity and become highly politicized. Whether it be climate change, ozone holes or the effects of tobacco, the various interest groups can assemble their science and scientists to defend their specific interests. At the other extreme, however, in a culture grounded upon particles and specialization, these interdisciplinary connections remain difficult to sustain in institutions and naturalists, environmentalists and scientists too often still remain in their respective solitudes.

The persistence of these solitudes helps us understand what is undoubtedly one of the most perplexing issue of our times, why is there so "...little causal relationship between our environmental education (knowledge acquisition, increased awareness, and changes in value orientation) and environmentally responsible behavior?"[13] How is it that when it comes to environmental issues we seem to have "...an infinite capacity for fantasy and denial."[14] This is what Wendell Berry calls the "human problem", the one that knowledge alone — the great hope of the modern Enlightenment — cannot solve.[15] For Berry, the "...split between what we think and what we do is profound.." and has its source in a "...corruption of character."[16] That corruption, I argue here, is a cultural and social rather than a personal or psychological problem, and it stems from our abandonment of a sensual world view that focused on humans as embedded in nature in favour first of a mechanistic model that fragmented nature and then an evolutionary model that set humanity above or even outside of nature.

What we derive from Schiller's account of the problem with modernity is the realization that not only our own internal alienation (Schiller's main concern), but also our alienation from nature itself is truly *our* problem. Civilization may have been the ultimate cause of both of these pathologies, but we are the actors, the agents that individually and collectively make up civilization. Neither civilization nor nature can summon a self-conscious will to bridge the dualities we necessarily live with. If we are to shed a mastery agenda rooted in dualism and destructive to both nature and human flourishing and develop instead a nurturing reciprocity both internally and externally, we have to make the first move individually and collectively. As a self-regulating set of mechanisms, nature, of course, acts as well and in response to our attempts at mastery it is visibly if unconsciously doing so now in ways that bode ill for us if we choose not to change.[17]

This takes us back *home* in a sense, to our attitudes, values and choices. Nature itself is ecologically sustainable, human civilization on its current trajectory is not. As the component of human civilization that has so far been able to dominate an increasingly globalized modernity, it seems largely up to Western culture to choose whether to abdicate its leadership role, to continue stubbornly leading human civilization to eco-collapse, or to shift its focus to ecologically sustainable beliefs, values, policies and practices. Cultures, of course, are not choosing beings. But a culture does embody and nurture sets of values and beliefs — a world view — that is (or should be) dynamic and subject to reform and re-definition by the social and political structures operative within it.

There are, of course, many impediments to the last choice, to taking the lead in a significant re-thinking of our relationship to the natural world. Despite mounting sensual and scientific evidence, there is still no popular or even professional consensus that a true ecological crisis is upon us or that human actions are the major cause of the various environmental problems we are encountering. As in so many similar cases, the scientists and other experts we turn to for the facts offer conflicting, contradictory and confusing responses. There are also powerful corporate, financial and industrial interest groups that profit from the status quo and at this point their political influence far outweighs that of emerging countervailing forces. In a market economy built on the foundation of these well-established private interests and their political and ideological support systems, the preferred response to environmental concerns has been to "...define problems according to the capacity of existing institutions to deal with them."[18] Thinking outside the box may be a popular marketing tool, but is not advice commonly followed by policymakers attuned to short-term goals and wedded to the status quo.

But systems like capitalism or socialism cannot really shoulder the blame for our current environmental dilemma. They are not "eternal verities" but are the current structural formations and ideologies through which our modernity operates and as such are subject to change. Based on the ideas of Norbert Elias, Stephen Quilley argues that despite a stubborn reliance on the ever-increasing complexity of our technologies and social systems to "cure" our ecological ills, it is possible that "..the species with the greatest capacity for destabilizing impacts on non-human nature, may yet prove to be the only species capable of exercizing evolutionary self-restraint."[19] Perhaps, but the jury is still out. Just recently the previous American President's Senior Environmental Advisor asserted that the solution to climate change is "the advancement of technology."[20]

I have argued throughout this book that there are more than just competing interests in modern culture, there are also competing or alternative world views embedded within modernity, world views that do not include a rejection of modernity, but rather a re-visioning. There are also increasingly influential voices from Buddhist, Taoist and Hindu traditions that are now entering modernity as potentially equal competitors in defining a modernity for the 21st century.[21] Fritjof Capra made this point in terms of Asian philosophy and Western science in his very influential book *The Tao of Physics*, and the same kind of kinship holds between many modern environmentalists and Asian conceptions of human relations with the natural world.[22] The

NEW INSIGHTS ABOUT CULTURE AND NATURE | 235

lesson that Schiller and Wollstonecraft learned in the 1790s was that changing structures via revolution or other means without a preceding or at least
corresponding revolution in thought and values accomplished little. That
lesson would be repeated after the 1917 Russian Revolution. So in crafting a
21st century response to the world's environmental crisis, I return to the issue of thought and action. To preserve a planet habitable for humans we must
change our ways, but to do that we must first change our minds. We should
start, as Jonathan Bate argues, by viewing nature "with wonder and reverence,
not rapaciousness."[23]

FOUR IMPORTANT CURRENT TRENDS

Lynn White in his classic essay on the "Historical Roots of Our Ecological
Crisis" reminds us that "...what people do about their ecology depends on
what they think about themselves in relation to the things about them."[24] Neil
Evernden opens his *Natural Alien: Humankind and the Environment* by asserting
that the environmental crisis is:

> Essentially a cultural phenomenon...the source of the environmental crisis
> lies not without but within, not in industrial effluent but in assumptions
> so casually held as to be virtually invisible.....our scarred habitat is not only
> of our doing, but our imagining, and it will take a profound re-creation of
> the social world to 'un-say' the environmental crisis and constitute a more
> benign alternative.[25]

He goes on to speculate that there are "alternate voices" in Western culture that might help us with a "fundamental re-thinking" of that relationship,
the *Shadow Modernity* I have been referring to. But these alternative voices,
some of which we have heard in this book (e.g. Epicurus and Lucretius, St.
Francis, Rousseau, Muir, Leopold), have been muted, marginalized or even
deliberately suppressed. And as we saw in the last chapter, alternatives such
as a return to a simpler life or advocating a slower or reduced rate of change,
while supported by many, have consistently failed to garner mass public support or the sustained attention of policy makers.

But we must not lose heart. As I suggested early on, there has been a consistent shadow to the modernity that emerged from the classical era. During
the Roman era and well into the Medieval millennium that followed there was
strong representation within Western culture of what we might call a holistic, sensual and humane ethos, with a companion politics in tow. There was
a consistent body of opinion that saw animals and humans in relationship,
that acknowledged a naturalist connection of human with landscape, and

236 | RE-THINKING CULTURE AND NATURE

advocated a light human step on the terrain of nature. By the beginning of what we consider modernity however, these views began slipping deeper into the shadows of the dominant culture. Renaissance Humanism began to solidify a consistent anthropocentric tendency and the rise of instrumental reason, science and an accompanying progressivist ideology soon—as we saw in Chapter Three—held sway. And we live with the consequences of its victory—positive and negative.

It is time, then, to heed the counsel of Coleridge's Ancient Mariner and work toward a unification of these two streams of modernity that Western culture has been so assiduous in both creating and keeping asunder. And since it was largely the issue of nature, and the relation of humans to nature, that divided this modernity, in order to create a more holistic 21st century modernity that human bond with nature must be re-kindled. The dualist temptation of a triumph of either stream of modernity at the expense of the other is not the answer. As the mastery agenda of our current modernity begins to shudder under the weight of its own victories we can begin to see the potential for the emergence of a new *mutualist modernity* that engages traditional human-centered notions of rights and obligations with the material interests of eco-systems and specific life-forms, and a new politics that integrates our modern technologies with nature. The modernity we need to create must rest on integration rather than separation, reciprocity rather than singularity, and a more pragmatic dualism tempered by a sensible holism. We can in effect employ our rational abilities to overcome our anthropocentrism and "...transcend the narrow limits of this humanism."[26]

On the surface there seems little evidence for optimism that these kinds of shifts in thinking, values and politics are imminent. While one might have anticipated that given the various dismal assessments of "the state we are in" that at least within the developed nations there would have been popular and political pressure to consider these kinds of changes. Instead, Green politics has stalled and "the environment" until quite recently has remained a consistent also-ran in public opinion polls about the concerns of voters. Social activism on behalf of the environment remains marginalized in groups like Greenpeace or the Sierra Club or in self-interested groups like Ducks Unlimited. Some more radical critics, of course, perceive these same environmental groups as hopelessly mainstream and hence too open to compromise.

Where might we look then for more optimistic trends in contemporary culture? Four developments that matured in the latter decades of the 20th century come to mind that may contribute in significant ways in changing how we think about and act upon the natural world in the 21st century:

NEW INSIGHTS ABOUT CULTURE AND NATURE | 237

- the revival of a persistent alienation from and doubts about material progress and consumption being the basis of human happiness;
- the development in modern feminism of a powerful challenge to hierarchical world views and dualist thinking;
- a new kind of scientific thinking that trumps the fragmentation characteristic of Baconian and Cartesean thinking and instrumental empiricism;
- the increasing influence of Asian philosophies in Western culture stemming from immigration and the dramatic economic growth in East and South Asia.

Following a discussion of these trends, I will explore in more detail in the Conclusion the two significant new directions that might flow from them, directions that build on the kind of alternative conceptions of modernity outlined earlier. One of these involves a radical extension of our assignment of moral value and legal standing to include components of the natural world. The other envisages shifting the control of technological change and the process of development itself from the private to the public sphere. But first, we need to establish how this proposed extension of rights and more democratic control of change would embody a modernity that integrates the romantic and the rational, the sensual and the empirical, passion and reason.

The modernity that that has come to dominate the globe is ecologically dysfunctional, not by nature but by choice. The choices made in the 17th and 18th centuries to privilege hierarchy, instrumental reason and growth at the expense of egalitarianism, sensuality and stability—qualities that could be equally at home in the modern—created a powerful but drastically imbalanced culture which reached its destructive apogee in the wars, alienation and ecological destruction of the 20th century. The extreme nature of this modernity's "successes" in turn provoked countervailing forces both within and outside of that modernity, forces that in the 21st century will need to assert their influence in order to create a more balanced and sustainable global modernity. Before suggesting what that modernity might look like we need to explore these newly emergent forces.

RE-ASSESSING HAPPINESS

In combination Freud and Rousseau got it right. Homo sapiens is at base (at heart?) a pleasure-seeking, pain-aversive creature. Socially these primal predispositions can translate into a sympathetic and even benevolent creature who shuns causing pain to others and actively seeks a general flourishing

RE-THINKING CULTURE AND NATURE

of both individuals and kinship or other affiliated groups. This is essentially the image we encountered in Chapter One with our leisure-loving, generally peaceful and self-sufficient hunter-gatherers. Then in the second chapter we saw how the veils of culture dropped down one by one to mask, divert or repress these fundamental predispositions. Benevolence came to be restricted to a chosen few and pleasure began to be displaced from the sensuality of the immediate to the delayed or distant pleasures stemming from externally derived cultural and economic norms.

But the river of nature runs deep and these originary pleasures still speak to us. We retain what are perhaps primal attractions to certain landscapes because they give us pleasure, both sensual and aesthetic. We aspire through reverie, meditation and other means to connect with that sentiment of existence that Rousseau described at the Isle St.Pierre. We anthropomorphize easily, ascribing feelings to trees and land as Aldo Leopold did on his Sand County farm and Mary Wollstonecraft did during her travels in Scandinavia. And we worry, with Wordsworth, about whether the culture we have created really brings us the kind of pleasure and completeness we sense in nature:

> The birds around me hopped and played:
> Their thoughts I cannot measure,
> But the least motion which they made,
> It seemed as thrill of pleasure.
>
> The budding twigs spread out their fan,
> To catch the breezy air;
> And I must think, do all I can,
> That there was pleasure there.
>
> If I these thoughts may not prevent,
> If such be of my creed the plan,
> Have I not reason to lament
> What man has made of man?[27]

But what do we really want? What would ease this lament? Much ancient wisdom from the likes of Lucretius, the Buddha, or Chang Tzu would insist that peace of mind, contentment, and balance are the particular keys to human pleasure. In this view the constant drive in modernity to increase the material standard of living only serves to decrease the actual quality of life.[28] Martha Nussbaum resurrects the ancient notion of *Eudaimona* or human flourishing in her account of basic human desire or need and argues that it has little to do with the accumulation of material goods. Instead, in Nussbaum's blending of ancient and modern usages, to flourish is to lead a "complete life", a life that allows for the full flourishing of both our moral and our social natures.[29] But, she says, in the cultures in which we live *what man has made of man* is often seen as an impediment to this flourishing, to this human yearning for some kind of completeness or authenticity.

Nussbaum's work in this area highlights for us the direct line of critique between Wordsworth in the 19th century and contemporary cultural critics like the Canadian political philosopher Charles Taylor—indeed the line stretches back at least to Epicurus and the Stoics and finds resonance in the likes of the poverty vows of medieval monks and nuns, the egalitarian politics of 17th century English Levelers, and in the celebration of simplicity in Rousseau and Thoreau. Always searching for a path to serene flourishing, Rousseau has the lead character in his novel *Julie, Or the New Heloise*, describe her own achievement of this end as one in which "..everything I see is an extension of my being, and nothing divides it; it resides in all that surrounds me, no portion of it remains far from me; there is nothing left for my imagination to do, there is nothing for me to desire; to feel and to enjoy are to me one and the same thing; I live at once in all those I love, I am sated with happiness and life."[30] But Rousseau knew that his modernity would have none of this primitivist, ecstatic holism devoid of imagination and desires and that his fictional heroine must therefore die.

Julie's temporary happiness is at once an affirmation of self and a diffusion of self into all its surroundings, an acknowledgement of an inevitable dualism and a necessary holism. Human cultures have over time responded to this paradoxical situation by favouring either one side or the other, seldom reaching the serenity of acceptance that Rousseau describes. The historian and cultural critic Christopher Lasch explores this paradox in a similar way:

> The achievement of selfhood, which our culture makes so difficult, might be defined as the acknowledgement of our separation from the original source of life, combined with a continuing struggle to recapture a sense of primal union by means of activity that gives us a provisional understanding and mastery of the world without denying our limitations and dependency. Selfhood is the painful awareness of the tension between our unlimited aspirations and our limited understanding, between our original intimations of immortality and our fallen state, between oneness and separation. A new culture—a postindustrial culture, if you like—has to be based on a recognition of these contradictions in human experience, not on a technology that tries to restore the illusion of self-sufficiency or, on the other hand, on a radical denial of selfhood that tries to restore the illusion of absolute unity with nature.[31]

The moral and social "complete life" that Nussbaum says we need to aspire to, the serenity of Rousseau's Julie, and the acceptance of the contradiction between unlimited aspirations and natural limits are the components of happiness, pleasure or flourishing that we need to move toward if we are to become a humanity that can create and nurture ecological sustainability.

Instead, as Hannah Arendt said, most of us live in cultures that base selfhood on a "tyranny of possibilities."[32] The feminist critic Kate Soper aptly describes the consumption and achievement possibilities that pervade our imaginations as making "plenitude the feast of sufficiency."[33] More is all we need and our imaginations are free to set aspirations far beyond our or nature's limitations.

Charles Taylor calls this the "malaise of modernity."[34] Pulling back from what turned out to be his perhaps naïve 1970s hopes for a steady state culture, Taylor retreated to an exploration of how we arrived at this out-of-control imaginative self. The result was his seminal text *Sources of the Self: The Making of Modern Identity*, which in turn formed the basis for his lecture-based book *The Malaise of Modernity*. Taylor identified three themes in modernity that he argues have resulted in a general sense of cultural decline and malaise:

1. a specific formulation of individualism that imposed more responsibility on each of us to define our needs, desires, choices and values without strong support from the older moral horizons framed by religion, family and tradition.
2. too great a focus on instrumental reason with its attendant twin engines of efficiency and progress, each driving to other toward ever faster rates of change.
3. a gradual loss of freedom with the rise of bureaucracy and reliance on technology which leads to a celebration of private life at the expense of community and conceptions of citizenship.

Taylor's diagnosis has obviously informed much of my thinking about ecological sustainability and is rooted in the long-standing critique within Western culture of excessive individualism, instrumental reasoning and the loss of an organic link with community and nature. While sharing a profound mistrust of the social and moral norms of modernity and fearful that they may negate the possibility of human flourishing, Taylor does not turn to nature for his solution, but instead to a re-formulation of culture. He argues that we are, by nature, persons-in-community and that we can flourish in community thanks to an innate moral sense that we can access in dialogical relations with others in that community. We are also, by nature as well, individuals with a powerful need for self-fulfillment which can only become authentic if we derive our individual values, desires and needs via a process of

NEW INSIGHTS ABOUT CULTURE AND NATURE | 241

negotiation between our selves, our community and the larger set of human values that sit above both. With Taylor we move beyond Wordsworth's lament and move toward a philosophical, cultural and social solution to the problem of culture, a problem of happiness and human flourishing.

Ironically, the global economic crisis that appeared rather too suddenly in 2008 has proved to be a stimulus to this process of re-thinking modernity. While the crisis of liquidity, employment and economic security is a genuine hardship for millions from Malaysia to New York, contrary to some expectations it has not caused a loss of interest in environmental issues but rather seems to have reinforced a desire to "get it right" this time. Thirty years of conspicuous consumption and de-regulation appears to be giving way to a new interest in lowering expectations and seeking a "green path" to economic recovery.

THE CONTRIBUTION OF ECOFEMINISM

Mother Nature. Women and Reproduction. Maternal Nurturing. These are just some of the common attempts to forge a special connection between women and the natural world. For many feminist thinkers, these connections are deeply problematic. One may bask in the glowing wonder of Mother Nature on a warm spring day, but curse her in the midst of a merciless storm or prolonged drought. The fact of human sexual reproduction is surely wondrous, but it also places a severe and sometimes fatal burden on the female. And nurturing is a fine and obviously natural thing, but if this natural disposition to nurture is assigned exclusively or even primarily to the female gender, the resulting social, economic and political inequities result in serious disadvantages.

These issues have been addressed at length by modern feminist thinkers and the suggestion that females have a special connection with nature remains a hotly contested issue, even within the feminist movement. But this debate is not the dimension of the feminist discussion of nature that I want to pursue. There is a large and growing ecofeminist literature and it is not my intention to attempt a summary. Rather I wish only to highlight the work of a few scholars who write within the ecofeminist tradition and whose insights and arguments contribute in a significant way to re-thinking the human relationship to nature.

In the wide-ranging discussions concerning ethics and the environment, many women contributors, whether feminist or not, have focused on the centrality of compassion and empathy as our primary "tools" in considering any re-thinking of our relations with nature, whether our own or nature *per se*.

242 | RE-THINKING CULTURE AND NATURE

There may be a hint of essentialism here, but several women writers seem to have been especially prominent in reinvigorating these classic human attributes—often at the expense of utility and reason—and employing them as the means to erode the strength of the mastery agenda.

Martha Nussbaum comes immediately to mind given the critical role that she assigns to compassion, empathy and imagination in her pursuit of cultural norms that might enhance human eudaimonia or well-being. In her magisterial *Upheavals of Thought* she cites compassion as the tool or attribute most central in our attempts to extend the boundaries of what we can imagine: "... only when we can imagine the good or ill of another can we fully and reliably extend to that other our moral concern."[35] An obvious extension of the well-established Kantian approach which privileges reason, Nussbaum's focus on the more subjective or at least affective side of human nature makes it much more possible to imagine the realm of animals, plants and eco-systems becoming subjects of genuine moral standing in their own right rather than merely in relation to human well-being.

The philosopher Annette Baier builds on the idea of natural sympathy or compassion which we explored in Chapter Four to make the point that of necessity we exist in a "climate of trust" and that a betrayal of that trust, rather than destroying it, "...just makes us more acutely aware of our dependence on it and vulnerability when it fails."[36] She is referring to a set of human relations but, once again, the idea of our relationship with the natural world being built on a climate of trust rather than a struggle for mastery is challenging. We trust that tsunamis, earthquakes and other "natural disasters" will in fact not occur, but when they do it is our vulnerability that is highlighted and our ultimate dependence on living with nature, not above it or outside it.

Moving to more direct feminist approaches to the natural world we can turn to Carolyn Merchant, historian and author of several books that focus on a feminist critique of conventional thinking on human relations with the natural world.[37] In her latest book she has proposed a "partnership ethic" based on the idea that "...people are helpers, partners, and colleagues and that people and nature are equally important to each other."[38] She suggests that thinking about ourselves and about nature in this way can bring about a "discourse of cooperation" based on values such as equity, respect, inclusion and management objectives based on health rather than efficiency which would bring humans and nonhuman nature into a "...dynamically balanced, more nearly equal relationship."[39]

Merchant's contribution to the task of re-thinking our relationship with the natural world is a prime example of the feminist insistence on linking cultural critique with a parallel critique of power, while at the same time sustaining a focus on the importance of ideas and historical patterns of thought. Her critique of the patterns of historical development in Western culture has focused in part on the importance of asserting an autonomous feminine voice in creating cultural norms, but she has also insisted on nature itself being seen as an "...autonomous actor that cannot be predicted or controlled except in very limited domains."[40] Assigning the role of autonomous actor to nonhuman nature goes a long way toward establishing nature a subject that warrants moral standing and respect instead of merely a resource or problem that merits only exploitation or control.

The last writer I will cite in this glimpse at the contribution of ecofeminist thought to our project of re-thinking nature is the Australian feminist philosopher Val Plumwood, often cited in earlier chapters. Capturing the sentiments of each of the feminist theorists discussed above, Plumwood launches a full-scale assault on the Cartesean/Kantean assumptions that have dominated the approach to nature in Western culture, above all the focus on rights:

> Rights seem to have acquired an exaggerated importance in ethics as part of the prestige of the public sphere and the masculine, and the emphasis on separation and autonomy, on reason and abstraction. A more promising approach for an ethic of nature, and also one much more in line with the current directions in feminism, would be to remove rights from the centre of the moral stage and pay more attention to some other less universalistic moral concepts such as respect, sympathy, care, concern, compassion, gratitude, friendship and responsibility.[41]

When applied to nature, reason and the focus on rights that it champions is, for Plumwood, analogous to the relation of husband to wife, master to slave. It is the dualism and its attending power imbalance that has been infused into reason that Plumwood objects to, not to reason *per se*. For her, the current ecological crisis is "...a crisis of the dominant culture and a crisis of reason...or of what the dominant global culture has made of reason."[42] And so with her as well as with the chain of voices with which we opened this book, the goal is not to move backward or shed the benefits that reason and science have generated, but rather to reject those versions of reason which promote dualistic accounts of otherness.[43] This will involve developing within ourselves and throughout our culture what Plumwood calls "stances of openness and attention" toward the other than human world. Only then, after we learn to be attentive to nature and sensitive to its requirements, can we begin what will of necessity be a long process of negotiation and mutual adaptation.[44]

NATURE AND THE NEW PHYSICS

If it is reasonable to speculate that there is an increasing skepticism about modernity's linkage of materialism, progress and human happiness, and that contemporary feminism has profoundly altered our comfort with long-standing dualist assumptions, it is less obvious that developments in science such as chaos theory, the Heisenberg principle and quantum theory have had a similar cultural impact. Despite the prevalence of attempts by various media to popularize these and other post-Newtonian developments in modern physics, it is fair to assume that an in-depth understanding of these theories and concepts is limited to a very few. Yet, their influence on contemporary culture can be quite powerful.

Where to start in describing this vague influence? Perhaps with the new sense of the fragility of scientific knowledge with its accompanying skepticism about even science being able to delineate truths. Most scientists following in the Baconian and Newtonian tradition knew that the truths they were pursuing always had a fragile hold on reality, but the wider public never really appreciated the subtleties involved with such caution. Instead, the popular view for the past 300 years has been that modern science comprises an ever-growing body of unchanging and unchallengeable truth.[45] Victor Frankenstein's science may have raised a caution, but few rallied to Hume's absolute skepticism, treating as a clever philosophical trick his warning that just because we have observed the sun rising in the East for millennia does not prove that it will do so tomorrow.

Paradigm stability has the charm of security, of knowing it is certain that two is the result of one plus one. Progress, the core belief of contemporary modernity in the West and increasingly across the globe, implies more or less constant change and hence the potential for disruption, even catastrophe. Science, from Bacon to Einstein, promised to keep one step ahead of the disruption and thereby mitigate most of the negative impacts of constant innovation and progress. A gradual loss of faith in the ability of science to accomplish this task—a loss triggered first by the carnage of 1914–1918, then later by 35 years of possible nuclear annihilation and currently by fears of a massive ecological collapse—undermines a culture's ability to tolerate such constant change.

Truth, the assuredness that there is an answer to be discovered by science, is the first casualty of this new fragility. The famous laws of science become tendencies, their supposed regularity flawed by the bias of the observer and hence now filled with uncertainties. Science now becomes a set of theories or interpretations of phenomena. John Danford in linking this fragility to the thought of the 18th century Scottish philosopher David Hume, explains the issue this way:

> Science produces different interpretations of what can in principle only be interpreted, never known in the sense reserved for truth. Scientific theories are thus capable of being judged on the basis of predictive power, but not as accurate accounts of what *is*, of the beings which together constitute the universe. What was once called scientific truth is in this view only interpretation, the suitability of which is to be judged according to the purpose for which we interpret (What's true is what works)......If scientific theories are only creations of the human mind, more or less acceptable, decorative, or powerful according to our particular goals at any given time, truth is a delusion.....all intellectual endeavour, including science, is merely imaginative construction. The methodological program of modern science, notwithstanding that for three centuries it has been a wonderfully successful tool in our efforts to master nature for the betterment of human life, has in our century managed to undermine all claims to genuine truth. If what's true is what works, then power is our only standard for determining if something is provisionally true.[46]

This view of science as only interpretation leads to a deep kind of subjectivism that in turn feeds into a renewed sensibility reminiscent of 19th century romanticism. Werner Heisenberg's argument that there is no such thing as an independent observer, that our method of questioning phenomena, our consciousness and behaviour are inextricably part of our experiments and observations significantly erodes the pretensions of objectivity or so-called value free science. As Heisenberg said, "What we observe is not nature itself, but nature exposed to our method of questioning."[47] And because that questioning tends necessarily to focus on the pragmatic dimensions of measurement and prediction, we are most interested in phenomena that can be measured and predicted — hence our obsession with the weather and our frustration at not being able to predict it.

There are, then, no independent observers, no objective views of nature or ourselves and the scientific hypotheses that we rely on are fragile constructions. In certain post-modern versions these new insights can lead to a questioning of the existence of nature *per se*. If all is constructed and dependent on the observer and on language, then perhaps concepts such as wilderness, sympathy or existence itself are mere social constructions? But a more important impact of Heisenberg's discoveries for our purposes is the crucial new insight that the subject, or what Morris Berman calls the "perceiving apparatus", and the object being perceived must then form one seamless web.[48]

RE-THINKING CULTURE AND NATURE

And this seamlessness leads us to another aspect of these new ideas about science, one that meshes well with feminist thought and with the romantic perceptions that we have been following throughout this book. Fritjof Capra cites the physicist David Bohm on the "...inseparable quantum interconnectedness of the whole universe.." being our fundamental reality, with the seemingly independent parts we observe being merely temporary and contingent parts of this whole.[49] This image of an inseparably connected sub-atomic world is easily linked to the 'web-of-life' images popular among ecologists and Romantics like Rousseau, Rolland and Shelley who seek some connection with existence *per se*. Going beyond Epicurean atoms, the realm of phenomena is still composed of particles, but we know of them only "...through their interactions with other systems."[50] The idea of the realm of the natural—the environment—as a "biospheric web" in which "...each entity draws its specific character from its relations, direct and indirect, to all others.." is now our common understanding of ecology.[51]

We have also come to understand that these interactions are not necessarily predictable. Like the weather systems we confront daily in a more visceral sense, they are too complex and hence appear to us as chaotic. But these non-linear systems that actually make up our reality are not chaotic, they are simply complex. They generate, on their own, emergent properties that "...cannot be predicted by examining just the parts."[52] Our tried and tested tool of mathematics can never describe for us a "real world" because that world—nature—is always subject to uncertainty and hence unpredictable. The historian William McNeill who writes sweeping books on the history of Western culture, sums up the impact of these changes on contemporary understandings:

> During the 20th century the physical sciences converged with biology in transforming the Newtonian world machine governed by eternal, universal, and mathematical laws into an evolving—indeed exploding—cosmos where uncertainty prevails, and human efforts at observation affect what is observed. This brings the mathematical sciences closer to the social sciences, and turns history into another kind of black hole from which no branch of knowledge can now escape.[53]

Our future, which only a few decades ago seemed a story of infinite technologies making the pursuit of happiness the dominant substance of being, now becomes as mysterious and filled with apprehension as a black hole in deep space.

And how have these insights affected our view of nature? It has undermined our long-cherished idea of a "balance of nature", that nature tends toward an equilibrium and stability which has been disturbed by human interventions, but which humans could correct if they wished to. Instead ecologists see long-term disequilibrium and flux with predictability just as likely as unpredictability and each independent of human action. In the face of the contingent and chance-riddled world that modern Chaos and Complexity theories describe, the human-as-steward seems an improbable dream. James Lovelock, whose Gaia Hypothesis portrays an ever-changing but self-managing planet, somewhat cynically observes that we "...are all too plainly failing even to manage ourselves and our own institutions. I would sooner expect a goat to succeed as a gardener than expect humans to become responsible stewards of the Earth."[54]

Nature, then, becomes a more powerful actor in its own right, not a passive object for manipulation or correction. Earthquakes, hurricanes, tsunamis, fires and tornadoes are nature's way of initiating dramatic change, but just as important are the constantly changing flora and fauna, climate shifts and the changing flows of air and water. Here was a new "...organic view of the Earth, a view in which we are part of a living and changing system."[55] This does not mean that nature is hopelessly "chaotic", but rather that chance and contingency now join the more obvious interventions by humans and other species and phenomena as forces that shape the environment we all share. In Daniel Botkin's aptly titled book, *Discordant Harmonies*, he acknowledges the surface confusion and chaos of nature but insists that there exists a harmony or unity that lies below the surface and argues that once we understand this ever-changing but still consistent nature we can learn to "...manage in terms of uncertainty.." and thereby to live with nature rather than above or apart from it and benefit both ourselves and life in general.[56]

The most important contribution, then, of these new scientific understandings of nature is their erasure of any clear and morally relevant dividing line between humankind and the rest of nature, that the nonhuman nature exists only to serve human interests.[57] Stripped of its pretensions Life, or Being in general is essentially only energy flowing to the earth in the form of solar photons, with individual forms of Being existing only as "...local perturbations in the universal energy flow....knots in the web of life."[58] We cannot help but be humbled by these ideas, but that may be the first essential step necessary to persuade us to focus on sustainability not just for us, but for the planet and all its ecosystems.

ASIAN PHILOSOPHIES

It has long been a comfort for participants in Western culture to think of themselves as the moderns, the source culture of modernity, with other cultures seen to be tribal, traditional, mired in primitive religiosities or in some other way backward and stalled. Surely that view can no longer hold. In a world of Asian Tigers, booming Chinese and Indian economies, and hypermodern cities in the Persian Gulf the boundaries between West and other are permanently blurred. And these regions are not, despite the dreams of the "westernizing" school of thought and some superficial appearances, becoming just like "us".

The vibrant cultures and economies of South and East Asia are contributing to the maturing of a global modernity with now more diverse ancient roots. One can imagine that 3,000 years ago the various cultures that occupied the long stretch of the Eurasian land mass from Cornwall to Korea shared and borrowed the early attributes of modernity in the form of tools, trade, ideas and myths. We know from the traveler tales of Ibn Khaldhun and Marco Polo that this belt of civilization was vibrant and cosmopolitan a thousand years ago, but then began to coalesce into more isolated centres of power and rivalry. Now, in the 21st century a new, expanded global modernity has surfaced and we begin the process of shifting from 400 years of the global dominance of one variant of modernity to once again shaping modernity collectively.

Throughout this book I have commented on the similarities between *shadow modernity* in Western culture and certain elements of Eastern traditions. Several of the individuals I have used as "voices" of this shadow modernity embody this connection. In his Introduction to the 1927 Brentano edition of Rousseau's *Reveries of a Solitary Walker*, John Fletcher notes that many passages in the book are reminiscent of a Hindu or Buddhist focus on a life of dissociated consciousness, with the feel of a waking dream or contemplative trance. Coleridge and Shelley both sought the kind of connection or intimacy with a larger, more transcendent awareness that finds a kindred spirit in the Tao. Both were more obsessed with "Being" than with beings, human or otherwise and shared with Wollstonecraft (in her less pragmatic moments) a sense of humans and nature harmoniously blended together—key aspects of Chinese culture.[59] These connections culminate in the 20th century with environmentalist like Muir and Leopold who shed the hostility toward wilderness endemic to Western culture, embracing instead a "...man-nature relationship marked by respect, bordering on love..." which Roderick Nash argues is similar to stances toward nature found in Jainism, Buddhism, Hinduism and Daoism.[60]

Fritjof Capra's *The Tao of Physics* awakened a wide readership in Western culture to the often uncanny similarities between contemporary post-Newtonian science and the fundamentals of Asian philosophical ideas about nature. In both idea systems, notions of fluidity and curvature trump right angles and progressivism; ever-changing energy flows negate the hubris of prediction; and a deep sense of the whole overwhelms any focus on the particular. There are other important points of convergence between Asian thought and a Western culture now more open to re-imaginings. Over thirty years ago E.F. Schumacher coined the phrase "Buddhist Economics" to describe what he saw as a necessary transition to obtaining given ends with the minimum means. Simplicity and non-violence were singled out by Schumacher as the qualities central to Buddhist thought that needed to become central to Western culture if it was to flourish.[61]

Equally interesting is the connection with the strain of romantic thought in Western culture. For the Dali Lama and for Shelley, it is the ability to participate intimately with the natural world via an emotional connection that allows one to imagine and act upon a new relationship with the natural world. As we saw earlier with the Romantics, within Buddhism this emotional connection requires that one "imitate the child-like mind", a mind uncluttered (as with Rousseau floating on Lake Bienne) by discursive reflection.[62] From Epicurus through St. Francis and Shelley, Leopold, Carson and Nussbaum it is the role of emotion, compassion and empathy that has provided the path to this more reciprocal connection with nature, and this is the same path charted by Buddhist and Daoist thought.

NATURE—MOTHER, COMRADE, QUEEN OR CITIZEN

These four themes (reassessment of happiness, contemporary feminism, post-Newtonian science, and the growing influence of Asian philosophies) provide, I argue, key building blocks for a major shift in humanity's relationship with the rest of the natural world.[63] The words "reciprocity", "mutuality", "compassion" and "sympathy" have figured large in the preceding pages as necessary components of this new relationship. I have argued that these concepts and dispositions are inherent in the human nature that evolved millennia ago and can be found in all human and many animal cultures, even if not dominant at all times. They are integral parts in what I have identified as the romantic strain in Western culture and are as well implicit in much contemporary scientific thought and explicit in much of Asian philosophy.

In thinking about how these qualities could become the more salient, even dominant, guidelines for relating to nature in contemporary Western

250 | RE-THINKING CULTURE AND NATURE

culture, I first imagined *comrade nature* might be the best replacement for the rather tired notion of *mother nature* which, in addition to being problematic from a feminist perspective, also has the feel of a relationship that one grows away from. The idea of being one with nature, linked in the truly organic way with nature that Deep Ecologists would have us be, seemed to mesh well with the meaning of "comrade". But the term seemed to have an already archaic feel to it and the failure of the environmental movement to really take hold in the latter decades of the 20th century seemed to argue against its usefulness.

Sometimes words and labels can be particularly important. One might imagine that during the 20th century in the West we were essentially preoccupied with a struggle between "comrades" and "citizens", two ways in which modern social systems could solidify the post-Enlightenment shift from all phenomena, including fellow humans, being seen as mere objects. Comrade implied a deep subjectivity, in its uncorrupted form a truly romantic line of social thought. Citizen, on the other hand, has a more formal, individualist, and legalistic as opposed to emotional attraction. But in its radical form, epitomized during the French Revolution, it had a powerful leveling quality, providing an indisputable social and political baseline for all who could claim the label. In today's world of constant human movement, it has the decisive role of providing essential rights and "standing" for those who have left or been driven from their homeland.

In these early decades of the 21st century the idea of citizenship and the related idea of civil society have been infused with new energy. The continued migration of populations, the emergence of large parts of the world from long-standing authoritarian regimes, the disillusionment with populist alternatives and the increasing acceptance that citizenship implies responsibilities as well as rights have all contributed to this renewed energy. While many may still limit their citizenship to voting (or even the "right" not to vote!), others assert that it implies a necessary engagement with community issues, peace issues, and environmental issues. And so we need to ask what role might this new, more energized and expansive notion of citizenship play in sorting out and, indeed, revolutionizing our relationship with nature? Would extending some aspects of citizenship into the other-than-human natural world enable us to reconcile human flourishing with planetary sustainability? Is it possible that a human-centered notion of rights, legal and moral, could have application in an other than human context? And if so would the necessary mutuality and reciprocity of this new relationship enable us to reclaim some of the more humane attributes of our "primitive" state that we have repressed over the millennia in the pursuit of mastery? These were the questions we started with and must now return to in the Conclusion to the idea of *Citizen Nature*.

NOTES

1 Wollstonecraft's eventual son-in-law Percy Shelley agreed, seeing the lure of power as distorting the ideals of men like Robespierre. He concluded that "What was required was a change in the mental attitude of the people, a mental revolution...". Alan Weinberg, *Shelley's Italian Experience* (New York: Macmillan, 1991), p 105.

2 Friedrich Schiller, "Letters on the Aesthetic Education of Man", *Essays* (New York: Continuum, 1993), p 106.

3 Schiller, p 95.

4 Schiller, p 99. Note that Schiller remains a utopian in that these "harmonious powers" are innate in humans.

5 Schiller, p 107.

6 Adorno and Horkheimer in their *Dialectic of Enlightenment* (1944) repeat Schiller's argument that while the instrumentalization of Nature freed mankind from its tyranny, the resulting disenchantment of Nature licenses its destruction and hence the destruction of humankind. And like Schiller, they saw poetry and art as the only means of breaking this cycle and re-enchanting the world.

7 See Plato's metaphor of the white and the black horse mediated by the charioteer in the *Phaedrus* (pp 61–63)

8 Kohak, p 92.

9 Schiller, p 121.

10 Carr

11 Schiller, p 125.

12 But we know as well that the generation of data is not free from passion and politics. Still, the empirically grounded understanding of how evolution and nature may "work" is crucial to an environmental politics and sensibility,

13 Thiele, p 258.

14 Caldwell, p 275. Catherine Roach reminds us that "...the way we understand nature and the way we respond to nature are not shaped at the conscious level alone, but are also shaped in complex ways by unconscious desire and affect; accordingly, they are not necessarily responsive to conscious adoption of new or 'better' models of nature...." Catherine Roach, "The Unconscious, Aggression and Mother Nature", *Journal of Feminist Studies in Religion* (v. 13:1, 1997, pp 105–117), p 107.

15 Wendell Berry, "People, Land and Community", in *The Graywolf Annual Five*, ed. by Rick Simonson, (St. Paul: Graywolf Press, 1988), p 42. Andrew Dobson addresses this issue directly in the final chapter of his book on *Citizenship and the Environment* (New York: Oxford University Press, 2003).

16 Wendell Berry, (1977), p 19.

17 This is a perspective made prominent by James Lovelock with his Gaia Hypothesis, the idea that "...the Earth is actively maintained and regulated by life on the surface...". Lovelock, (1990).p 3. For Lovelock's dire warnings about what will happen if we do not acknowledge the planet's ability to protect itself against us, see James Lovelock, *The Revenge of Gaia: Why the Earth is Fighting Back — And How We Can Still Save Humanity* (London: Penguin, 2005).

18 Leiss, (1990), p 96.

19 Stephen Quilley, "Ecology, 'human nature' and civilizing processes: biology and sociology in the work of Norbert Elias" (pp420–58) in Quilley & Loyal, p 55.

20 Reuters, "Technology is key on global warming: Bush Advisor", 14 August 2007. www.reuters.com/articlePrint?articleId=USPEK985120070814.

252 | RE-THINKING CULTURE AND NATURE

21 Of course there are other powerful traditions such as a rejuvenated fundamentalist Islam that define their mission as one of resisting or radically altering that modernity.

22 Capra, (1991).

23 Bate, (1993). p 161.

24 White, in Gruen and Jamieson, p 10.

25 Evernden, p xii.

26 Collier, p 85.

27 William Wordsworth, from "Lines Written in Early Spring", *William Wordsworth* (New York: Oxford University Press, 1984), p 81.

28 Capra, p 106.

29 Nussbaum, (2001) p 32.

30 Rousseau, *Julie, or the New Heloise*, trs by Philip Stewart ands Jean Vache (Hannover: University Press of New England, 1997), p 566.

31 Lasch, (1984), p 20.

32 In Bauman, (1998), p 25.

33 Kate Soper, "A Difference of Needs", *New Left Review*, #152, 1985 (pp 109–119), p 117. "We live in a society that offers so little diversification in the way of work and creative activity that consumption comes to figure as the main vehicle of self-expression."

34 Taylor, (1991). When Harvard University Press issued a U.S. edition of Taylor's book in 1992 the title was changed to *The Ethics of Authenticity*. One suspects that this was a response to the fact that Americans reacted so strongly to then President Carter's 1979 "Malaise Speech" (though the actual word was not used in the speech), quickly proceeding to dump Carter and his warnings about a potential cultural (and natural) crisis in favour of the "feel good politics" of Ronald Reagan and Bill Clinton.

35 Nussbaum (2001), p 388.

36 Annette Baier, *Moral Prejudices: Essays on Ethics* (Cambridge: Harvard University Press, 1994), p 352.

37 These include *The Death of Nature: Women, Ecology and the Scientific Revolution* (New York: Harper and Row, 1983); *Radical Ecology: The Search for A Livable World* (New York: Routledge, 1992); and *Reinventing Eden: The Fate of Nature in Western Culture* (New York: Routledge, 2003)

38 Merchant, (2003), p 223.

39 Carolyn Merchant, "Reinventing Eden: Western Culture as a Recovery Narrative" (pp 132–159) in William Cronon, p 158.

40 Merchant (1996), p 158.

41 Plumwood, (1993), p 173.

42 Plumwood, (2003) p 5.

43 Plumwood, (1993) p 42.

44 Plumwood, (2003) p 169.

45 Colin Russell, *The Earth, Humanity and God* (London: UCL Press, 1994), p 6. Carolyn Merchant describes logical positivism at the base of this science as a theory that "...assumes that valid, verifiable, hence positive, knowledge of the world derives ultimately from experience obtained through the senses or by experiment and interpretation via the conventions and rules of mathematical language and logic. Scientific knowledge is rule-governed, context-free, and empirically verifiable and as such claims to be objective, that is, independent of the influence of particular historical times and places.....The basic dichotomy is between subject and object; indeed objectivity, the hallmark of logical positivism, depends on it. .....The dualism between activity and passivity hypothesizes an active subject—man—who

NEW INSIGHTS ABOUT CULTURE AND NATURE | 253

receives, interprets, and reacts to sense data supplied by a passive object—nature." In *Earthcare: Women and the Environment* (New York: Routledge, 1996), p 60.

46 John Danford, *David Hume and the Problem of Reason* (New Haven: Yale University Press, 1990), pp 19–20.

47 Cited in Capra, p 140. Thus as Capra points out, "...the observer decides how he is going to set up the measurement and this arrangement will determine, to some extent, the properties of the observed object. If the experimental arrangement is modified, the properties of the observed object will change in turn."

48 Berman, (1981) p 137.

49 Capra, p 138.

50 Neils Bohr, cited by Capra, p 138.

51 Abrams, p 85.

52 Tom Wessels, *The Myth of Progress: Toward a Sustainable Future*, (Burlington: University of Vermont Press, 2006), p 9

53 William McNeill, "A Short History of Humanity", *New York Review of Books*, 29 June 2000, p 9

54 James Lovelock, "Planetary Medicine: Stewards or partners on Earth?" *TLS* 13 Sept 1991.

55 Botkin, p 189. Referring to the region around Lake Superior, Botkin notes that following the last glaciation period the region was comprised of tundra (low shrubs, moss & lichen). The tundra was replaced by a forest of spruce. About 9000 years ago the spruce forest was replaced by a forest of jack pine and red pine, trees characteristic of warmer climates. Paper birch and alder immigrated into the area around 8300 years ago. White pine appeared about 7000 years ago and then there was a return of spruce, jack pine and white pine suggesting a cooling of the climate. "Which of the forests represented the natural state? p 59.

56 Botkin, p 155.

57 Eckersley, p 51.

58 Harold Horowitz, "Biology of a Cosmological Science" (pp 37–49) in Callicott and Ames.

59 Francis Cook, "The Jewel Net of Indra" (pp 213–229) in Callicott and Ames, eds., p 216.

60 Nash (2001), p 20. "...wilderness, in Eastern thought, did not have an unholy or evil connotation but was venerated as the symbol and even the very essence of the deity. As early as the fifth century B.C., Chinese Taoists postulated an infinite and benign force in the natural world....Freed from the combined weight of Classicism, Judaism, and Christianity, Eastern cultures did not fear or abhor wilderness."

61 Schumacher, p 61.

62 David Shaner, "The Japanese Experience of Nature" (p 163–182) in Callicott and Ames,

63 One assumes that a similar shift may be looming in humanity's relationship with it's own built world.

CONCLUSION

HARMONIZING SENSE AND SENSIBILITY— TOWARD CITIZEN NATURE

"We have abandoned the bond that connects us to the world"

MICHEL SERRES

But seriously, how can trees or dogs or river basins be thought of as citizens? Surely it is a human term, specifically designed for beings who can participate in a polity, who are conscious actors with language and interests. And citizenship involves more than just rights—it requires responsibility as well. Isn't the idea just another well-meaning anthropomorphism?

As we saw in Chapter Three, the tradition of anthropocentrism, or what came to be called humanism in Western culture, is powerfully embedded. The human capacities for freedom, autonomy, sociality and language were singled out by Kant and other humanists as definitively separating humanity from all other phenomena in nature and that tradition remains powerful. For the contemporary critical theorist Jurgen Habermas, the unique nature of human communicative powers must mean that "our relation to nature, insofar as nature is taken as an object of cognition, can only be one of instrumental control.[1] All else is simply a nostalgic appeal to a "distant pastoral scene."[2] Andrew Dobson, writing on the issue of citizenship and the environment, agrees that it would be a serious "mistake" to extend any notion of citizenship to animals since they "...can have no awareness as it is usually thought of."[3] Citizenship for Dobson must remain a "fundamentally anthropocentric notion."[4] The philosopher Bernard Williams insists that we must come to "... accept the implications of our Promethean potential for dominating nature and as part of that come to "respect" nature as a potentially dangerous *other*."[5] Perhaps the most rigorous critique of any move to extend rights to animals

CONCLUSION

or other non-human phenomena is that of Luc Ferry who argues that such a move implies the "passing of the humanist era" and must end in an "ecological dictatorship."[6]

Philosophers are not alone in asserting this importance of human uniqueness. Within science, a new kind of strong anthropocentrism based on a renewed social Darwinism is lurking, insisting that there is a logic to evolution, a trend toward increasing complexity and "encephalization" and that we humans are "…the leading edge, the cambium of nature.." and that "humankind is what nature has been trying, all these millennia, to be."[7] This is, in a sense, Luc Ferry's argument from Kant and from his reading of Rousseau. For whatever reason (God or evolution) for Ferry Homo sapiens is the only "perfectible" species, the only one with a brain large enough to allow reason to surpass the reliance on instinct and thus allow for unlimited personal development and creativity. Thus "…man is the anti-natural being par excellence…", independent of natural cycles and living according to laws derived from reason.[8]

Aldo Leopold might question this common sense view of the issue. On his Sand County farm in Wisconsin Leopold envisaged a powerful bridge spanning the great divide between humans and nature. That divide, so perfect an example of the dualism lodged at the heart of Western culture may, as we have seen, be hard-wired by evolution, forcing us to see virtually all phenomena in dualist terms. In this era of the planet's history Homo sapiens seems to have evolved to such a point that separations too profound to ignore have opened up between humanity and the rest of nature. The holistic frame of the Paleolithic and the deep connections with the land that were present in the Neolithic and persisted for millennia after have eroded almost without hope of recall.

But the various environmental crises that stem from this divide still loom and Leopold identified the source of the problem clearly in 1948. The dualist insistence on rigidly viewing nature as the other—as a collection of objects—has wrought terrible havoc and threatens to so radically alter the biosphere and atmosphere that the species Homo sapiens itself appears at risk. We cannot will ourselves into a remote primitive holism and we dare not proceed blindly down a path that may lead only to a cliff edge. But Leopold's vision of a bridge between humans and nature remains as a means of at least managing this dilemma.

Aldo Leopold's *Sand County Almanac* is about this modern divide between human culture and the rest of the natural world and the price both we and nature have paid in allowing it to reach a point of virtual rupture. But he does not just focus on the threat, offering as well a naturalist's love of the particulars of nature, an ecologist's understanding of the connections that bind us

HARMONIZING SENSE AND SENSIBILITY—TOWARD CITIZEN NATURE | 257

to nature, and a conservationist's prescription for a path towards a cure. And so he is the perfect final *voice* for us to consider in this search for a path to human and ecological flourishing. Leopold calls for us to re-unite sense and sensibility by extending the reach of our ethical universe to include all of nature, creating what he calls the "land ethic".

> All ethics so far evolved rests upon a single premise: that the individual is a member of a community of interdependent parts...the land ethic simply enlarges the boundaries of the community to include soils, waters, plants and animals.[9]

And within this matrix called land we humans must use this ethic to make the transition from conquerors to "biotic citizens,"[10] citizens of the land community. The ethic Leopold urges us to apply, as we saw earlier, is one that sees something as "right when it tends to preserve the integrity, stability and beauty of the biotic community....wrong when it tends otherwise."[11]

But where is the sense and where the sensibility? *Sand County Almanac* opens with the sensibility, because for Leopold that is where the change has to begin. We first have to feel something for this land community if we are to have any hope of using our sense to adopt and adhere to a new ethic. Look again at Leopold's tale of the mouse and the hawk, an ecological Aesop's Fable:

> A meadow mouse, startled by my approach, darts damply across the skunk track. Why is he abroad in daylight? Probably because he feels grieved about the thaw. Today his maze of secret tunnels, laboriously chewed through the matted grass under the snow, are tunnels no more, but only paths exposed to public view and ridicule...The mouse is a sober citizen who knows that grass grows in order that mice may store it as underground haystacks, and that snow falls in order that mice may build subways from stack to stack: supply, demand and transport all neatly organized. To the mouse, snow means freedom from want and fear.
>
> A rough-legged hawk comes sailing over the meadow ahead. Now he stops, hovers like a kingfisher, and then drops like a feathered bomb into the marsh. He does not rise again, and so I am sure he has caught, and is now eating, some worried mouse-engineer who could not wait until night to inspect the damage to his well-ordered world. The rough-leg has no opinion why grass grows, but he is well aware that snow melts in order that hawks may again catch mice.....for him a thaw means freedom from want and fear.[12]

Lots of food for thought in this little vignette. Feelings, thoughts and plans are ascribed to the little rodent and his fellows, and he is a "citizen" of his community—of our community. The hawk, a predator for sure, is only

258 | CONCLUSION

doing what comes naturally and, as Leopold's later essay on the importance to the mountain and its grasses of the wolf's culling of the deer, the hawk's predatory predisposition is crucial for the on-going sustainability of the land community.

> I now suspect that just as a deer lives in mortal fear of its wolves, so does a mountain live in mortal fear of its deer. And perhaps with better cause, for while a buck pulled down by wolves can be replaced in two or three years, a range pulled down by too many deer may fail of replacement in as many decades.[13]

Throughout the observations that Leopold makes his romantic, sensual and appreciative immersion within the biotic communities he participates in is salient, as is his understanding of the unique role of humans within them.

The central insight in Leopold's understanding of his Sand County farm is that it is in effect a community. In this community not only animals but even the various forms of plant life assume an active presence, participating as it were in the vibrant life that Leopold sees and—given his specifically human talents—oversees and stewards. Leopold is aware of the trap of anthropomorphism and its companion the "pathetic fallacy", the ascribing human qualities to other species. He confronts it directly in observing the return of the geese to the Wisconsin marshes.

> In...watching the daily routine of a spring goose convention, one notices the prevalence of singles—lone geese that do much flying about and much talking. One is apt to impute a disconsolate tone to their honkings, and to jump to the conclusion that they are broken-hearted widowers, or mothers hunting lost children. The seasoned ornithologist knows, however, that such subjective interpretation of bird behavior is risky. I long tried to keep an open mind on the question.
>
> After my students and I had counted for half a dozen years the number of geese comprising a flock, some unexpected light was cast on the meaning of lone geese. It was found by mathematical analysis that flocks of six or multiples of six were far more frequent than chance alone would dictate. In other words, goose flocks are families, or aggregations of families, and lone geese in spring are probably what our imaginings had suggested. They are bereaved survivors of the winter's shooting, searching in vain for their kin. Now I am free to grieve with and for the lone honkers.[14]

This is more than the pity of grieving for a seemingly injured beast, but grieving with a member of one's fellow community. Just as humans within their political culture allow for vast differentials among citizens in terms of access to the resources needed for flourishing and occasionally even existence itself, so Leopold accepts predation in the form of humans hunting geese while acknowledging deep fellow-feeling with these members of his community. It is this fellow-feeling, Adam Smith's empathy for the man on the rack, that must provide a starting point for re-thinking the human-nature relationship. The horrors of the factory farm, the beakless chickens crammed into their life-long prisons need not force us to abandon consuming beef or fowl (though it might!), but rather should persuade us to address the horror—just as a rejection of concentration camps need not lead us to prison abolition.

In the *Sand County Almanac* is clear that Leopold is aware that something dramatic has happened to the land community since the end of the Paleolithic era; the most important variable being changes in the behaviours of humans.

> For the first time in the history of the human species, two changes are now impending. One is the exhaustion of wilderness in the more habitable portions of the globe. The other is the world-wide hybridization of cultures through modern transport and industrialization...To the laborer in the sweat of his labor, the raw stuff on his anvil is an adversary to be conquered. So was wilderness an adversary to the pioneer. But to the laborer in repose, able for a moment to cast a philosophical eye on his world, that same raw stuff is something to be beloved and cherished, because it gives definition and meaning to his life.[15]

Here is the essence of Leopold's message; the need for each of us to cast a philosophical eye on our world, our biotic community.

For Leopold, such a casting of the eye is an ecological necessity as well as an ethical obligation. He is no misanthrope complaining about humanity as a scourge on the land—quite the opposite. He places humans at the apex of a moral hierarchy.

> For one species to mourn the death of another is a new thing under the sun. The Cro-Magnon who slew the last mammoth thought only of steaks. The sportsman who shot the last pigeon thought only of his prowess. The sailor who clubbed the last auk thought of nothing at all. But we, who lost our pigeons, mourn the loss. In this fact...lies objective evidence of our superiority over the beasts.[16]

It is we humans who because of our power and prowess must be the citizens of the land community who employ an ethic as a limitation on our freedom of action in the struggle for existence.[17] Plumwood proposes an approach to nature based on "respect, care and love", referring to all creation as "our planetary partners."[18] Leopold affirms that the creation and nurturing of such

260 | CONCLUSION

an ethic of care requires first "...an internal change in our intellectual emphasis, loyalties, affections, and convictions."[19]

Many contemporary voices echo Leopold's concern for a fundamental value change as our first priority. Vaclav Havel, the prototypical modern humanist, insists on the necessity of a change in the "sphere of spirit, in the sphere
of human conscience", leading to a "...new understanding of the true purpose of our existence on this Earth."[20] In my home place, Vancouver, British
Columbia, a prominent entrepreneur/philanthropist, Milton Wong, states
Leopold's case in a particularly incisive way:

> I believe intuitively that the notion of a civil society hinges on a finely balanced
> notion of interdependency among all living organisms, such that each is free
> to optimize its full potential without significantly infringing on another's
> ability to do the same......all living beings should enjoy the opportunity to live
> the most meaningful lives possible—whatever that actually means for each
> particular creature.[21]

With these sentiments and with Leopold, then, we return to the opening
premise of this book; that a politics of ecological sustainability must be accompanied by a set of values that supports and nurtures it. On the one hand,
acknowledging the practical wisdom of Murray Bookchin and others who focus on issues of structure and power, we dare not reduce our relationship to
the natural world to simplistic psychological terms. Simply changing thinking patterns or values, individual by individual, will not do the job.[22] But neither can we ignore the crucial sustaining and nurturing contributions that
changes in individual and cultural values can provide to structural changes
that must occur.

It is the balance or relationship between these two necessities—structural change and value change—that we must get right. In Jane Austen's novel *Sense and Sensibility* she draws for us classic portraits of these two qualities
(often known as reason and passion) in the persons of Elinor Dashwood and
her younger sister Marianne Dashwood. The latter is all sensibility. She pro-
claims on several occasions that she "must feel" and is only too aware that
others whom she encounters are too often oblivious to feeling. Or, as in the
case of her more pragmatic and rational sister, may choose not to.[23] In an exchange between the sisters while walking on a tree-lined path covered with
dead leaves, Marianne, thinking of their recent expulsion from a childhood
home now the possession of those she imagines do not feel, exclaims:

> Oh...with what transporting sensations have I formerly seen them fall! How
> I have delighted, as I walked, to see them driven in showers about me by the
> wind! What feelings they have, the season, the air altogether inspired! Now

there is no one to regard them. They are seen only as a nuisance, swept hastily off, and driven as much as possible from the sight".[24]

Elinor replies that "It is not everyone who has your passion for dead leaves". Elinor, of course, does feel but has chosen to repress feelings in order to get on with the task of acting in the world, in her case responding to a very real family crisis. And in the end Marianne too must do the reasonable thing and in her case marry out of duty and respect, not passion. And so Austen brings about a reconciliation of the two ways of being, but it is a reconciliation weighted in favour of a very pragmatic, instrumental reason. The new bride Marianne Brandon had been persuaded to "...counteract by her conduct, her most favourite maxims."[25] Marianne has been persuaded to dominate her "inner nature" via a "repression and renunciation of the instinctual, aesthetic, and expressive aspects of [her] being".[26]

But what has this to do with our search for a means of ecological sustainability? Marianne's emotional connection to the dead leaves and Aldo Leopold's empathy for the mouse and the hawk point to a crucial quality of human nature. I have noted throughout this exploration of humans and nature that we have within us a strong element of this sensibility toward the other, what E.O. Wilson calls our "...innate tendency to focus upon life and lifelike forms, and in some instances to affiliate with them emotionally."[27]

Elinor Dashwood's powerful dedication to sense finds a theoretical home in Freud's celebration of repression as a key to the success of civilization. In Austen's novel Elinor's stern, pragmatic instrumental reason "works", but it exacts a cost. For her the dead leaves on the ground embody no life force but are mere decayed matter, pragmatism must govern social relations, and disappointments in life must submit to stoic self-command. She stands in for an emerging materialist and technocratic culture that was, in its mature 20th century form, to be the "first responder" to the environmental crisis of the latter decades of that century. Rejecting any so-called romantic retreats to more simplistic or primitive economic models, the focus instead was to be on a sustainable development built on high standards of environmental protection—finding a more sustainable means to preserve existing ends.[28]

This "ecological modernization" approach stressed working to "...control rigorously the workings of capitalism in the interests of the preservation of the biosphere [involving] a mixture of state and international regulations, taxation policies, licensing regimes and the increased democratic accountability of capitalist corporations."[29] The core belief was that a better understanding of technology and science would lead to discovering the means of lowering resource depletion, pollution and excessive consumption.[30] From

262 | CONCLUSION

the Brundtland Commission to the Kyoto Accord with its carbon credits, this pragmatic reliance on an Adam Smith inspired combination of market incentives and stoic self-command has dominated contemporary policy responses to the environmental crisis.

But is this really enough? As I prepare to bring this book to a conclusion media reports announce new alarming evidence of a dramatic increase in the rate of global warming. *Science* magazine notes that recent satellite photos show that Antarctica is losing ice at the pace of about 36 cubic miles each year. At the northern pole the same process is occurring with predictions that by the end of the century the Arctic ocean will be ice-free in the summer months. Elizabeth Kolbert, who recently completed a book on global warming based on her series of articles in the *New Yorker* in 2005, reports that the response of governments to this alarming news is to focus on responding to the now seemingly inevitable problems caused by this trend while still refusing to address in any meaningful way the root causes.[31] Closer to home in Canada, a series of warmer winters have made it possible for the mountain pine beetle to devastate huge areas of British Columbia — and it is heading East.

So even if there is now a glimmer of understanding that human actions are playing a major role in global warming and that even a hyper-modern technological culture cannot really master it, the power of that glimmer is seemingly insufficient to cause any substantial change in values or behaviour within the institutions of modernity. Instead, one imagines that risk management specialists are already drawing up plans for shifting populations from low-lying areas, developing new seed strains that will withstand higher temperatures and resist new pests, and trying to anticipate the geo-political realities of an "...Earth so warm as to be practically a different planet."[32]

But even such draconian planning will produce no stability. The ever-increasing complexity of human cultures and societies means that there is no way of ever getting the social/nature relations *right*. As Botkin's research demonstrates, there can be no stable or sustainable situation. Instead we face the necessity of on-going negotiations.[33]

What, then, is to be done? The promotion of environmental citizenship on the part of both individuals and institutions seems a reasonable starting point and this idea has provoked a wide range of debate within academic and policy-making circles. Thinking and acting Green has turned out to be more complex than it might seem and once again a dualist frame has come to dominate debate. Brian Baxter describes the two dominate approaches as Green Rationalism and Green Romanticism; one focusing on politics and structure the other on culture and values. Andrew Light expands on this distinction, using the terms environmental materialism versus environmental ontology:

Environmental materialists argue that the appropriate human response to environmental problems must primarily involve an analysis of the causes of those problems in the organization of human society through the material conditions of capitalist (or state capitalist) economies, and the social and political systems which sustain those societies. Material conditions, such as who owns and controls the technological processes that are used to stimulate economic growth, expand markets and consume natural resources, are for these thinkers the starting points for unpacking the complex web of environmental problems. In contrast, "environmental ontologists" see more potential in diagnosing environmental problems as primarily involving individual human attitudes toward nature. For environmental ontologists, social, political, and material problems are the symptom of a larger crisis involving the relation of the self with nature, not the root cause. The primary cause of environmental problems involves some assessment of our disconnection from nature, spiritual or otherwise. As such, the principle location of solutions to environmental problems for ontologists is to be found in changing the "consciousness" (for lack of a better term) of individual humans in relation to the non-human natural world.[34]

A growing body of literature is now dedicated to exploring these and related approaches.[35] Echoing the social ecology approach of Bookchin, those associated with Light's "environmental materialist" approach argue that our current problems stem not from a culture of anthropocentrism but rather "...from hierarchical and authoritarian power structures that allow *some* humans to dominate others while simultaneously misusing the natural world."[36] Likewise Barry Commoner has argued that our difficulties with the environment stem from social causes rather from any natural limits or foundational cultural flaws. Going far beyond populist campaigns for re-cycling and re-using, the focus here is on making the necessary political and social changes in order that a transformed technology can be directed toward meeting the needs of both humans and the environment. Thus Commoner argues that technology choice must be driven by public policy rather than private interests.[37]

I have tended to focus on the position that the difficulties inherent in the contemporary relationship between humanity and the natural world are essentially a questions of values and that an ecologically sustainable solution requires a cultural critique of the mastery agenda that has dominated Western culture since at least *Genesis*. I have argued that this historic commitment to mastery is not the fulfillment an inherent evolutionary drive within Homo sapiens nor is it necessarily an inevitable component of the Western cultural

264 | CONCLUSION

tradition. Certainly it is not universal phenomenon across human cultures. Hence the need to focus on a cultural critique of mastery as an approach to the natural world — and indeed to each other! This involves, as I have stressed throughout, finding the means to bring about changes in attitudes, values and beliefs — we are closer here, then, to Light's "environmental ontologist" and Baxter's Green Romanticism.

The twain, however, must meet. If a cultural shift persuades us to link human flourishing with notions of interdependence and seeing nature as consisting of fellow moral beings or phenomena with interests, proponents of that new world view must still have the political means to operationalize the change and that requires the commitment of economic and political institutions to ecological sustainability. Hence the old 20th century argument about whether political change or value change should come first in attempts to transform human behaviour must give way to a 21st century position that the changes must be concurrent, not consecutive.[38]

Perhaps the best example of an attempt to link these two approaches is what has come to be called "collective ecological management" — sometimes referred to as "ecological stewardship" or, to use Dobson's formulation, "ecological citizenship". The goal is to enhance the compatibility of human and non-human interests by a focus on prudence in consumption and production, enhancing civic virtue via informed citizen participation in environmental decision-making, and finding a mean between "arrogant humanism" on the one hand and "ecological quietism" on the other.[39]

Andrew Feenberg, a self-described environmental optimist who rejects the suggestion that we must opt for "small is beautiful" or other regressive steps, calls for a "...critical, democratic politics of technology within and not against the general project of modernity."[40] For this to happen, though, Feenberg does insist that there must be a cultural shift in which "...a clean and healthful environment is considered not as an exogenous dumping ground but as a component of individual well-being." Such a cultural shift, Feenberg says, would obviate the need to have new environmental practices imposed by market incentives or coercion and technology could be changed to serve the interests of the changed culture.[41]

Here once again is our dilemma: we need to act immediately to address the very real damage that modern industrial/technological systems are inflicting on the natural systems upon which we all depend, but for that action to be effective a significant cultural shift is required, one which strengthens our personal and cultural ties with the very natural systems we are threatening. Dualism thus rears its head again. Feenberg, Leiss and others argue that new technologies can be developed that will assist people in maintaining the ecosystems in which they exist, but admit that to persuade people to create and employ those technologies requires a set of supportive beliefs and values.[42]

It is possible, then, that an ecologically enlightened environmental/ ecological citizenship might be the answer to our civilizational dilemma. Instead of going "back to the future" via a romanticized primitivism, we can try to use the technologies of modernity to in effect undo the damage caused by their use in a culture dominated by an uncritical anthropocentrism and an instrumentalized reason. In this vision technology would become a source of "...innovation that can lead us to less consumption, less pollution, less depletion of resources, and lower rates of population growth."[43] In one of the more imaginative versions of this approach, Roderick Nash suggests an "Island Scenario" for humanity, based on using high technology to create several hundred "habitats" containing closed circle technologies for producing food, water and energy and disposing of waste. "Boundaries would be drawn around the technological human presence, not around wild nature, permit[ting] humans to fulfill their evolutionary potential while not compromising or eliminating the chances of other species fulfilling theirs."[44]

Andrew Dobson has worked this out in more detail in stressing the distinction between environmental and ecological citizenship. The former, manifesting itself in the politics of liberal political cultures, would be the informed citizenry that would demand new environmentally sustainable political and technological choices. But for Dobson these environmental citizens remain state-bound and act only in the public sphere. Ecological citizenship, on the other hand, provides the universal force of beliefs and values that Feenberg and others see as essential to sustain these new policy choices. In the true sense of the word, these ecological or "earth citizens" would become the stewards of life envisaged in Genesis.

While there are those who argue that only a more complete, efficient and thoughtful domination and mastery of nature will suffice, the voices I have focused on take the contrary view that domination is the problem, not the solution. Most of them reject any dramatic reversal in which humans submit to the erratic whims of nature via a return to some pre-modern or primitive culture. Instead, living in *harmony with nature* is set as the goal. But this too is slippery ground. Bill Leiss notes the "seductive charm" of the phrase, warning that it is often employed as an "...apotheosis of primitivism and a proscription of all mechanization."[45] And we know that the linking of harmony with nature does not reflect the dynamism and constant change that characterizes natural systems.

But there is no easy consensus about what living in harmony with nature

266 | CONCLUSION

would actually look like. For the likes of Feenberg, Leiss, or Bookchin this harmony retains a strong focus on the centrality of the human, a weakened but nonetheless still vibrant anthropocentrism. For Leiss, human dominion over nature need not be abandoned but only tempered, since it has proven useful in undermining earlier practices of projecting human social forms onto the natural order; i.e., giving nature purpose, intention or will.[46] There is a certain naturalness or inevitability to human anthropocentrism and, given the abilities and talents acquired by Homo sapiens via evolution, a certain logic to an assertion of dominion over less able creatures and perhaps over nature and evolution as well.

These are all very persuasive arguments that humans and nature must stand in a relation of subject and other, if not always subject and object. Fundamental to this position is a deep moral ambiguity surrounding the standing—moral and legal—of these "others". As we have seen with Habermas and Ferry, for many any attribution of moral or legal standing to non-human entities poses a mortal threat to the dignity of the human. Others might concede a moral relationship between human and non-human, but deny the possibility of any legal or "fellow citizen" relationship. Others still might concede legal standing but deny any moral obligations. For many, any concession of moral or legal standing is the first step on a slippery slope leading to nostalgic primitivism.

All these competing prevarications leave us in a muddle. The global environment continues to erode all around us, technical and sometimes fantastical solutions are proposed, doomsayers abound, pragmatists counsel slow and steady change and politicians seem structurally unable to cope with the long-term nature of the problem. Can a culture built on such a foundation ever generate the means and the will to address the severity and depth of the clash with nature that now seems just over the horizon?

TOWARD CITIZEN NATURE

The great divide in the prescriptions for addressing our environmental crises remains the distinction between those who are anthropocentric in some way and those who claim to be biocentric or ecocentric. There are well known conflicts between Deep Ecologists and Ecofeminists and between each of these and Social Ecologists, and the "Reds" and the "Greens" are still often at odds. The weakly anthropocentric Ecological Management approach, rooted as it is in the incisive critique of modernity by the likes of Adorno and Marcuse, has perhaps the greatest persuasive power and is also one of the more pragmatic options on the table. Acknowledging as it does the need for a fundamental shift in values resting on an "ethics of care" and accepting the reality of ecological limits, it seems to come closest to an enlightened stewardship approach to the relationship between humans and the environment.

Ecocentrists, however, remain doubtful, suspicious that people will "..keep slipping their opposable thumbs on the scales of justice. They worry that the very process of balancing values leaves environmental Goods too vulnerable to continued exploitation."[47] For many ecologists, environmentalists and naturalists even ecologically sound management is too fragile a foundation upon which to construct a new and more sustainable relationship with nature. Instead it is argued that the ".. real solution to problems in environmental policy lies in a specific transformation of values — the transcendence of human-based systems of ethics, and the development of an ecological ethic. Humanity must acknowledge that moral value extends beyond the human community to the communities within natural systems."[48] Mary Midgley insists that we are faced here with a moral, not a scientific choice of whether we view the natural world through what she describes as our modern lens of resentment, terror or predatory greed or "...with a mixture of awe, fear, respect and gratitude" that is characteristic of most past and many contemporary human cultures.[49] This call for a radical paradigm shift in the realm of ethics is based on the conviction that we are the animal that is free to choose the means and the moral frame through which we address the natural world within which we live.

Having choices is no doubt a good thing, but with choice comes the responsibility to make informed choices. Being able to choose how or to whom to apply value, or how to act in the face of troubling circumstances would have little meaning if the choices were simply random. Our 21st century choices centre on finding a way to combine in some way the pragmatic approach of the Green Rationalists with the more transformational agenda of the Green Romantics and thereby to move toward an ecologically sustainable culture that re-acquaints us with lost or repressed holistic and communitarian assets from our prehistoric and historic past while preserving the humane benefits of modernity.

Creating more environmentally enlightened citizens along with more democratic and critical policies and practices in overseeing science and technology is an important part of this task, but it has what can be seen as a core flaw in its focus on doing things for the environment primarily in order to ensure human well-being. To move toward a more holistic understanding of the human/nature relationship we should consider supplementing this approach with a focus on reciprocity, solidarity, and even mutuality with the natural world. In addition, then, to nurturing good environmental and ecological

268 | CONCLUSION

citizens, we should consider assigning natural phenomena—life forms and ecological systems—moral and even legal standing as fellow citizens in our shared biotic community with all the mutuality that this implies. This would in effect put humans in sync with other species for whom mutuality with each other and with 'home places' is the norm.[50] Just as granting such moral and legal rights to other humans who had previously been denied them raised fears of social chaos and unworkable procedures and the similar fears attendant on this extension will no doubt prove to be exaggerated. As I have illustrated by exploring alternative voices within Western culture from the Greeks to the present, there is nonetheless a strong tradition of receptivity for this kind of mutualism.

Still, the suggestion that one might want to mix notions of citizenship with other-than-human phenomena is often greeted with incredulity or even hostility. For some environmentalists it is hopelessly atomistic, assigning rights to individual animals, trees or other forms instead of addressing the broader issue of ecosystems.[51] Plumwood is very critical of this approach, seeing it as an outdated attachment to Kantian ideas about rights and reason and perpetuating "..a process of (masculine) universalisation, moral abstraction, detachment and disconnection, involving the discarding of the self, emotions and special ties (all of course associated with the private sphere and femininity) [and] is highly problematic."[52] And, of course, there is the argument that citizenship demands voice, consciousness and the ability to reciprocate which most animals and all other aspects of the natural world seem to lack.

If "do rocks have rights?" is for many a ridiculous question posed by deep ecologists and fringe environmentalists, the notion of rocks (or armadillos or watersheds) having responsibilities must seem even more bizarre. But before we consider how such an extension of citizenship into the natural world might work in practice we should explore further the rationale for such a dramatic development. We may reasonably assume that given current activities in the biosphere an ecologically sustainable earth inclusive of humankind requires changes in behaviour by humans; the current path of human development epitomized by Western industrial/technological culture is simply not sustainable globally.

We can further assume that changes in behaviour that are more than mere short-term responses to immediate crises require simultaneous changes in thinking, new understandings about the relationship between humans and their environments. These shifts in understandings are in turn, as Mary Midgley argued earlier, linked inextricably to changes or modifications of values and beliefs which provide the crucial foundations of any future changes in behaviour.

New understandings, then, of ecosystems, sustainability, genetic connections between humans and other animals or the intricacies of climate will have a decisive impact on our behaviour only when they are paralleled with changes in our values, in our moral understandings of nature. And we have every reason to be optimistic that the moral sentiments and ethical systems that have evolved over time in relation to human-to-human interactions may very well find a more universal and ecological application, a history of extension that James Q. Wilson identifies as the "...chief feature of the moral history of mankind."[53] Roderick Nash calls this idea "...one of the most extraordinary developments in recent intellectual history....the most dramatic expansion of morality in the course of human thought."[54]

An interesting rendition of Wilson's and Nash's optimism can be found in Michael Schermer's "Bio-Cultural Evolutionary Pyramid", an attempt to model the origin and evolution of human ethical systems. In the 1.5 million years portrayed in Schermer's model, we start with the individual's concern for survival of self and move through successive stages in which this concern for self preservation acquires a moral dimension as it extends to family, extended family, community, society, species and, finally, the biosphere.[55] Nash has a similar model that has ethics being extended from animals to plants, life *per se*, rocks, ecosystems, the planet and then the universe.

There is a clear logic to this notion of extending moral standing and moral rights. Considering the dimension of citizenship that is associated with basic rights, we have seen the status associated with the word move from white males with property to all white males, to women, to all adults, to children and the disabled. The United Nations Universal Declaration of Human Rights and the various tribunals and global bodies linked to the UN are further evidence that we are gradually extending the legal and moral standing linked to "citizen" beyond specific societies toward humanity as a whole—Schermer's notion of "belongingness to a species". At the same time various scientists and naturalists are making claims for moving beyond this species-border and extending legal and moral standing—both important dimensions of citizenship—to various mammals, ranging from dolphins to dogs. According to UN Resolution A-RES-37: "Every form of life is unique, warranting respect regardless of its worth to man, and, to accord other organisms such recognition, man must be guided by a moral code of action."[56]

The extensionist position has the logic of history behind it, but the seeming qualitative distinctions between humans and 'everything else' does give one pause. How might we come to know what non-speaking phenomena desire or need? Kimberly Smith suggests that we need to hone our listening skills, opening our decision-making to communications from nature,

270 | CONCLUSION

especially feedback from natural systems: "we need better, more creative ways to communicate with the non-human world."[57] Matthew Calarco sets the task as being "...ethically attentive and open to the possibility that anything might take on a face."[58] We should avoid drawing strict boundaries, since our past attempts to do so have more often than not been failures. Still, as Derek Bell warns, we most likely will need to prioritize relationships, perhaps by placing value on something like a capacity for "richness of experience" when conflicts arise (e.g. the moral claims of a lake or stream in conflict with a beaver).[59]

But as we have seen, we remain doggedly anthropocentric and there are powerful arguments against what Bookchin calls the possible "animalization of humanity."[60] But it is that very focus on humans as "just another animal" instead of a being "other than animal", what Plumwood calls a celestial instead of terrestrial self-image, that may be necessary if we are to re-think our relation to nature.[61] This stubborn anthropocentrism exists side-by-side with the knowledge of the evolutionary history that we explored in Chapter One and the increasing awareness that in ecosystems organisms cannot be separated from their physical milieus. As Stan Rowe tells us, ecology, derived from *oikos* or "house" or "home", means knowledge of home — home wisdom.[62] And that home or eco-system is above all an interdependent system in which organisms like us are only a part and in which "everything is connected to everything else.[63] One might imagine that we are on the cusp of one of those transitional points on schemes like Schermer's or Nash's pyramid, between being species-centered and biosphere centered — with ecocentrism still far in the distance. David Abrams, in his ground-breaking book *The Spell of the Sensuous*, insists that these interdependencies are now widely understood in the more industrialized cultures, even if their institutions lag behind:

> We have at last come to realize that neither the soils, the oceans, nor the atmosphere can be comprehended without taking into account the participation of innumerable organisms, from the lichens that crumble rocks, and the bacterial entities that decompose organic detritus, to all the respiring plants and animals exchanging vital gases with the air. The notion of an earthly nature as a densely interconnected organic network — a 'biospheric web' wherein each entity draws its specific character from its relations, direct and indirect, to all the others — has today become commonplace.... an intertwined, and actively intertwining, lattice of mutually dependent phenomena, both sensorial and sentient, of which our own sensing bodies are a part.[64]

Must this new ecological understanding, while insisting in language similar to much of the Asian philosophic traditions that everything is connected and that even humans are less singular objects than sets of relationships or processes in time, in effect *flatten* all nature and humanity into sameness? Can it really be expected that we would accord equal moral, legal and political standing to the AIDS virus, to the Grand Canyon and to Nelson Mandela?

Here we can turn again to Val Plumwood and her thoughts on ecofeminism for guidance. While pursuing her attack on dualism and the rigid subject/object relationship that seems to flow inevitably from it, she rejects an "oceanic holism" or the ecocentrism of the deep ecologists as the better choice. Instead she argues for a paradox, that continuity and difference must exist together. That a human sees a "table" while a dog sees a "thing to lie under" does not mean one image is superior to the other -"...every animal has its own excellence of interpretation."[65] Self and other, organism and land form, human and animal must, according to Plumwood, be in the kind of relation that allows them to exist but not be oppressed or denied agency by being labeled the "other" or by losing identity through immersion in some oceanic fullness.[66] Again, it is the hierarchical dimension of dualism that is rejected and the insistence that agency not be denied due to the seeming superiority of one over another. For Plumwood, this means recognizing "...continuity and hybridity between the human and the natural. It does not require us to deny nature's otherness or separateness, or to deny or submerge human distinctness from other species..... we need a de-polarizing recognition of nonhuman nature which recognizes the *denied space* of our hybridity, continuity and kinship, and is able to recognize, in suitable contexts, the difference of the nonhuman in a non-hierarchical way."[67]

Plumwood identifies here an important aspect of our problem with nature, namely the linking of difference with hierarchy. This is perhaps the most objectionable dimension of dualism and it is, in her view, unnecessary. Arne Naess and other so-called Deep Ecologists base their philosophy (or ecosophy as Naess calls it) on a strong rejection of dualism or a fixation on otherness, choosing instead a more or less oceanic, romantic identification with all of nature.[68] Plumwood and other environmental ethicists (such as Holmes Rolston) accept difference, indeed they accept the rather profound differences that set Homo sapiens apart from even very genetically similar species. But they insist that different species can still stand in solidarity with one another despite barriers of language, skills and reasoning. They reject, in other words, the necessity of identification as a requirement for solidarity. And they reject the 'humanist' argument that these differences are decisive.[69]

There is much splitting of hairs by those who debate these issues, but the stakes are indeed significant. There is an apparent anti-human strain in much of deep ecology and even James Lovelock's Gaia theory can sometimes reduce humankind to an alarming inconsequentiality. The attack on modernity in its various forms that is at the heart of much Green Romanticism implies a kind of social revolution that has little appeal for most of humanity. There is also a too easy absolutism in many of the critiques of dualism, a conviction that a focus on difference (e.g., humans have language, tool-making, consciousness, etc. while other species and natural formations do not) must lead to a subject-object dichotomy and thus a denial of any moral claims or obligations. Finally, there is an unacceptable harshness in the response of many environmentalists to the persistent material inequalities within and among human cultures, a reluctance to acknowledge that the issue of global equity for humanity must be addressed before severe limits on development can be imposed. Garrett Hardin, for instance, in making a strong case for independent, self-sustaining bioregions argues that this principle over-rules any moral claim to relieve famine-related human suffering in neighbouring and perhaps poorly managed bioregions.[70]

But these issues do not negate the serious flaws in the mastery agenda, the idea that our cultural goal should be the conquest and control of our external environment and mastery or repression of our inner drives. And despite the continued existence of economic disparities, we cannot avoid the truth of Leiss's conclusion that there is "...simply no real possibility that the entire world's population, at any time in the future, can obtain the material standard of living now possessed by the majority of inhabitants in the industrialized nations."[71] Our persistent attempts to either deny this or to overcome these disparities by more growth through more mastery will, Leiss argues, lead to "catastrophic environmental degradation." In the face of the global economic crisis of 2008, our response should thus be one of re-assessing materialism, progress and mastery rather than 're-starting' the global economic system.

And so if "staying the course" is not possible and voluntary or imposed radical reversals both undesirable and unlikely, we are left with Feenberg's suggestion that there must be a cultural change in which "...a clean and healthful environment is considered not as an exogenous dumping grounds but as a component of individual well-being...", Leiss's insistence that our task is to "...find adequate political forms for an appropriate representation of the relation between humanity and nature", and Plumwood's demand that we must create "...sustainable social institutions and values which can acknowledge deeply and fully our dependence on and ties to the earth."[72] Nurturing an ecologically sustainable planet and humankind must, then, engage us simultaneously with the cultural, political and social dimensions of human life—with the body, mind and spirit.

EXTENDING OUR REACH

I want to bring this exploration to a close by once again referring to the work of Neil Evernden whose insights have been so crucial in the evolution of my own understanding of the human/nature relationship. Despite the warnings of its "leveling" implications for human well-being vis-a-vis other forms of being, I can see no way of avoiding the acceptance of what Evernden calls "...the intrinsic worth of life, of human beings, of living beings, ultimately of Being itself."[73] And I agree with the environmental philosopher Michael Fox in seeing all forms of life as equally striving for their optimum well-being or the most beneficial functioning for other members of their species.[74] As to how we come to internalize this notion of the intrinsic value of all aspects of Being and respecting if not always conceding the right to attain optimal well-being, I think Callicott is correct in highlighting the importance of ecological knowledge but that Abrams may be closer to the mark in stressing the importance of a "...rejuvenation of our carnal, sensorial empathy with the living land that sustains us."[75] The poets, philosophers and environmentalists that I have referred to throughout this book would all agree with Friedrich Schiller that the path to the head must be via the heart.[76]

In all kinds of inchoate ways humans worldwide are taking steps in these directions, both individually and through various institutional frameworks. One thinks of the growing awareness of re-cycling and re-using, a new interest in vegetarianism, tree-planting programs in Africa and myriad other conservation projects, population planning, animal rights activities, and international initiatives like the Kyoto Accord and the Brundtland Commission. At the same time, as noted earlier, despite our new awareness of its pathological implications there remain powerful forces of resistance to any substantive change in the project of virtually unlimited economic growth that has been the key feature of modernity.

Returning once again to the insight offered by Neil Evernden that the environmental crisis of our times is essentially a cultural phenomenon, that "...how we act toward the non-human is a consequence of our beliefs about how we should act and about what we are acting on," I am persuaded that three *extensions* are required if we are to re-think our relationship with nature in a way that leads to mutuality and reciprocity: extending the realm of our ethical concern; extending our natural capacity for empathy; and extending our field of legal rights and obligations.[77]

As we have seen, the idea of extending ethical and legal standing to the other-than-human is logical in an historicist or evolutionary sense, and deeply controversial. We have a clear history of basing such extensions not only on

274 | CONCLUSION

the basis of fairness or as acknowledgment of a natural right, but as Robert Goodin says also on the basis of interest. Women, the poor, aliens, children and the variously disabled have "interests" and therefore a claim to certain rights.[78] The natural world clearly has interests as well and thus some kind of claim for representation and the "...standard way to include excluded interests, politically, is simply to enfranchise them."[79] But, like infants and a future human generation, nature cannot speak or articulate in any clear way those interest claims. If one sets aside the humanist chauvinism argument, this issue of articulation remains the most difficult aspect of the extensionist case.

Moral Extension

Humankind's moral universe has undergone a steady expansion in the past 300 years, albeit with several tragic retrogressive episodes. It was not that long ago that those considered barbarians, primitives, enemies, and humans with different skin tones were excluded from moral consideration — in effect denied their standing as humans. As we saw in Chapter Four, this extension of moral standing has not been a linear process. Now in the 21st century we are considering the idea that the human-nature relationship should be considered a moral issue "...conditioned or restrained by ethics."[80] There are a variety of forms that this extension of ethics has taken, ranging from the utilitarian argument that as sentient beings animals at least deserve the same moral consideration as sentient humans, to the "reverence for life" position espoused by Fox, Albert Schweitzer and others in which all living beings have a *telos* or an excellence toward which they strive, to the more holistic and ecocentric focus on the complete interdependence of the biotic community including animals, plants, oceans and landforms.[81]

An important model of such an extension of moral standing is Martha Nussbaum's Capabilities Approach as outlined in her most recent book *Frontiers of Justice*.[82] Agreeing with Nash and other proponents of the need for an extension of moral standing beyond humans, Nussbaum considers "...the possession of sentience as a threshold condition for membership in the community of beings who have entitlements based on justice....if a creature has *either* the capacity for pleasure and pain *or* the capacity for movement from place to place *or* the capacity for emotion and affiliation *or* the capacity for reasoning...(we might add play, tool use, and others), then that creature has moral standing."[83] She is focusing here on the individual being, asking us to assess its capabilities and judge how important each is to the flourishing of the being in question, thereby avoiding applying moral standing to ecosystems, issues of biodiversity or toward non-sentient phenomena. This is not "nature-worship", but rather the extension to beings with human-like qualities the ethical norms applicable to humans. There is no oceanic holism here, but rather a practical first step in the extension of the ethical pyramid beyond Homo sapiens.[84]

Nussbaum's Capabilities Approach is designed to provide a much thinner edge to the wedge that environmentalists use to pry apart the anthropocentric wall between human and animals. By awarding animals moral standing that has been derived from the experience of humans she expands the existing social contract and thereby opens the door for even further extensions. There is a kind of progression implied in this spread of ethical approaches. The focus on sentience or feelings of pleasure and pain is perhaps the easiest for humans to consider since most of us are a long way from thinking of our fellow animals as mere Cartesean machines and we are, as Nussbaum says, more convinced now "...that there is something wonderful and wonder-inspiring in all the complex forms of life in nature."[85]

This limited extension of moral standing has drawn strong criticism from those more inclined to an ecocentric view. Carey Wolfe, for instance, acknowledges that while short-term gains in animal welfare can come from an approach like Nussbaum's, in the end this inhibits "...a more ambitious and more profound ethical project...a more inclusive form of ethical pluralism."[86] Critics of the extensionist position from the environmental "left" like Wolfe and Plumwood see it as too limited, based too much on the possible human-like qualities of the species being "cared for".

Holmes Rolston celebrates the fact that wild animals can never "enter culture" and bemoans the fact that when we domesticate or shelter them they "lose their wildness" and hence their "nature."[87] And, of course, this moderate extensionism excludes awarding such moral standing to eco-systems, bioregions or inorganic matter. Plumwood concludes that it merely expands and validates Cartesean dualism.[88]

This illustrates how difficult it is to break from what James Lovelock calls the "...apartheid of Victorian biology and geology."[89] While the sentience and even animal rights approaches can be firmly grounded in established moral frames such as Kant's categorical imperative or even the Golden Rule, the more holistic approach of environmentalists like Val Plumwood, Aldo Leopold, J. Baird Callicott and Erazim Kohak require a more personal or subjective sense of the not-necessarily-living other. Kohak frames this in a version of Martin Buber's I-Thou imperative, arguing that all reality must be seen in personalistic terms, independent of attributes such as language, mind or even sensation. Each aspect or manifestation of the natural world would therefore be encountered as a "thou", thereby giving it moral content and an

276 | CONCLUSION

intrinsic claim to respect—as humans have with each other. "For a person, ultimately, is not just a being who possesses a psyche or manifests certain personality traits as much as a being who stands in a moral relation to us, a being we encounter as a Thou."[90]

Such an extension or moral relationships to the natural world, whether just to sentient beings or to all of being, returns humankind to a reconstructed animism or holistic world view that we presume was prevalent in our hunter-gatherer past and persists today in many aboriginal cultures. It implies not a passive helplessness in the face of a world of thous, but rather consumption and manipulation by humans of the other—animal, vegetable and mineral—from a stance of respect and acknowledgement of mutuality. Hence humans may choose to "take no enemies" as Bernard Williams counsels in reference to the AIDS virus, just as they may choose to imprison or even execute other humans, but if this more universal notion of moral standing is in effect these actions would be done with respect and even a dose of reverence [91].

This extension of moral standing is then the base from which further extensions can be envisaged. It is perhaps the easiest starting point, except for those of us wedded too strictly to anthropocentrism. In some cultures, especially those with a tradition of keeping animals as 'pets', the first stages of moral extension have been taken and in recent years many have pushed that further, beyond pets to endangered species, to food species and to species crucial to ecosystem health. Those in the *avant garde* of the environmental movement over the past fifty years have been gradually extending the area of moral concern to include land formations, bodies of water and the atmosphere. Western culture, the prime offender in terms of environmental degradation for the past 300 years, is also the home of many of these groups.

Extension of Empathy

An ethics that depends upon establishing a personal, even intimate relationship with the other brings us to the importance of extending our sympathetic imagination to the natural world. In the great debate between Thomas Hobbes and Jean-Jacques Rousseau over whether humans are naturally good and empathetic or nasty and egocentric—a debate that continues to enliven our times—the Rousseauean position appears to be gaining strength. Anthropologists and archaeologists continue to find evidence of a "caring nature" from past and present and psychologists in "tit for tat" games and in theories of reciprocal altruism make a strong case for sympathy, sharing and caring being the more successful human option. In philosophy, Paul Taylor says that our ability to imaginatively look at the world from the animal's standpoint — our empathetic ability — is the crucial first step in extending moral standing to animals.[92] The fact that animals (or plants or rivers or the earth itself) do not reciprocate and are not by nature themselves sympathetic beings, does not excuse us, he argues, from exercising this unique dimension of our consciousness. Certainly from the holistic perspective we are beings that have co-evolved with a much larger biotic community toward which we "...ought to feel sympathy or benevolence toward our fellow members and loyalty and respect toward the community as such."[93]

It would seem, given Taylor's comment, that the extension of empathy should be the first step, the base from which an ethics might emerge. Obviously ethics and empathy are closely connected, but I see the latter, interpreted as more than mere sympathy, as more difficult than developing a moral grounding for an environmental ethic. An ethic is a rule, a set of rational arguments concerning how we ought to behave toward these new others'. Empathy is an internal, deeply subjective phenomenon that is linked tour conscience more than our reason; affective more than cognitive. Empathy, I argue, is more difficult as a guide to action. Truly internalized, it is less subject to evasion or tendentious objections or counter arguments. It is a truer test, if you will, of a genuine change in our relations with nature.

By far the most fully developed expression of this perspective is the "Partnership Ethic" proposed by Carolyn Merchant. Deeply antipathetic to the ecocentric views of deep ecology which she feels levels value hierarchies to the point that humans become "...morally equivalent to a bacterium or a mosquito", Merchant sees two communities in operation, one human and the other nonhuman.[94] Her partnership Ethic sees "...the human community and the biotic community in mutual relationship, [in] mutual living interdependence."[95] The key here is that the relationship between the two communities is no longer one of mastery and subordination, but rather is characterized by a dynamic balance, with a recognition by humans that each has power over the other. The human side of this partnership would be bound to exercize sound ecological management (the enlightened stewardship discussed earlier), would have "moral consideration" for other species, and would acknowledge nonhuman nature as "...an autonomous actor that cannot be predicted or controlled except in very limited domains."[96]

Merchant is careful that this partnership be based on a "more nearly equal" relationship with the biotic community, thereby avoiding the homocentric tradition of mastery and the ecocentric tradition of oceanic holism. This new "earthcare ethic" is — and of course must be — generated by humans, but only after "...listening to, hearing, and responding to the voice of nature."[97] Central to its success is the potential for humans to develop feelings

278 | **CONCLUSION**

of compassion for nonhumans as well as for humans and perhaps with a "new consciousness" replace a discourse of environmental conflict with a "...discourse of cooperation."[98] Merchant's Partnership Ethic seems a progressive, thorough and practical proposal designed to find a middle path between the extreme positions. It borrows from the Collective Environment Management position, goes beyond Nussbaum's more limited focus on sentience and is cautious in dealing with issues of intrinsic value and rights. And it acknowledges the importance of a "new consciousness", a new way of thinking and feeling about the nonhuman, one that relies on an extension of our sympathetic imaginations.

But Merchant's earthcare ethic may have difficulty getting to the heart of the matter, its seeming focus on sympathy and compassion avoiding a broader and more profound development of empathetic feelings for all of the natural world. Her work sits astride, as she acknowledges, the gap between a homocentric world view and one based on a more ecocentric holism. As has been clear throughout this exploration of human relations with the natural world, this is a difficult gap to try and bridge.

Stan Rowe, echoing the work of Kohak, Abrams and many deep ecologists, insists that our problems with the environment all stem from a lack of "...sympathy with and care for the land and water ecosystems that support life."[99] Rowe wants us to internalize in our everyday life the interests (which we determine via reason and science) of these systems and make their well-being prominent "in our hearts and imagination." This will require a movement from sympathy or caring to empathy. Adam Smith in the famous opening section "On Sympathy" in his *Theory of Moral Sentiments* (1759) warned us of the difficulty we face in such a transition: "Though our brother is upon the rack, as long as we ourselves are at our ease, our senses will never inform us of what he suffers....it is by the imagination only that we can form any conception of what are his sensations."[100] So it is to something more than reason or sympathy that we must turn. Plumwood seems to side here more with Rowe than with Merchant, her ethic of care firmly grounded in respect, care and love.[101]

The introduction of the requirement that we cultivate a relationship of love not just with beings but the systems of nature raises the bar considerably. Environmental ethicists of whatever persuasion do seen to base the extension of moral standing on rules, on reasoned arguments and critiques of existing boundaries. By working toward extending our empathetic abilities (which I have argued earlier is a core element of our human nature) we shift the source of an ethics of care, respect and love from law, duty or self-imposed rule following toward disposition, an internalized inclination or "habit of the heart". This becomes a core "civic virtue", a dramatic expansion of our range of significant others. Such civic virtue is "...a second nature, a predisposition to act voluntarily in some wider interests.[102] Ian Hacking speaks to this when he urges us to become more "Humean and first worry about how to enlarge our sympathies. Rights and utilities will fall into place much later."[103] In the context of our discussion of shadow modernities and alternate voices, in assessing the cultural roots that ground us we need to elevate Rousseau's concerns about achieving a balance among self, community and nature and, contrary to our dominant Kantian heritage with it's celebration of reason. We need to listen more carefully to David Hume's conviction that: "Reason is, and ought only to be the slave of the passions, and can never pretend to any other office than to serve and obey them".[104]

Extended to its full potential, this ability to use our imaginative powers to achieve a sympathetic understanding of and benevolence toward not only other humans but also to other beings and possibly nature as whole is central to the 21st century task of nurturing a human culture that is ecologically sustainable. The now global enterprise of modern science is demonstrating for us the dangers we face and our ultimate dependency of the health and viability of ecosystems. Sadly, we now realize that this knowledge alone is still not persuasive enough or present enough in our lives to alter our pattern of mastery and control which remains powerfully embedded in our institutions and patterns of consumption. When, however this scientific understanding is combined with a "...rejuvenation of our carnal, sensorial empathy with the living land that sustains us.." and with an ecocentric holism from which we derive a "...feeling of an aesthetic and material oneness with all things...", we can begin to sense the possibility of the re-thinking that Evernden wants and the cultural change that Feenberg, Leiss and Plumwood urge us toward.[105]

Extension of Rights

But how might these changes in thinking and in cultural preferences find social expression? As individuals we may choose to celebrate oneness, engage in sensuous communion with nature, become vegans or adopt a radical reverence for life in conducting daily affairs, but the ecological problems we face require a more concerted and enforceable set of social, political and economic changes. It is necessary, therefore, that our third "extension"—legal rights—be considered. In the Western political tradition and in many others as well, legal rights are accorded to citizens, who in return undertake certain responsibilities toward the whole. Originally, as defined by Aristotle in the 4th century BCE, the title citizen was limited to "...a man who shares in the administration of justice and the holding of office."[106] This was a "contract" in

280 | CONCLUSION

a very real sense, awarding the citizen significant rights and privileges in return for an obligation to reserve parts of one's self for service to the community. This ancient contract, theorized in the 18th century by Hobbes, Rousseau and Locke, had of course no signatory moment but simply evolved out of the understanding by individuals of the practical wisdom of collective action.

Given the Rousseauean position that for humans, freedom and individuality are the primary natural goods, entering into a social contract is justifiable only because it is assumed that one's self-interest (freedom and individuality) are in fact best advanced under the protection of a society of like-minded individuals. Thus while the United States Declaration of Independence insists that a "Creator" has endowed each individual with an intrinsic (i.e. natural) right to "...life, liberty and the pursuit of happiness", it is in fact the temporal social contract that allows humans to enjoy this Creator's gift. As Rousseau argued, when individuals submit themselves to the social contract they "... find themselves in a situation preferable in real terms to that which had prevailed before ...they have exchanged natural independence for freedom."[107] In its modern form a product of the 18th century European Enlightenment, this notion of the citizen with rights and responsibilities has spread in scope from property-holding males to all males, to males and females and beyond. It has become a global force via initiatives such as the United Nations' Universal Declaration of Human Rights. It is important to note that this central category of human cultures, the rights bearing and responsibility-assuming citizen, is ours to define and to decide whom to enfranchise.

In our legal world citizens are "persons" and have standing in our courts. In recent years in parts of the world children, prisoners, aliens and other once disenfranchised non-citizen beings have been made "persons" in this sense. Likewise corporations, trusts, municipalities and other beings in other-than-human form are often given standing as persons-in-law.[108] So despite the derision of some at the prospect of rocks having rights, the idea that forests, watersheds, streams or oceans might gain a kind of citizenship via becoming "persons-in-law" is likely only a matter of time.[109]

But beyond including some or all natural phenomena in a human framework of law and legal rights, could there ever be a *politics* of nature? Humans, as citizens, envisage a good life and argue, mobilize and negotiate to achieve it. Surely this is beyond the potential of landscapes, animal species or ecosystems. Whatever moral or contractural arrangement we may imagine between humans and nature would surely have to rest on human subjects treating natural objects in a more equitable manner. We will, after all, because of our superior powers "...necessarily remain in some sense the rulers of the earth."[110]

Here we re-enter the issues discussed in Chapter Three. Who can be a subject? Must the perceived world be divided into subjects and objects? For adherents to the Kantian view we know that only rational agents can be subjects and thereby have moral and legal standing. They "...stand out against the background of nature, just in that they are free and self-determining... Everything else in nature conforms to laws blindly."[111]

There are, as we know, strong arguments opposed to this anthropocentric tradition. It is contested by the writers of the Romantic era and by anthropologists and psychologists informed by Darwinian evolutionary ideas. Here humans are very much like most other animals, made unique by historical circumstances and evolutionary pressures and it is impossible to establish any essential distinction between humans and other animals.[112] Evidence continues to mount that animals reason, have language, make tools, reflect, have memory and experience emotional attachments. "Of course animals know" insists Erazim Kohak. "Their knowing may be of a different order from human reflection, conceptualization, and articulation, but their purposeful behavior in the context they treat as meaningful testifies both to a highly differentiated cognition and to an ability to store and transmit information."[113]

But Kohak goes further. Falling back on the work of Martin Buber, Kohak insists that a person is a being who "stands in a moral relation to us, a being we encounter as a thou."[114] And he argues that insects, trees and plants share a dimension of "knowing" that he takes as a given in animals and hence can qualify for personhood as well.

What rationale, aside from sheer power, do we have to deny moral and legal standing to the various components of the natural world? Robyn Eckersley admits that this extension could very well come to pass, but like so many when confronted by this possibility he sees something strained and ungainly in the attempt to extend to the nonhuman world political concepts that have been especially tailored over many centuries to protect *human* interests."[115] He suggests instead that nature would be better served by a more ecologically enlightened human culture than by any extension of human-centered rights.

Christopher Stone, famous among environmentalists for his essay *Should Trees Have Standing*, agrees that such a shift in ecological consciousness would make for better humans, but he doubts that this would in the long run be the best protection for a natural world faced with ever increasing human demands for development and access to resources. Stone responds to the seemingly radical nature of his proposal that trees and other aspects of nature should have legal rights by noting that "...each time there is a movement to confer rights onto some new *entity*, the proposal is bound to sound odd or frightening or laughable. This is partly because until the rightless thing receives its rights, we cannot see it as anything but a *thing* for the use of

282 | CONCLUSION

us– those who are holding rights at the time."[116]

Stone rejects the idea that because nature has no *voice* it cannot speak on its own behalf. Lawyers, he says, can speak for nature just as they do for corporations, universities, infants and estates. Advocacy groups working on behalf of species such as chimpanzees are already working toward this end, seeking to allow non-profit groups to petition courts to act as guardians for chimps being used in medical or other experimental projects.[117] Martha Nussbaum supports Stone's position by arguing that in order to extend the contract idea to disadvantaged groups and to nonhuman species we will need to abandon the notion in contract theory that the parties must be roughly equal in power and authority and can agree on the issue of mutual advantage.[118]

Stone qualifies the notion that rocks might have rights by stressing the difference between "rights" and "legal recognition", arguing that there is no need to conflate the two. Animals are in many ways given legal protection without implying that they have any rights in law. He offers a similar example for a non-sentient aspect of nature in the case of a lake which by law is meant to be kept in a "certain condition" and punishments are meted out to those who alter that condition.[119] The lake is not a moral agent nor can it exercize rights, but a right to exist in its ideal form (as deemed by humans) may nonetheless be codified in law.

The issue of rights, responsibilities and citizenship becomes very complex when one moves beyond the boundaries of *ideal-types*. Even the prototypical adult resident of a given polity often fails to qualify. Beyond imprisonment, emigration or blatant non-conformity, there are problems with including children, the disabled or mentally challenged in the 'contract community'. These complexities increase tenfold when we move beyond humans. Speciesism is not as easy to address as were racism and sexism. One way of confronting this dilemma is by differentiating between moral subjects and moral agents:

> Moral subjects are worthy of moral consideration; they have moral standing. Moral agents are capable acting morally; they have moral responsibilities and duties. All moral agents are moral subjects but not all moral subjects are moral agents. Both moral agents and moral subjects have interests. Blacks and women are moral agents. Nonhuman nature is a moral subject; non human individuals are moral subjects.[120]

This is a "best of all possible worlds" solution—moral agents are not responsible, but still deserve our respect and consideration. We seem to have returned here to a version of Merchant's Ethics of Care. The components of nature that are limited to being moral and legal subjects are at the mercy of the moral agents' internalized understanding and compassion. Still, we have come a long way from the Kantian claims of social theorists such as Jurgen Habermas or Luc Ferry and the insistence that without language and self-consciousness non-human entities are merely objects to be controlled and cannot as a result be considered subjects in any way.

But for many this may not be enough. Plumwood sees it as a "highly problematic" in its opting for rights as opposed to the unique emotional obligations humans have (and feel) toward nature—the biophilia that Wilson champions. Here once again the demarcation lines are clearly drawn in the struggle between reason seen as abstract, universal and detached and the passions with their grounding in the particular and the emotional in guiding human values and behaviours. Passionate though advocates of a biophilia approach that relies on a deeply emotional connection with nature may be, the fears that this is too vulnerable a foundation to rest a 'case for nature' on persuades most that it must be combined with arguments based on reason as well. Carey Wolfe argues that recent animal studies as well as studies of humans with disabilities "..pose fundamental challenges" to civil procedures drawn from a tradition that bases inclusion in civic life solely on rationality, autonomy and agency.[121] The case of Washoe, the "signing chimp", further blurred the boundaries when a court in New York agreed that chimps could be considered "persons" in law because they "...are capable of rational thought, communications and other higher cognitive functions...", making them the equivalent in law of minors or disabled humans.[122] So here the rights of humans were extended to animals—or one type of animal—but with the subject/agent distinction in place. Some animals, then, may acquire "...a kind of civic personality which demands recognition from other civic persons."[123]

Are we, then, close to the idea of 'citizen nature'? We may reasonably conclude that a case can be made for an extension of moral standing to include nature, for an extension of our identification with or sympathetic understanding of the various parts of nature and for an extension of legal rights in some form. We are, then, essentially looking at an expansion of the social contract. And these extensions are warranted because we can imagine that the nonhuman world would comprise what Aldo Leopold saw as a "land community"—comparable in its complexity and its interdependence with the social community of humans. It remains perhaps more a metaphor than a practical idea to consider this land community to be made up of citizens. But since it is a word we all understand, a word that verifies membership and a certain minimal level of equality as members of a community, it seems useful to apply it in thinking about our fellow members of Leopold's community.

The case for thinking of members of the land community in this way

284 | CONCLUSION

finds powerful support in the work of Michel Serres. An historian of science trained in mathematics and skilled in philosophical discourse, Serres is the author of *The Natural Contract*, a work that provided the inspiration for my work in this area and he deserves to be a summative voice at the conclusion:

> ...we must add to the exclusively social contract a natural contract of symbiosis and reciprocity in which our relationship to things would set aside mastery and possession in favor of admiring attention, reciprocity, contemplation, and respect; where knowledge would no longer imply property, nor action mastery, nor would property and mastery imply their excremental results and origins. An armistice contract in the objective war, a contract of symbiosis, for a symbiont recognizes the host's rights, whereas a parasite—which is what we are now—condemns to death the one he pillages and inhabits, not realizing that in the long run he's condemning himself to death too.[124]

Serres has no illusions that nature is comprised of phenomena identical to human persons or that a natural contract could be a negotiated agreement any more than the social contract was a signed document. But he does consider nature a subject. Maria Assad, a commentator on Serres, interprets him as accepting much of what the anthropocentric position insists on, that humans, thanks to a complex series of biological and psychological changes over millennia, have essentially stepped outside of nature. It is this outsider situation—one thinks of Camus' *Outsider* or *Stranger*—that Serres argues gives us the ability to "partner" with nature—or not.[125] This is a very different contractual arrangement than that implied by the tentative granting of civic personhood to a chimp! Nature seen collectively, as a single phenomenon, provides humankind with all its material needs, but has the power—one might imply will—to refuse such provisions if we abuse them, much as Coleridge implied in his story of the Ancient Mariner. And the earth for Serres has a voice which warns us when that abuse becomes dangerous. We do not know that language well, but its forces, bonds and interactions are enough for Serres to make a contract a real possibility.[125] Serres insists we learn to translate this voice because nature is now able to "burst in on our culture."[127]

The natural contract is essential and possible because, as implied in the earlier discourse of the Romantics and Deep Ecologists, we have "...abandoned the bond that connects us to the world." [128] Like James Lovelock's dark prognostications in his more recent books on Gaia, Serres sees a dark future ahead if we do not re-open a connection with the earth, with the land community. This bond, at once spiritual and visceral, was abandoned according to Serres when the hold of tradition and religion on the speculative drive of science was lost. In the modern world the sciences rule and speak for nature, largely independent of the realm of the social contract. But for Serres the nature the sciences pretend to speak is changed forever. It is now a global subject, a subject whose actions and intentions "remain in large part unpredictable."[129]

For Serres there are three contracts possible, but only two in operation. The demand to love one another led to a social contract and the demand to love the truth led to a scientific contract. The natural contract would, if implemented, fulfill the third demand for a universal love of the physical earth:

> ...I mean by natural contract above all the precisely metaphysical recognition, by each collectivity [i.e. the social contract & the scientific contract], that it lives and works in the same global world as all the others....It is as global as the social contract and in a way makes the social contract enter the world, and it is as worldwide as the scientific contract and in a way makes the scientific contract enter history........the natural contract recognizes and acknowledges an equilibrium between our current power and the forces of the world....The natural contract leads us to consider the world's point of view in its totality.[130]

We return here, in a round-about way, to Plumwood's demand that we must love nature, that we must put into practice the biophilia claimed to be at our very centre, that we must accept the centrality to culture of our 'home place', and that we must regain that 'sentiment of existence' that bonded us with the whole rather than just the atomistic self. And for Serres it is the "whole" writ large, the Earth *per se* that we must love. We must now master mastery, the goal as well of the Ecological Management position discussed earlier. To do so will involve not only the increasing communal overview of scientific research and the diffusion of technology that William Leiss and many others counsel, but also the granting of rights to at least some nonhuman things as Christopher Stone suggests. In a recent work Serres goes much further, insisting that "...all living beings and all inert objects, in short, all of Nature.." should be considered legal subjects.[131]

Serres is not alone in this call for us to turn from mastery to respect, from control to reciprocity. We have heard a similar message from Erazim Kohak. Roderick Nash has called for an "ecological contract" with wild nature which would allow it to exist simply out of respect for its intrinsic value. And J. Baird Callicott, echoing Lovelock, reminds us that we need to enfranchise nature as a whole as well as the various individual nonhuman natural entities.[132] Serres warns us that this will entail a "...harrowing revision of modern natural law..", one which sees objects as legal subjects, no longer mere material for appropriation, individually or collectively. But, he insists, "if objects themselves become legal subjects, then all scales will tend toward an equilibrium."[133]

Have we answered the poet Shelley's three questions that opened this exploration? We ought to live in a reciprocal and sympathetic relationship with each other and with the land community, aware of our extraordinary powers but aware of its voices and powers as well. What is our relation with the rest of the land community? Certainly not mastery but neither is it slavery. We live in an interdependent world the parts of which must negotiate a means of survival and flourishing. To reconcile human flourishing with planetary sustainability we will need to acknowledge and respect both empirically and sympathetically the aspirations of all members of the land community to achieve as much of their life project as the conditions of their existence and ours allows. Above all we need to acknowledge that our planet is a community of communities and that each contains members/citizens with rights, possibilities and crucial functions to perform for the well-being of the whole.

NOTES

1 Joel Whitebook, "The Problem of Nature in Habermas", (pp 283–317) in Macauley, p 299.

2 Jane Bennett, *Unthinking Faith and Enlightenment* (New York: New York University Press, 1987), p 111.

3 Dobson (2003), p 109.

4 Dobson (2003), p 111.

5 Bernard Williams, "Must a Concern for the Environment be Centered on Human Beings?" (pp 46–52), in Gruen and Jamieson, p 51.

6 Luc Ferry, *The New Ecological Order* (Chicago: University of Chicago Press, 1995).

7 Pollan, p 48.

8 Ferry, p xxviii.

9 Leopold, p 204.

10 Leopold, p 223.

11 Leopold, p 224. We may reasonably critique Leopold's apparent assumption of stability, knowing now as Botkin and others have stressed that fluidity and change are more the case in nature. Likewise his focus on beauty may give us pause since that term is highly culturally dependent.

12 Leopold, p 4

13 Leopold, p 21

14 Leopold, p 132

15 Leopold, p 188.

16 Leopold, p 110.

17 Leopold, p 202.

18 Plumwood (2002), p 142.

19 Leopold, p 210.

20 Speth, p 200.

21 *Vancouver Sun*, 5 April 2008, p C6.

22 "The impact of this personalistic view of the ecological crisis and its sources has—like sociobiology and ecomysticism—significantly shifted public attention from the social roots of our ecological dislocations to a psychological level of discussion...". Bookchin, p 109.

23 Jane Austen, *Sense and Sensibility* (New York: Oxford University Press, 1990). p 164.

24 Austen, p 76.

25 Austen, p 333.

26 Eckersley, p 102,

27 Jon Turney, "Of Mites and Men", review of E.O. Wilson, *The Future of Life*, in *New York Times Book Review*, 17 Feb. 2002, p 11. Terry Glavin in his *Waiting for the Macaws* says that a "deep empathy for and identification with non-human forms of life can be found throughout human history, and not just in the mystical traditions of Eastern spirituality....", p 178.

28 John Barry, *Rethinking Green Politics* (London: Sage, 1999), p 114.

29 David Goldblatt, *Social Theory and the Environment* cited in Baxter, p 191.

30 George E. Brown (Chairman, House of Rep, Committee on Science, Space and Technology, "Reorienting Scientific and Technological Inquiry to Tackle the Global Crisis Facing Humanity" *Chronicle of Higher Education*, 22 April 1992.

31 See Elizabeth Kolbert, "The Climate of Man", *The New Yorker* 25 April 2005 (pp 56–71); 2 May 2005 (pp 64–73); 9 May 2005 (pp 52–64) and Elizabeth Kolbert, "Chilling", *The New Yorker* 20

CONCLUSION

March 2006 (pp 67–68).

32   Kolbert quoting James Hansen, head of the Goddard Institute. *The New Yorker*, 20 March 2006, p 68

33   Bert deVries, et. al. "Understanding: Fragments of a Unifying Perspective" in Goudsblom and deVries, p 273.

34   Andrew Light, "Democratic Technology, Population and Environmental Change", in T. Veak, ed., *Philosophy & Technology: New Debates in the Democratization of Technology* (Albany: SUNY Press, 2004), p 12.

35   See for example Dobson (2003); M. Smith, *Ecologism: Towards Ecological Citizenship* (Minneapolis: University of Minnesota Press, 1998); D. Bell, "Liberal Environmental Citizenship", *Environmental Politics* (v. 14:2, 2005, pp 179–194).

36   Zimmerman, p 76.

37   Brian Baxter cites David Goldblatt's work as an example of this focus on technology: "...only when industrialism was organized within the capitalist system which it had itself unleashed did the potential created by the discovery of coal-based forms of energy-production to release economic activity from previous ecological limits become a reality.....the aim of environmentally concerned people should be to control rigorously the workings of capitalism in the interests of the preservation of the biosphere. This will involve a mixture of state and international regulations, taxation policies, licensing regimes and the increased democratic accountability of capitalist corporations. It may also involve encouraging capitalist firms to engage in the activities which go under the heading of 'ecological modernisation'. This involves, among other things, the manufacture of environmentally beneficial technologies,by processes which are themselves made minimally damaging to the environment." Baxter, p 191.

38   This arguments is perhaps best put by Baxter: "It is certainly plausible to suggest that changes in ideas are not a sufficient condition of changes in people's behaviour. But it seems equally plausible to suggest that they are a necessary condition." Baxter, p 5

39   Barry, p 72.

40   Andrew Feenberg, "Values and the Environment" (unpub mss) March 2004, p 1.

41   Andrew Feenberg, "The Commoner-Ehrlich Debate: Environmentalism and the Politics of Survival" (pp 257–282)) in Macauley, p 272. Feenberg is seconding here the views of one of his mentors, Herbert Marcuse, who called for "...change, not only in the basic institutions and relationships of an established society, but also in individual consciousness in such a society. Radical change may even be so deep as to affect the individual unconscious....In other words, radical change must entail both a change in society's institutions, and also a change in the character structure predominant among individuals in that society." Marcuse, p 30.

42   Light (2004), p 22 (mss)

43   George E. Brown (Chairman, House of Rep, Committee on Science, Space and Technology, "Reorienting Scientific and Technological Inquiry to Tackle the Global Crisis Facing Humanity" *Chronicle of Higher Education*, 22 April 1992. Edward Wilson has generated a set of steps that could be taken to "save the environment" which reflect this approach:
- salvage immediately the world's hotspots
- keep intact the five remaining frontier forests (Amazon Basin, Congo, New Guinea, Canada, Russia
- Cease logging old-growth forests everywhere
- concentrate on lakes and river systems

- identify and focus on marine hotspots
- complete the mapping of the world's biodiversity
- ensure that the full range of the Earth's ecosystems are included in the conservation strategy
- make conservation profitable
- use biodiversity more effectively to benefit the world economy
- initiate restoration projects to increase the share of Earth allotted to Nature
- increase zoo capacity
- support population planning.

Edward Wilson (2002), p162-4.

44   Nash (2001), p 381. Nash sees this as a middle ground between a wasteland scenario and nature as "garden".

45   Leiss (1994), p 173.

46   Leiss (1990), p 85

47   Kerry Whiteside, (MIT Press, 2002) p 10.

48   Katz, p 166.

49   Mary Midgley (2003), p 24.

50   "Mutualism is a relationship between two species of organisms in which both benefit from interacting and their coexistence is essential fir the survival of one or, most commonly, both species." Wessels, p 67.

51   Eckersley, p 45

52   Plumwood (1993), p 171

53   James Q. Wilson, (1993), p 194.

54   Nash (1989), p 4-7. The philosopher Andrew Brennan sees the extension of rights moving from children, elderly and infirm humans to the dead, the senile, the insane, embryos, sentient animals, non-sentient animals, artifacts, groups, villages, and landscapes. Andrew Brennan, "Ecological Theory and Value in Nature", *Philosophical Inquiry* v. 8:1-2, 1986 (pp 66-95), p 67.

55   Michael Schermer, *The Science of Good and Evil: Why People Cheat, Gossip, Care, Share and Follow the Golden Rule* (New York: Henry Holt, 2004), p 48. The Canadian environmentalist John Livingston devised a similar potential progressive model moving from a focus on the individual self to a biospheric self but concluded that humankind was hopelessly stuck at the level of the "...infantile and placeless self." In Raymond Rogers, *Nature and the Crisis of Modernity* (Montreal: Black Rose Books, 1994), p 93.

56   Glavin, p 282.

57   Kimberley Smith, "Natural Subjects: Nature and Political Community", *Environmental Values*, v. 15, 2007 (pp 343-53), p 348.

58   Calarco, in Cavalieri, p 81.

59   Derek Bell, "Political Liberalism and Ecological Justice", *Analyse and Kritik*, v. 28, 2006 (pp 206-22), p 209.

60   Bookchin, p x.

61   Plumwood (1993), p 92. She accuses our religious traditions of creating an existential homelessness for humanity in which the earth is "...not a home to be cherished but a trial, a place of temporary passage and little significance compared to the world beyond."

62   Rowe (1990), p 46.

63   Commoner (1972), p 29.

64   Abrams, p 85.

290 | CONCLUSION

65   Peter Scheers, "Human Interpretation and Animal Excellence" (pp 56–64) in Drees, p 63.

66   Plumwood (1993), p 125–6.

67   Val Plumwood, "Nature as Agency and the Prospects for a Progressive Naturalism", *CNS*, vol. 12:4, 2001 (pp 3–32)

68   Arne Naess, "The Deep Ecological Movement: Some Philosophical Aspects" (pp 193–212) in Michael Zimmerman, et. al., eds. *Environmental Philosophy: From Animal Rights to Radical Ecology* (Englewood Cliffs: Prentice-Hall, 1993) p 195.

69   Plumwood (2002), p 202. "Solidarity requires not just the affirmation of difference, but also sensitivity to the difference between positioning oneself *with* the other and positioning oneself *as* the other, and this requires in turn the recognition and rejection of oppressive concepts and projects of unity or merger."

70   Vanderheiden, p 56.

71   Leiss, (1994) p xxvi.

72   Feenberg (1996), p 272; Leiss (1994), p xxvi; Plumwood (1993), p 71.

73   Evernden, p 13.

74   Michael Fox, "Thinking Ethically About the Environment", *Dalhousie Review* v. 73, Winter 1993–94 (pp 493–511), p 503.

75   Abrams, p 69.

76   Schiller, p 107.

77   Evernden, p35.

78   Robert Goodin, "Enfranchising the Earth, and Its Alternatives", *Political Studies* v. 44, 1996 (pp 835–849), p 837.

79   Goodin, p841.

80   Nash (2001), p4.

81   Fox, p495. Fox has subsequently proposed a Universal Bill of Rights for Animals and Nature. http://tedeboy.tripod.com/drmichaelwfox/id54.html

82   Nussbaum (2006).

83   Nussbaum (2006), p 362.

84   Nussbaum confronts directly the issue of the eating of animals., asking us to focus initially on good treatment of animals during their lives and painless killing, "...setting the threshold there, at first, where it is clearly compatible with securing all the human capabilities, and not very clearly in violation of any major animal capability, depending on how we understand the harm of a painless death for various types of animals." p 402. She also acknowledges that these are "...very slippery issues [and] that we are likely to be self-serving here, and biased toward our own form of life." (2006) p 387.

85   Nussbaum (2006), p 347.

86   Carey Wolfe, "Learning from Temple Grandin, Or, Animal Studies, Disability Studies and Who Comes after the Subject", v. 64, 2008 (pp 110–123), p 118.

87   Rolston in Gruen and Jamieson, p 67.

88   Plumwood (2002). "The fact that the rights concept is difficult to apply to nonhumans beyond the case of certain animals that resemble humans means that it tends to support neo-Cartesian moral dualism.", p 152.

89   Lovelock (1990), p 244.

90   Kohak, p 129.

91   Bernard Williams, "Must a Concern for the Environment be Centered on Human Beings?" (pp 46–52), in Lori Gruen and Jamieson, p 49. Robyn Eckersley points out that the AIDS virus issue is a bit of a red herring: "....ecocentrism merely seeks to cultivate a prima facie

orientation of non-favoritism; it does not mean that humans cannot eat or act to defend themselves or others from danger or life-threatening diseases....A nonanthropocentric perspective is one that ensures that the interests of nonhuman species and ecological communities...are not ignored in human decision making *simply* because they are not human or because they are not of instrumental value to humans....When faced with a choice...those who adopt an ecocentric perspective will seek to choose the course that will minimize...harm and maximize the opportunity of the widest range of organisms and communities—*including ourselves*—to flourish in their/our own way." Eckersley, p 57.

92   Paul Taylor, p 67.

93   J. Baird Callicott. "Can A Theory of Moral Sentiments Support A Genuinely Normative Environmental Ethic?", *Inquiry* (vol. 35, 1992, pp 183–98), p 196.

94   Merchant (1996), p 8.

95   Merchant (1996), p 216.

96   Merchant (in Cronon), p 158.

97   Merchant (1996), p xix.

98   Merchant (2003), p 229.

99   Rowe, (1990), p 26.

100 Smith, *The Theory of Moral Sentiments*, p9.

101 Plumwood 2002), p 142.

102 Iseult Honohan, (London: Routledge, 2002), p 160.

103 Ian Hacking, "Our Fellow Animals", *New York Review of Books*, 29 June 2000 (pp 20–26), p 26.

104 David Hume, *A Treatise of Human Nature* (South Bend: University of Notre Dame Press, 1965), p 179

105 Abrams, p 69; Shaner, "The Japanese Experience of Nature", (pp 163-182) in Callicott and Ames, p 179.

106 Aristotle, *Politics* (New York: Oxford University Press, 1962), p 93.

107 Jean-Jacques Rousseau, *The Social Contract* (London: Penguin, 1968), p 77. Notoriously, he also felt that this social freedom was so superior to natural freedom that recalcitrant individuals should be "forced to be free".

108 Christopher Stone, (Los Altos, CA: William Kaufmann, Inc., 1972), p 5.

109 Stone argues that components of nature that are granted rights would also have to assume responsibilities: if 'rights' are to be granted to the environment, then for many of the same reasons it may bear 'liabilities' as well—as inanimate objects did anciently. Rivers drown people, and flood over and destroy crops; forests burn, setting fire to contiguous communities. Where trust funds had been established, they could be available for the satisfaction of judgments *against* the environment, making it bear the costs of some of the harms it imposes on other right holders." p 34.

110 Collier, p 86.

111 Taylor (1992), p 365.

112 Megarry, p 122 In his review of the history of humankind, Fernandez Armesto concludes that "most of the attributes on which we humans congratulate ourselves seem...to be evolutionary compensations for physical feebleness." p 54.

113 Kohak (1984), p 185.

114 Kohak (1984), p 129.

115 Eckersley, p 59.

116 Stone, p 8.

117 David Bank, "Reformers Press to Give Chimpanzees Legal Standing", www.aegis.com/news/ap/2002/AP020454.html

292 | CONCLUSION

118 Nussbaum (2006), p 67.

119 Christopher Stone, *Earth and Other Ethics* (New York: Harper and Row, 1977),p 45.

120 Bell (2006), Robert Goodin agrees, hoping that nature's interests will be "internalized" by a sufficient number of voting humans to assure its protection. (p 844).

121 Wolfe (2008), p 110.

122 Jeffrey Toobin, "Rich Bitch: The Legal Battle over Trust Funds for Pets", *New Yorker*, 29 September 2008 (pp 38–47), p 40.

123 Kimberly Smith, "Animals and the Social Contract: A Reply to Nussbaum", *Environmental Ethics* v. 30, Summer 2008 (pp 195–207), p 206.

124 Serres (1995).

125 Maria Assad, *Time and Earth: Reading Le Contrat naturel* (Albany: SUNY Press, 1999), p 160.

126 Serres (1995), p 39.

127 Serres (1995), p 3.

128 Serres (1995), p 48.

129 Stephanie Posthumus, "Translating Ecocriticism: Dialoguing with Michel Serres", *Reconstruction: Studies in Contemporary Culture*, vol. 7:2 2007 (pp 1–30), p 7.

130 Serres (1995), p 46.

131 Michel Serres, Trs. Anne-Marie Feenberg—Speech in Vancouver 2006—Institute for the Humanities.

132 Callicott and Ames, p 3.

133 Serres (1995), p 37.

BIBLIOGRAPHY

Abrams, David. *The Spell of the Sensuous*. New York: Vintage, 1996.

Aeschylus. *Prometheus Bound*, Trs T.G. Tucker. Melbourne: Melbourne University Press, 1935.

Alexander, Donald. "Bioregionalism: Science or Sensibility?" *Environmental Ethics* 12:2, 1990.

Aristotle. *Politics*. New York: Oxford University Press, 1962.

——. *The Nicomachean Ethics*. London: Penguin, 2004.

Ashcraft, Richard. "Hobbes's Natural Man: A Study in Ideology Formation." *Journal of Politics* 33, 1971.

Assad, Maria. *Time and Earth: Reading Le Contrat naturel*. Albany: SUNY Press, 1999.

Attfield, Robin. "Christian Attitudes to Nature." *Journal of the History of Ideas* 44:3, 1983.

Austen, Jane. *Sense and Sensibility*. New York: Oxford University Press, 1990.

Bacon, Francis. *The Great Instauration and New Atlantis. Trs J. Weinberger*. Arlington Heights: Crofts Classics, 1980.

Baier, Annette. *Moral Prejudices: Essays on Ethics*. Cambridge: Harvard University Press, 1994.

Bank, David. "Reformers Press to Give Chimpanzees Legal Standing", www.aegis.com/news/ap/2002/AP020454.html.

Barkow, Jerome, Leda Cosmides and John Tooby, Eds. *The Adapted Mind* (New York: Oxford University Press, 1992.

Barnes, Frances. "The Biology of Pre-Neolithic Man." in Boyden 1987.

Barry, John. *Rethinking Green Politics*. London: Sage, 1999.

Batchelor, John. *John Ruskin: No Wealth but Life*. London: Chatto & Windus, 2000.

Bate, Jonathan. "Romantic Ecology Revisited." *The Wordsworth Circle* 24:3, 1993.

——. *The Song of the Earth*. London: Picador, 2000.

Battan, Alan. "Is there an absolute plan — or just a desire for one?" *Toronto Globe and Mail*, June 7, 2003.

Bauman, Zygmunt. *Modernity and the Holocaust*. Ithaca: Cornell University Press, 1989.

——. "Parvenu and pariah: heroes and victims of modernity." in James Good 1998.

——. *Community: Seeking Safety in an Insecure World*. London: Polity Press, 2001.

Baxter, Brian. *Ecologism*. Edinburgh: Edinburgh University Press, 1999.

Bell, Derek. "Liberal Environmental Citizenship." *Environmental Politics* 14:2, 2005.

——. "Political Liberalism and Ecological Justice." *Analyse and Kritik* 28, 2006.

Bennett, Jane. *Unthinking Faith and Enlightenment*. New York: New York University Press, 1987.

Bentham, Jeremy. "Introduction to the Principles of Morals and Legislation." *The Collected Works of Jeremy Bentham*. Eds. Burns, J.H. and H.L.A. Hart. New York: Oxford University Press, 1996.

Berlin, Sir Isaiah. *The Crooked Timber of Humanity*. Princeton: Princeton University Press, 1991.

——."The Magus of the North. *New York Review of Books*. 21 October, 1993.

Berman, Morris. *The Re-enchantment of the World*. New York: Bantam, 1981.

——. *Coming To Our Senses*. New York: Bantam, 1989.

——. *Wandering God*. Albany: SUNY Press, 2000.

Berry, Thomas. *The Dream of the Earth*. San Francisco: Sierra Books, 1988.

Berry, Wendell. "People, Land and Community." *The Graywolf Annual Five*. Ed. Rick Simonson. St. Paul: Graywolf Press, 1988.

——. *The Unsettling of America*. San Francisco: Sierra Club Books, 1977.

294 | BIBLIOGRAPHY

Bird-David, Nurit, "Beyond The Original Affluent Society: A Culturalist Reformulation. *Limited Wants, Unlimited Means.* Ed. John Gowdy. Washington, D.C.: Island Press, 1998.

Blackburn, Simon. "A not-so common good." *Times Literary Supplement,* May 5, 2000.

Boesche, Roger. "Why Did Tocqueville Fear Abundance?" *History of European Ideas,* 9:1, 1988.

Bolgar, R.R. Ed. *Classical Influences on Western Thought.* Cambridge: Cambridge University Press, 1979.

Bookchin, Murray. *Re-enchanting Humanity: A Defense of the Human Spirit Against Antihumanism, Misanthropy, and Primitivism.* London: Cassell, 1995.

Botkin, Daniel. *Discordant Harmonies.* New York: Oxford University Press, 1990.

Bourdeau, Pierre et al. *Environmental Ethics: Man's Relationship with Nature.* Luxembourg: Commission of the European Communities, 1990.

Boyden, Stephen. *Western Civilization in Biological Perspective.* Oxford: Clarendon Press, 1987.

Brennan, Andrew. "Ecological Theory and Value in Nature." *Philosophical Inquiry,* 8:1-2, 1986.

Brooke, John. "Improvable Nature?" *Is Nature Ever Evil?* Ed. Willem Drees. New York: Routledge, 2003.

Brown, Lester and Christopher Flavin. "A New Economy for a New Century." *State of the World.* Ed. Linda Starke. New York: W.W. Norton & Company, 1999.

Buber, Martin. *I and Thou.* New York: Charles Scribner's Sons, 1958.

Buell, Laurence. *The Environmental Imagination: Thoreau, Nature Writing and the Formation of American Culture.* Cambridge: Belknap Press of Harvard University Press, 1995.

Butler, Marilyn. "The First *Frankenstein* and Radical Science." *Times Literary Supplement,* April 9, 1993.

Calarco, Matthew. "Thinking through Animals: Reflections on the Ethical and Political Stakes of the Question of the Animal in Derrida." *Oxford Literary Review* 29, 2007.

Caldwell, Lynton. "Is Humanity Destined to Self-Destruct?" *Politics and the Life Sciences.* March 1999.

Calhoun, John. "Ecological Factors in the Development of Behavioral Anomalies." *Psychopathology: Animal and Human.* Eds. Joseph Zubin and Howard Hunt. New York: Grune and Stratton, 1967.

Callicott, J. Baird and Roger Ames. Eds, *Nature in Asian Traditions of Thought.* Albany: SUNY Press, 1989.

Callicott, J. Baird. "The Metaphysical Implications of Ecology." *Nature in Asian Traditions of Thought.* Eds. J. Baird Callicott and Roger Ames. Albany: SUNY Press, 1989.

——. "Can A Theory of Moral Sentiments Support A Genuinely Normative Environmental Ethic?" *Inquiry* 35, 1992.

Capra, Fritjof. *The Tao of Physics.* Boston: Shambhala, 1991.

Carolan, Trevor. "The Wild Mind of Gary Snyder." *Shambhala Sun.* May 1996.

Carr, Mike. *Bioregionalism and Civil Society.* Vancouver: University of British Columbia Press, 2004.

Carson, Rachel. *Silent Spring.* Boston: Houghton Mifflin,1962.

Cartmill, Matt. *A View to a Death in the Morning: Hunting and Nature through History.* Cambridge: Harvard University Press, 1993.

Cassirer, Ernst. *Rousseau, Kant, Goethe.* Princeton: Princeton University Press, 1945.

Cavalieri, Paola. *The Death of the Animal.* New York: Columbia University Press, 2009.

Christiansen, Rupert. *Romantic Affinities.* London: Sphere, 1988.

Cladis, Mark. "Rousseau and the Redemptive Mountain Village." *Interpretation* 29:1, 2001.

Clark, David. Ed. *Shelley's Prose.* New York: New Amsterdam Books, 1988.

Clark, Stephen. "The Moral Animals.",*Times Literary Supplement,* September 6, 1996.

BIBLIOGRAPHY | 295

———. "Have Biologists Wrapped Up Philosophy?" *Inquiry* 43, 2000.

Clastres, Pierre. *Society Against the State*. New York: Zone Books, 1987.

Coetzee, J.M. *The Lives of Animals*. Princeton: Princeton University Press, 1999.

Cohen, Michael. *The Pathless Way: John Muir and the American Wilderness*. Madison: University of Wisconsin Press, 1984.

Collier, Andrew. *Being and Worth*. London: Routledge, 1999.

Colletti, Lucien. *From Rousseau to Lenin*. New York: Monthly Review Press, 1972.

Commoner, Barry. *The Closing Circle*. New York: Bantam, 1972.

Cook, Michael. *A Brief History of the Human Race*. New York: Norton, 2003.

Corning, Peter. "The Sociobiology of Democracy: Is Authoritarianism in Our Genes." *Politics and the Life Sciences*. March 2000.

Cook, Francis. "The Jewel Net of Indra." *Nature in Asian Traditions of Thought*. Eds. J. Baird Callicott and Roger Ames. Albany: SUNY Press, 1989.

Cottingham, John. "A Brute to the Brute? Descartes' Treatment of Animals." *Philosophy* 63, 1988.

Cranston, Maurice. *Jean-Jacques: The Early Life and Work of Jean-Jacques Rousseau*. London: Allen Lane, 1983.

—.*John Locke: A Biography*. New York: Oxford University Press, 1985.

Crist, Eileen. "Against the Social Construction of Nature and Wilderness." *Environmental Ethics* 26:1, 2004.

Cro, Stelio. *The Noble Savage: Allegory of Freedom*. Waterloo: Wilfrid Laurier University Press, 1990.

Crombie, A.C. *Medieval and Early Modern Science*. Volume II. New York: Doubleday, 1959.

Cronon, William. Ed. *Uncommon Ground: Re-Thinking the Human Place in Nature*. New York: Norton, 1996,

Crosby, Alfred. *Ecological Imperialism*. Cambridge: Cambridge University Press, 1986.

Cunliffe, Barry. *The Oxford Illustrated Prehistory of Europe*. New York: Oxford University Press, 1994.

Daly, Herman and John Cobb. *For the Common Good: Redirecting the Economy toward Community, the Environment, and a Sustainable Future*. Boston: Beacon Press, 1989.

Damasio, Antonio. *The Feeling of What Happens*. New York: Harcourt, Brace & Co., 1999.

Danford, John. *David Hume and the Problem of Reason*. New Haven: Yale University Press, 1990.

Daniel, Carey. "Reconsidering Rousseau: sociability, moral sense and the American Indian from Hutcheson to Bartram." *British Journal of Eighteenth Century Studies* 21, 1998.

Dart, Gregory. *Rousseau, Robespierre and English Romanticism*. Cambridge University Press, 2005.

Darwin, Charles. *The Origin of Species*. London: Penguin Books, 1985.

De Las Casas, Bartolome. *A Short Account of the Destruction of the Indies*. London: Penguin Books, 1992.

Denielou, Guy. "Developing a Company Outlook in Universities." *European Journal of Education* 20:1, 1985.

Descartes, Rene. *Discourse on Method*. London: Penguin, 1968.

DeVries, Bert. "Understanding: Fragments of a Unifying Perspective." *Mappae Mundi: Humans and Their Habitats in a Long-Term Socio-Ecological Perspective*. Eds. Johan Goudsblom and Bert deVries. Amsterdam: Amsterdam University Press, 2002.

DeVries, Bert and R.A. Marchant. "Environment and the Great Transition: Agrarianization." *Mappae Mundi: Humans and Their Habitats in a Long-Term Socio-Ecological Perspective*. Eds. Johan Goudsblom and Bert deVries. Amsterdam: Amsterdam University Press, 2002.

DeWaal, Frans. *Good Natured: The Origins of Right and Wrong in Humans and Other Animals*. Cambridge: Harvard University Press, 1997.

Diamond, Stanley. *In Search of the Primitive*. (New Brunswick: Transaction, 1974/1993.

296 | BIBLIOGRAPHY

Dickens, Charles. *Hard Times*. London: Penguin, 1995.

Dobson, Andrew. *Citizenship and the Environment*. New York: Oxford University Press, 2003.

Dobson, Andrew and Paul Lucardie, eds. *The Politics of Nature*. London: Routledge, 1993.

Dore, Elizabeth. "Debt and Ecological Disaster in Latin America." *Race & Class* 34:1, 1992.

Douthwaite, Richard. *The Growth Illusion*. Bideford, Devon: Resurgence, 1992.

Drees, Willem, ed. *Is Nature Ever Evil?*, New York: Routledge, 2003.

Dudley, Donald R. Ed. "Introduction." *Lucretius*. New York: Basic Books, 1965.

Dudley, Edward and Maximillian Novack. *The Wild Man Within: An Image in Western Thought from the Renaissance to Romanticism*. Pittsburgh: University of Pittsburgh Press, 1972.

Dyson, Freeman. "What A World!" *New York Review of Books*, 15 May 2003.

Eckersley, Robyn. *Environmentalism and Political Theory*. Albany: SUNY Press, 1992.

Eisenberg, Evan. "Back to Eden." *Atlantic Monthly*. November 1989.

——. "The Call of the Wild." *The New Republic*. April 30, 1990.

Elias, Norbert. "The Sciences." *The Norbert Elias Reader: A Biographical Selection*. Eds. Johan Goudsblom and Stephen Mennell. Oxford: Blackwell, 1998.

Ellingson, Ter. *The Myth of the Noble Savage*. Berkeley: University of California Press, 2001.

Ely, John. "Ernst Bloch, Natural Rights, and the Greens." *Minding Nature: The Philosophers of Ecology*. Ed. David Macauley. New York: The Guilford Press, 1996.

Emery, Leon. "Rousseau and the Foundations of Human Regeneration." *Yale French Studies* 28, 1961.

Evernden, Neil. *The Natural Alien*. Toronto: University of Toronto Press, 1993.

Fairchild, Hoxie. *The Noble Savage: A Study in Romantic Naturalism*. New York: Columbia University Press, 1928.

Falk, W.D. *Ought, Reason and Morality*. Ithaca: Cornell University Press, 1986.

Feenberg, Andrew. "The Commoner-Ehrlich Debate: Environmentalism and the Politics of Survival." *Minding Nature: The Philosophers of Ecology*. Ed. David Macauley. New York: The Guilford Press, 1996.

——. "Values and the Environment." unpub mss, March 2004.

Fernandez-Armesto, Ernesto. *Civilizations*. London: Pan, 2001.

Ferry, Luc. *The New Ecological Order*. Chicago: University of Chicago Press, 1995.

Finlayson, Clive. *Neanderthals and Modern Humans*. Cambridge: Cambridge University Press, 2004.

Flannery, Tim. *The Weather Makers*. New York: Harper Collins, 2005.

Fleck, Richard. *Henry Thoreau and John Muir Among the Indians*. Hamden, Conn: Archon Books, 1985.

Foley, Robert. *Humans before Humanity*. Oxford: Blackwells, 1995.

Fox, Michael. "Thinking Ethically About the Environment." *Dalhousie Review* 73, Winter, 1993-94.

Fox, Stephen. *The American Conservation Movement: John Muir and His Legacy*. Madison: University of Wisconsin Press, 1981.

France, Peter. *Politeness and Its Discontents: Problems in French Classical Culture*. Cambridge: Cambridge University Press, 1992.

Freud, Sigmund. *Totem and Taboo*. London: Routledge & Kegan Paul, 1960.

——. *The Future of an Illusion*. New York: Norton, 1961.

——. *Civilization and Its Discontents*. New York: Norton, 1961.

Frost, Robert. "The Road Not Taken", *Mountain Interval*. 1920 (http://www.bartleby.com/119/1.html)

Gailey, Christine. *Civilization in Crisis*. Vol. 1 Gainesville: University of Florida Press, 1992.

Gamble, David. "Loren Eisley: Wilderness and Moral Transcendence." *The Midwest Quarterly* 33:1, 1991.

BIBLIOGRAPHY | 297

Gander, Eric. *On Our Minds*. Baltimore: Johns Hopkins, 2003.

Gay, Peter, *Freud: A Life for Our Time* (New York: Norton, 1988

Geoghegan, Vincent. "A Golden Age: From the Reign of Kronos to the Realm of Freedom." *History of Political Thought*. 12:2, 1991.

Glavin, Terry. *Waiting for the Macaws*. Toronto: Viking Canada—Penguin, 2006.

Goethe, Wolfgang von. *The Sufferings of Young Werther*. New York: Ungar, 1957.

Golding, William. *Lord of the Flies*. New York: Perigree Books, n.d.

Good, James, ed. *The Politics of Postmodernity*. Cambridge: Cambridge University Press, 1998.

Goodin, Robert. "Enfranchising the Earth, and Its Alternatives." *Political Studies* 44, 1996.

Goody, Jack. *The Domestication of the Savage Mind*. Cambridge: Cambridge University Press, 1977.

Gore, Rick. "Neanderthals: The Dawn of Humans." *National Geographic Magazine*, January 1996.

Goudsblom, Johan and Bert deVries , eds. *Mappae Mundi: Humans and Their Habitats in a Long-Term Socio-Ecological Perspective*. Amsterdam University Press, 2002.

Goudsblom, Johan and Stephen Mennell, eds. *The Norbert Elias Reader: A Biographical Selection*. Oxford: Blackwell, 1998.

Gowdy, John, ed. *Limited Wants, Unlimited Means*. Washington, D.C.: Island Press, 1998.

——. "Hunter-Gatherers and the Mythology of the Market." *Cambridge Encyclopedia of Hunters and Gatherers*. Eds. Richard Lee and Richard Daly. Cambridge: Cambridge University Press, 1999.

Grayson, Donald. "The Archaeological Record of Human Impacts on Animal Populations." *Journal of World Prehistory*, 15:1, 2001.

Gruen, Lori and Dale Jamieson, Eds. *Reflecting On Nature*. New York: Oxford University Press, 1994.

Guignon, Charles. *On Being Authentic*. New York: Routledge, 2004.

Gunter, Pete, "The Disembodied Parasite and Other Tragedies; Or: Modern Western Philosophy and How to get out of it." *The Wilderness Condition*. Ed. Max Oelschlaeger. San Francisco: Sierra Books, 1992.

Hacking, Ian. "Our Fellow Animals." *New York Review of Books*, 29 June 2000.

Hadot, Pierre. *The Veil of Isis*. Cambridge: Harvard University Press, 2006.

Harkin, Michael and David Lewis, eds. *Native Americans and the Environment: Perspectives on the Ecological Indian*. Lincoln: University of Nebraska Press, 2007.

Harlan, Jack. *Crops and Man*. Madison: American Society of Agronomy, 1992.

Harrison, Robert. *Forests, the Shadow of Civilization*. Chicago: University of Chicago Press, 1992.

Hayden, Brian and Robert Gargett. "Big Man, Big Heart? A Mesoamerican view of the emergence of complex society." *Ancient Mesoamerica*, 1, 1990.

Hayles, N. *Chaos and Order: Complex Dynamics in Literature and Science*. Chicago: University of Chicago Press, 1991.

Hefner, Philip. "Nature Good and Evil." *Is Nature Ever Evil?* Ed. Willem Drees. New York: Routledge, 2003.

Hesiod. "Works and Days." *Primitivism and Related Ideas in Antiquity*. Eds. Arthur Lovejoy and George Boas. Baltimore: Johns Hopkins, 1935/1997.

Hinchman, Lewis and Sandra. "Should Environmentalists Reject the Enlightenment?" *Review of Politics*, 63:4, 2001.

Holmes, Stephen. *The Young John Muir: An Environmental Biography*. Madison: University of Wisconsin Press,1999.

Horkheimer, Max and Theodor Adorno. *Dialectic of Enlightenment*. New York: Continuum, 1995.

Horowitz, Asher. "Laws and Customs Thrust Us Back into Infancy: Rousseau's Historical Anthropology." *Review of Politics* 52:2, 1990.

Horowitz, Harold. "The Metaphysical Implications of Ecology." and "Biology of a Cosmological

298 | BIBLIOGRAPHY

Science." *Nature in Asian Traditions of Thought.* Eds. J. Baird Callicott and Roger Ames. Albany: SUNY Press, 1989.

Hossenfelder, M. "Epicurus—hedonist malgre lui." *The Norms of Nature: Studies in Hellenistic Ethics.* Eds. M. Schofield and G. Striker. Cambridge: Cambridge University Press, 1986.

Houellebecq, Michel. *The Elementary Particles.* New York: Vintage International, 2000.

Hoy, Terry. *Towards a Naturalistic Political Theory: Aristotle, Hume, Dewey, Evolutionary Biology, and Deep Ecology.* Westport, Conn: Praeger, 2000.

Hume, David. *Treatise of Human Nature,* Book 2, Section 3. (http://www.class.uidaho.edu/mickelsen/texts/Hume%20Treatise/hume%20treatise2.htm#PART%20III

Hutchinson, Thomas, ed. *The Complete Poetical Works of Percy Bysshe Shelley.* London: Oxford University Press, 1948.

Hutchinson, D.S. "Introduction." *The Epicurus Reader.* Indianapolis: Hackett Publishing, 1994.

Ignatieff, Michael. *The Needs of Strangers.* London: Penguin, 1986.

Illathuparampil, Matthew. "Normativity of Nature." *Is Nature Ever Evil?* Ed. Willem Drees. New York: Routledge, 2003.

Ingold, Tim. "On the Social Relations of the Hunter-Gatherer Band." *Cambridge Encyclopedia of Hunters and Gatherers.* Eds. Richard Lee and Richard Daly. Cambridge: Cambridge University Press, 1999.

Inwood, Brad Trs. *The Epicurus Reader.* Indianapolis: Hackett Publishing, 1994.

Johnson,W.R. *Lucretius and the Modern World.* London: Duckworth, 2000.

Jolly, Allison. *Lucy's Legacy: Sex and Intelligence in Human Evolution.* Cambridge: Harvard University Press, 1999.

Jung, Hwa Yol. "The Ecological Crisis: A Philosophic Perspective, East and West." *Bucknell Review,* 20:3, 1972.

Kalter, Susan. "The Path to *Endless:* Gary Snyder in the Mid-1990's." *Texas Studies in Literature and Language* 41:1, 1999.

Kant, Immanuel. "Conjectures on the Beginning of Human History". *Kant: Political Writings,* Ed. Hans Reiss. Cambridge: Cambridge University Press, 1991.

——. "Conjectural Beginning of Human History." *On History.* Ed. Lewis Beck. Indianapolis: Bobbs-Merrill, 1963.

Katz, Eric. *Nature as Subject: Human Obligation and Natural Community.* New York: Rowman and Littlefield, 1997.

Kenyon-Jones, Christine. *Kindred Brutes.* Aldershot: Ashgate Press, 2001.

Kingdon, Jonathan. *Self-Made Man: Human Evolution from Eden to Extinction?* New York: John Wiley, 1993.

Kohak, Erazim. *The Embers and the Stars.* Chicago: University of Chicago Press, 1984.

Kolbert, Elizabeth "Why Work: A Hundred Years of The Protestant Ethic." *The New Yorker,* November 29, 2004.

——. "The Climate of Man." *The New Yorker,* April 25, 2005; May 2 2005 (pp. May 9, 2005; and, "Chilling.", *The New Yorker,* March 20, 2006.

Kretch, Shepard. *The Ecological Indian: Myth and History.* New York: Norton, 1999.

Lafleur, William. "Buddhist Value of Nature." *Nature in Asian Traditions of Thought.*Eds. J. Baird Callicott and Roger Ames. Albany: SUNY Press, 1989.

Lamontagne, Maurice. "The Loss of the Steady State." *Beyond Industrial Growth.* Ed. Abraham Rotstein. Toronto: University of Toronto Press, 1976.

Larmore, Charles. "Romanticism and Modernity." *Inquiry* 34, 1991.

Lasch, Christopher. *The Minimal Self: Psychic Survival in Troubled Times.* New York: Norton, 1984.

BIBLIOGRAPHY | 299

Latour, Bruno. *Conversations on Science, Culture and Time*. Ann Arbor: University. of Michigan Press, 1995.

Lauck, Joanne. *The Voice of the Infinite in the Small: Re-Visioning the Insect-Human Connection*. Boston: Shambala Press, 2002.

Lawson, John Howard. *The Hidden Heritage*. New York: Citadel Press, 1950.

Leacock, Eleanor. "Women's Status in Egalitarian Society: Implications for Social Evolution." *Limited Wants, Unlimited Means*. Ed. John Gowdy. Washington, D.C.: Island Press, 1998.

Leader, Zachary and Michael O'Neill, eds. *Percy Bysshe Shelley: The Major Works*. Oxford: Oxford University Press, 2003.

Lee, Richard. "Demystifying Primitive Communism." *Civilization in Crisis*, Vol. 1. Ed. Christine Gailey. Gainesville: University of Florida Press, 1992.

——. "Demystifying Primitive Communism." *Civilization in Crisis*, Vol. 1. Ed. Christine Gailey. Gainesville: University of Florida Press, 1992.

——. "What Hunters Do for a Living, or, How to Make Out on Scarce Resources." *Limited Wants, Unlimited Means*. Ed. John Gowdy. Washington, D.C.: Island Press, 1998.

——. "Hunter-Gatherer Studies and the Millennium: A Look forward (and Back)." Keynote Address, 8th Conference on Hunting and Gathering Societies, National Museum of Ethnology, Osaka, Japan 1998.

Lee, Richard and Richard Daly, eds. *Cambridge Encyclopedia of Hunters and Gatherers* Cambridge: Cambridge University Press, 1999.

Lee, Richard and I. Devore, Eds. *Man the Hunter*. Chicago: Aldine, 1968.

Lee, Keekok. "To De-Industrialize—Is it so irrational?" *The Politics of Nature*. Eds. Andrew Dobson and Paul Lucardie. London: Routledge, 1993.

Leigh, R.A., "Jean-Jacques Rousseau and the Myth of Antiquity in the 18th Century." *Classical Influences on Western Thought*. Ed. R.R. Bolgar. Cambridge: Cambridge University Press, 1979.

Leiss, William. *Under Technology's Thumb*. (Montreal: McGill-Queen's University Press, 1990

——. *Domination of Nature*. Montreal: McGill-Queens University Press, 1994.

Leopold, Aldo. *A Sand County Almanac*. New York: Oxford University Press, 1949.

Levi Strauss, Claude. "Rousseau: Father of Anthropology." UNESCO *Courier*, March 1963.

——. *Tristes Tropique*. New York: Atheneum, 1978.

Levin, Harry. *The Myth of the Golden Age in the Renaissance*. Bloomington: Indiana University Press, 1969.

Lewis-Williams, D. and D. Pearce. *Inside the Neolithic Mind*. London: Thames and Hudson, 2005.

Light, Andrew. "Democratic Technology, Population and Environmental Change." *Philosophy & Technology: New Debates in the Democratization of Technology*. Ed. T. Veak. Albany: SUNY Press, 2004.

Livingston, John. *The John A. Livingston Reader*. Toronto: McClelland and Stewart, 2007.

Locke, John. *Second Treatise of Government. Reflecting On Nature*. Eds. Lori Gruen and Dale Jamieson. New York: Oxford University Press, 1994.

Lockridge, Laurence. *The Ethics of Romanticism*. Cambridge: Cambridge University Press, 1989.

Lovejoy, Arthur and George Boas. *Primitivism and Related Ideas in Antiquity*. Baltimore: Johns Hopkins, 1935/1997.

Lovelock, James. "Gaia and the Balance of Nature." *Environmental Ethics: Man's Relationship with Nature*. Ed. Pierre Bourdeau. Luxembourg: Commission of the European Communities, 1990.

——. "Planetary Medicine: Stewards or partners on Earth?" *TLS* Sept. 13, 1991

——. *The Revenge of Gaia: Why the Earth is Fighting Back—And How We Can Still Save Humanity*. London: Penguin, 2005.

300 | BIBLIOGRAPHY

——. "Gaia must go nuclear. *Toronto Globe and Mail*, March10, 2005.

Lowey, Erich. *Suffering and the Beneficent Community*. Albany: SUNY Press, 1991.

Lowy, Michael and Robert Sayre. *Romanticism Against the Tide of Modernity*. Durham: Duke University Press, 2001.

Lucretius. *The Nature of Things*. New York: Norton, 1977.

——. *On the Nature of the Universe*. London: Penguin, 1994.

Luke, Tim. "Re-Reading the Unabomber Manifesto." *Telos* 107, Spring 1996.

Macauley, David, Ed., *Minding Nature: The Philosophers of Ecology*. New York: The Guilford Press, 1996.

Machiavelli, Niccolo. *The Prince*. London: Penguin, 1999.

MacIntyre, Alastair. *Dependent Rational Animals*. Chicago: Open Court, 1999.

Macnaghten, Phil and John Urry. *Contested Natures*. London: Sage 1998.

Manuel, Frank & Fritzie Manuel. *Utopian Thought in the Western World*. Cambridge: Harvard University Press, 1979.

Marcuse, Herbert. "Ecology and the Critique of Modern Society." *Capitalism, Nature, Socialism* 3:3, 1992.

Marshall, Peter. *William Godwin*. New Haven: Yale University Press,1984.

Marx, Karl and Friedrich Engels. *On Historical Materialism*. Moscow: Progress Publishers, 1972.

Masson, John. *Lucretius: Epicurean and Poet*. London: John Murray, 1909.

McIlroy, Anne. "World's Oldest Necklace Discovered.", *Toronto Globe and Mail*, April 17, 2004.

McKusick, James. "A Language that is ever green: The Ecological Vision of John Clare."*University of Toronto Quarterly*, 61:2, 1991.

McLaughlin, Andrew. *Regarding Nature*. Albany: SUNY Press, 1993.

McNeil, William. "A Short History of Humanity." *New York Review of Books*, June 29, 2000.

Meek, Ronald. *Social Science and the Ignoble Savage*. Cambridge: Cambridge University Press, 1976.

Meeker, Joseph. "The Assisi Connection."*Wilderness*, Spring 1988.

Megarry, Tim. *Society in Prehistory: The Origins of Human Culture*. London: Macmillan, 1995.

Mehring, Franz. *Karl Marx*. Ann Arbor: University of Michigan Press, 1959.

Meier, Heinrich. "The Discourse on the Origins and Foundations of Inequality Among Men." *Interpretation*, 16:2, 1988.

Mellars, Paul. "The Upper Paleolithic Revolution." *The Oxford Illustrated Prehistory of Europe*. Ed. Barry Cunliffe. New York: Oxford University Press, 1994.

Mendes-Flohr, Paul. *Martin Buber: A Contemporary Perspective*. Syracuse: Syracuse University Press, 2002.

Merchant, Carolyn. *The Death of Nature*. New York: Harper and Row, 1983.

——. *Radical Ecology: The Search for A Livable World*. New York: Routledge, 1992.

——. *Earthcare: Women and the Environment*. New York: Routledge, 1996.

——. "Reinventing Eden: Western Culture as a Recovery Narrative." *Uncommon Ground: Re-Thinking the Human Place in Nature*. Ed. William Cronon. New York: Norton, 1996.

——. *Reinventing Eden*. New York: Routledge, 2003.

Midgely, Mary. "Criticizing the Cosmos." *Is Nature Ever Evil?* Ed., Willem Drees. New York: Routledge, 2003.

Miller, Sally. *John Muir: Life and Work*. Albuquerque: University of New Mexico Press, 1993.

Mitchell, Alanna. "European Fishing linked to Extinction of African Animals." *Toronto Globe & Mail*, November 12, 2004.

Mitford, Mary. *Our Village: Sketches of Rural Character and Scenery*. London: G. Bell & Sons, 1877.

Mizzoni, John. St. Francis, "Paul Taylor and Franciscan Biocentrism.", *Environmental Ethics* 26:1,

2004.

Moran, Francis. "Between Primates and Primitives: Natural Man as the Missing Link in Rousseau's Second Discourse."*Journal of the History of Ideas* 54:1, 1993.

——. *My First Summer in the Sierra*. London: Penguin, 1978.

——. *The Story of My Boyhood and Youth*. Edinburgh: Canongate, 1987.

——. *The Yosemite*. San Francisco: Sierra Club Books, 1988.

Naess, Arne. "The Deep Ecological Movement: Some Philosophical Aspects." *Environmental Philosophy: From Animal Rights to Radical Ecology*. Ed. Michael Zimmerman. Englewood Cliffs: Prentice-Hall, 1993.

Nash, Roderick. *The Rights of Nature*. Madison: University of Wisconsin Press, 1989.

——. *Wilderness and the American Mind* New Haven: Yale University Press, 2001.

——. "Island Civilization: A Vision for Human Occupancy of Earth in the Fourth Millennium." *Confluence*, 14:2, 2009.

Neiman, Susan. *Evil in Modern Thought*. Princeton: Princeton University Press, 2002.

Nesse, Randolph. "Natural Selection and the Capacity for Subjective Commitment." Ed., R. Nesse. *Evolution and the Capacity for Commitment*. New York: Sage, 2001.

Neumann, Harry. "Philosophy and Freedom: An Interpretation of Rousseau's State of Nature." *The Journal of General Education*. 27:4, 1976.

Nicholson, Linda. *Gender and History: The Limits of Social Theory in the Age of the Family*. New York: Columbia University Press, 1986.

Nicholson, Peter. "Aristotle: Ideals and Realities." *Political Thought from Plato to NATO*. Ed. Brian Redhead. Chicago: Dorsey Press.

Nussbaum, Martha. *The Therapy of Desire*. Princeton: Princeton University Press, 1994.

——. *Upheavals of Thought: The Intelligence of Emotions*. Cambridge: Cambridge University Press, 2001.

——. *Frontiers of Justice*. Cambridge: Harvard University Press, 2006.

Oelschlager, Max. *The Idea of Wilderness*. New Haven: Yale University Press, 1991.

——. *The Wilderness Condition*. San Francisco: Sierra Books, 1992.

Oerlemans, Onno. *Romanticism and the Materiality of Nature*. Toronto: University of Toronto Press, 2002.

Pagden, Anthony. *The Fall of Natural Man: The American Indian and the Origins of Comparative Ethnology*. Cambridge: Cambridge University Press, 1982.

Passmore, John. "The Treatment of Animals." *Journal of the History of Ideas* 36:2, 1975.

Pinker, Stephen. *How the Mind Works*. New York: Norton, 1997.

Plant, Christopher and Judith Plant, Eds. *Turtle Talk: Voices for a Sustainable Future*. Santa Cruz: New Catalyst, 1990.

Plato. *Phaedrus and Letters VII and VIII*. London: Penguin, 1973.

——. *Timaeus and Critias*. London: Penguin, 1977.

Plumwood, Val. *Feminism and the Mastery of Nature*. London: Routledge, 1993.

——. "Nature as Agency and the Prospects for a Progressive Naturalism." *CNS*, 12:4, 2001.

——. *Environmental Culture*. New York: Routledge, 2002.

Pollan, Michael. "Only Man's Presence Can Save Nature." *Harper's Magazine*, April 1990.

Posthumus, Stephanie. "Translating Ecocriticism: Dialoguing with Michel Serres." *Reconstruction: Studies in Contemporary Culture* 7:2 2007.

Quilley, Stephen and Steven Loyal, Eds. *The Sociology of Norbert Elias*. Cambridge: Cambridge University Press, 2004.

Quilley, Stephen. "Ecology, 'human nature' and civilizing processes: biology and sociology in the work of Norbert Elias." *The Sociology of Norbert Elias*. Eds. Stephen Quilley and Steven Loyal. Cambridge: Cambridge University Press, 2004.

Quilley, Stephen and Steven Loyal. "Towards a central theory: the scope and relevance of the sociology of Norbert Elias." *The Sociology of Norbert Elias*. Eds. Stephen Quilley and Steven Loyal. Cambridge: Cambridge University Press, 2004.

Quinton, Anthony. "The Right Stuff." *The New York Review of Books* 32, December 5, 1985.

Raskin, Paul, ed. *The Great Transformation*. Stockholm: The Global Scenario Report, 2002.

Readings, Bill. *The University in Ruins*. Cambridge: Harvard University Press, 1996.

Redhead, Brian. *Political Thought from Plato to NATO*. (Chicago: Dorsey Press, 1984.

Reed, Peter. "Man Apart: An Alternative to the Self-Realization Approach." *Environmental Ethics* 11:1, 1989.

Regan Tom. *The Case for Animal Rights*. Berkeley: University of California Press, 1983.

Rice, Daryl. "Teaching Rousseau: Natural Man and Present Existence." *Teaching Political Science* 16:4, 1989.

Ridley, Mark. "Do we love nature?" Review of Stephen Kellert and Edward Wilson, eds. *The Biophilia Hypothesis*, TLS, September 9, 1994.

Rist, J.M. *Epicurus: An Introduction*. Cambridge: Cambridge University Press, 1972.

Roach, Catherine. "The Unconscious, Aggression and Mother Nature." *Journal of Feminist Studies in Religion* 13:1, 1997.

Roberts, Hugh. *Shelley and the Chaos of History*. University Park, PA: Pennsylvania State University Press, 1997.

Roe, Nicholas. *The Politics of Nature: William Wordsworth and Some Contemporaries*. London: Palgrave, 2002.

Rogers, Raymond. *Nature and the Crisis of Modernity*. Montreal: Black Rose Books, 1994.

Rolston, Holmes. "Environmental Ethics: Values in and Duties to the Natural World." *Reflecting on Nature*. Eds. Lori Gruen and Dale Jamieson. New York: Oxford University Press, 1994.

———. "Naturalizing and Systemizing Evil." *Is Nature Ever Evil?* Ed. Willem Drees. New York: Routledge, 2003.

Rorty, Richard. *Contingency, Irony, and Solidarity*. Cambridge: Cambridge University Press, 1989.

Rotstein, Abraham. *Beyond Industrial Growth*. Toronto: University of Toronto Press, 1976.

Rousseau, Jean-Jacques. *Confessions*. London: Penguin, 1953.

———. *Emile*. Chicago: University of Chicago Press, 1979.

———. *Reveries of the Solitary Walker*. London: Penguin, 1979.

———. *Discourse on the Origins of Inequality*. London: Penguin, 1984.

———. *Rousseau, Judge of Jean-Jacques: Dialogues*. Ed. Roger Masters and Christopher Kelly. Dartmouth: University Press of New England, 1990.

———. *Julie, or the New Heloise*. Trs by Philip Stewart and Jean Vache. Hannover: University Press of New England, 1997.

———. "Letter to Voltaire on Optimism." *Candide and Related Texts*. Ed. David Wooten. Indianapolis: Hackett, 2000.

Rowe, Stan. *Home Place*. Edmonton: NeWest, 1990.

Rowe, Stan. "The Changing World: An Ecological Perspective." un pub mss, "Nature and Human Community" series, Simon Fraser University, December 1992.

Russell, Colin. *The Earth, Humanity and God*. London: UCL Press, 1994.

Ruston, Sharon. *Shelley and Vitality*. Basingstoke: Palgrave, 2005.

Ryan, Alan, "Reasons of the Heart", *New York Review of Books*, 23 September 1993

Ryder, Richard. *Speciesism: The Ethics of Vivisection*. Edinburgh: Scottish Society for the Prevention of Vivisection, 1974).

Sahlins, Marshall. *Stone Age Economics*. New York: Aldine de Gruyter, 1972.

——. "The Original Affluent Society." *Limited Wants, Unlimited Means*. Ed. John Gowdy. Washington, D.C.: Island Press, 1998.

Sale, Kirkpatrick, "Preface." E.F. Schumacher. *Small Is Beautiful*. New York: Harper, 1989/1973

——. "Foreward." *Turtle Talk: Voices for a Sustainable Future*. Eds. Christopher Plant and Judith Plant. Santa Cruz: New Catalyst, 1990.

——. *Dwellers In the Land*. Philadelphia: New Society Publishers, 1991.

Scarry, Elaine. *On Beauty and Being Just*. Review by Stuart Hampshire. *New York Review of Books*, 18 November 1999.

Scheers, Peter. "Human Interpretation and Animal Excellence." *Is Nature Ever Evil?* Ed. Willem Drees. New York: Routledge, 2003.

Schelling, Thomas. "Commitment: Deliberate Versus Involuntary." *Evolution and the Capacity for Commitment*. Ed. R. Nesse. New York: Sage, 2001.

Schermer, Michael. *The Science of Good and Evil: Why People Cheat, Gossip, Care, Share and Follow the Golden Rule*. New York: Henry Holt, 2004.

Schiller, Friedrich, "Letters on the Aesthetic Education of Man." *Essays*. New York: Continuum, 1993.

Schofield, M. and G. Striker. *The Norms of Nature: Studies in Hellenistic Ethics*. Cambridge: Cambridge University Press, 1986.

Schumacher, E.F. *Small Is Beautiful*. New York: Harper, 1989/1973.

Scott, John. "The Theodicy of the Second Discourse: The Pure State of Nature and Rousseau's Political Thought." *American Political Science Review* 86:3, 1992.

Segal, Charles. *Lucretius on Death and Anxiety*. Princeton: Princeton University Press, 1990.

Seneca. *Letters from a Stoic*. London: Penguin, 2004.

Serres, Michel. *The Natural Contract*. Ann Arbor: The University of Michigan Press, 1995

——. *The Birth of Physics*. Manchester: Clinamen Press Ltd, 2000.

Shaner, David. "The Japanese Experience of Nature." *Nature in Asian Traditions of Thought*. Eds. J. Baird Callicott and Roger Ames. Albany: SUNY Press, 1989.

Sharpe, Lesley. *Friedrich Schiller*. Cambridge: Cambridge University Press 1991.

Shelley, Mary. *Frankenstein or, The Modern Prometheus*. London: Penguin, 1994.

Shelley, Percy Bysshe. *Letters of Percy Bysshe Shelley*, vol. 2. London: Pitman and Sons, 1909.

——. "Essay On Life." *Shelley's Prose*. Ed. David Clark. New York: New Amsterdam, 1988.

——. "A Defence of Poetry." *Shelley's Prose*. Ed. David Clark. New York: New Amsterdam, 1988.

——. "Vindication of Natural Diet." *Shelley's Prose*. Ed. David Clark. New York: New Amsterdam, 1988.

——. "Julian and Maddalo." *The Complete Poetical Works of Percy Bysshe Shelley*. Ed. Thomas. Hutchinson. London: Oxford University Press, 1948.

Shepard, Paul, *Man in the Landscape*. College Station: Texas A&M Univ. Press, 1991/1967.

——. *Nature and Madness*. Athens: University of Georgia Press, 1982.

——. "A Post-Historic Primitivism." *The Wilderness Condition*. Ed. Max Oelschlaeger. San Francisco: Sierra Books, 1992 .

——. *Coming Home to the Pleistocene*. Washington, D.C.: Island Press, 1998.

Sheriff, John. *The Good-Natured Man: The Evolution of a Moral Ideal, 1660-1800*. Montgomery: University of Alabama Press, 1982.

Sherratt, Andrew "The Transformation of Early Agrarian Europe: The Later Neolithic and Copper Ages 4500-2500 BC." *The Oxford Illustrated Prehistory of Europe*. Ed. Barry Cunliffe. New York:

Oxford University Press, 1994.

Shi, David. *The Simple Life* . New York: Oxford University Press, 1985.

Shklar, Judith. "Rousseau's Two Models: Sparta and the Age of Gold." *Political Science Quarterly*, 81, 1966.

Shaner, David. "The Japanese Experience of Nature." *Nature in Asian Traditions of Thought*. Eds. J. Baird Callicott and Roger Ames. Albany: SUNY Press, 1989.

Silberstein, Laurence. *Martin Buber's Social and Religious Thought*. New York: New York University Press, 1989.

Simmons, Alan. *The Neolithic Revolution in the Near East*. Tucson: The University of Arizona Press, 2007.

Singer, Peter. "Practical Ethics." *Reflecting on Nature*. Eds. Lori Gruen and Dale Jamieson. New York: Oxford University Press, 1994 (2nd ed. Cambridge University Press, 1993 in Gruen and Jamieson 1994

——. "A Response to Martha Nussbaum.", http://www.petersingerlinks.com/nussbaum.htm,

Skakoon, Elizabeth. "Nature and Human Identity." *Environmental Ethics* 30, Spring 2008.

Slater, Candace. "Amazonia as Edenic Narrative." *Uncommon Ground: Re-Thinking the Human Place in Nature*. Ed. William Cronon. New York: Norton, 1996.

Smith, Adam. *The Theory of Moral Sentiments*. New York: Oxford University Press, 1976.

——. *The Wealth of Nations*. London: Penguin, 1986.

Smith, Andrew. "Archaeology and evolution of hunters and gatherers." *Cambridge Encyclopedia of Hunters and Gatherers*. Eds. Richard Lee and Richard Daly. Cambridge: Cambridge University Press, 1999.

Smith, Kimberly. *Wendell Berry and the Agrarian Tradition*. Lawrence: University Press of Kansas, 2003.

——. "Natural Subjects: Nature and Political Community." *Environmental Values* 15, 2007.

——. "Animals and the Social Contract: A Reply to Nussbaum." *Environmental Ethics* 30, Summer 2008.

Smith, M. *Ecologism: Towards Ecological Citizenship*. Minneapolis: University of Minnesota Press, 1998.

Snyder, Gary. "Baby Jackrabbit." *Danger on Peaks*. Washington, D.C. : Shoemaker Hoard, 2004.

Soper, Kate. "A Difference of Needs." *New Left Review* 152, 1985.

——. *What Is Nature?* Oxford: Blackwell, 1995.

——. "Looking At Landscape." *CNS* 12:2, 2001.

Sophocles. *Antigone*. Indianapolis: Hackett Publishing, 2001.

Sorrell, Roger. *St.Francis of Assisi and Nature*. New York: Oxford University Press, 1988.

Specter, Michael."The Extremist." *The New Yorker*, April 14, 2003.

Speth, James. *The Bridge at the Edge of the World*. New Haven: Yale University Press, 2008.

Stanley, Millie. *The Heart of John Muir's World*. Madison, Wisconsin: Prairie Oak Press, 1995.

Starobinski, Jean. "The Antidote in the Poison." in *Blessings in Disguise*. Cambridge: Harvard University Press, 1993.

Stegner, Wallace. *Wolf Willow*. New York: Viking, 1963.

Stock, Gregory. *Redesigning Humans*. Boston: Houghton Mifflin, 2003.

Stone, Christopher. *Should Trees have Standing?* Los Altos, California: William Kaufmann, 1972.

——. *Earth and Other Ethics*. New York: Harper and Row, 1977.

Stuart, Tristram. *The Bloodless Revolution: A Cultural History of Vegetarianism from 1600 to Modern Times*. New York: Norton, 2006.

Studer, Heidi. "Strange Fire at the Altar of the Lord: Francis Bacon on Human Nature." *The Review* of *Politics* 65:2, 2003.

Sulloway, Frank. "Darwinian Virtues." *New York Review of Books*, April 9, 1998.

Sutcliffe, F.E. "Introduction." Rene Descartes, *Discourse on Method and the Meditations*. London: Penguin, 1968.

Symcox, Geoffrey. "The Wild Man's Return: The Enclosed Vision of Rousseau's Discourses." *The Wild Man Within: An Image in Western Thought from the Renaissance to Romanticism*. Eds. Edward Dudley and Maximillian Novack. Pittsburgh: University of Pittsburgh Press, 1972.

Tainter, J.A. *The Collapse of Complex Societies*. Cambridge: Cambridge University Press, 1988.

Taplin, Kim. *Tongues in Trees: Studies in Literature and Ecology* . Bideford, Devon: Green Books, 1989.

Taylor, Charles. *The Malaise of Modernity*. Concord, Ontario: Anansi Press, 1991.

———. *Sources of the Self*. Cambridge: Harvard University Press, 1992.

Taylor, Paul,.*Respect for Nature* . Princeton University Press, 1986.

Terborgh, John. "The Age of Giants." review of Tim Flannery. *The Eternal Frontier: An Ecological History of North America and Its Peoples. New York Review of Books*, Sept 20 2001.

Thacker, Christopher. *The Wildness Pleases*. New York: St. Martin's, 1983.

Thiele, Leslie. "Learning the Lesson of Interdependence." *Politics and the Life Sciences*. Sept 1999.

Thiele, Leslie *Environmentalism for a New Millennium: The Challenge of Coevolution*. New York: Oxford University Press, 1999.

Thomas, Keith. *Man and the Natural World*. London: Penguin, 1984.

Thompson, E.P. *The Making of the English Working Class*. New York: Vintage, 1963.

Tobias, Philip. "Twenty Questions about Human Evolution." *Human Evolution* 18:1-2, 2003.

Todd, Janet. *Mary Wollstonecraft: A Revolutionary Life*. London: Phoenix Press, 2000.

Todorov, Tzvetan. *Imperfect Garden: The Legacy of Humanism*. Princeton: Princeton University Press, 2002.

Toobin, Jeffrey. "Rich Bitch: The Legal Battle over Trust Funds for Pets." *The New Yorker*, September 29, 2008.

Torgovnick, Marianna. *Gone Primitive: Savage Intellects, Modern Lives*. Chicago: University of Chicago Press, 1990.

Tuan, Yi-fu. *Topophilia: A Study of Environmental Perception, Attitudes and Values*. Englewood Cliffs, NJ: Prentice-Hall, 1974.

Tucker, Mary and John Berthrong, eds. *Confucianism and Ecology: The Interrelation of Heaven, Earth, and Humans*. Cambridge: Harvard University Press, 1998.

Turner, Frederick. *Rediscovering America: John Muir and His Times*. San Francisco: Sierra Club, 1985.

Turney, Jon. "Of Mites and Men.", review of E.O. Wilson, *The Future of Life. New York Times Book Review*, Feb. 17, 2002.

Vanderheiden, Steve. "Two Conceptions of Sustainability." *Political Studies* 56, 2008.

Vanelli, Ron. *Evolutionary Theory and Human Nature*. Boston: Kluwer Academic Publishers, 2001.

Veak, T. Ed. *Philosophy & Technology: New Debates in the Democratization of Technology*. Albany: SUNY Press, 2004.

Vickers, Brian. "Francis Bacon, Feminist Historiography and the Dominion of Nature." *Journal of the History of Ideas* 69:1, 2008.

Von Rad, Gerhard. *Genesis: A Commentary*. Westminster; John Knox Press, 1972.

Wackernagel, Mathis and William Rees. *Our Ecological Footprint*. Philadelphia: New Society Publishers, 1996.

Washburn, S.L. and C.S. Lancaster. "The Evolution of Hunting." *Man the Hunter* Eds. R.B. Lee and I. DeVore. Chicago: Aldine, 1968.

Weatherford, Jack. *Savages and Civilization: Who Will Survive?* New York: Crown Publishers, 1994.

306 | BIBLIOGRAPHY

Webb, David ."Introduction" to Michel Serres, *The Birth of Physics*. Manchester: Clinamen Press, 2000.

Weiming, Tu. "Beyond the Enlightenment Mentality." *Confucianism and Ecology: The Interrelation of Heaven, Earth, and Humans* Eds. Mary Tucker and John Berthrong. Cambridge: Harvard University Press, 1998.

Weinberg, Alan. *Shelley's Italian Experience*. New York: Macmillan, 1991.

Weisman, Alan. *The World Without Us*. New York: Harper Collins, 2007.

Wessels, Tom. *The Myth of Progress: Toward a Sustainable Future*. Burlington: University of Vermont Press, 2006.

West, David. *The Imagery and Poetry of Lucretius*. Bristol: Classical Press, 1969.

Weston, Peter. "The Noble Primitive as Bourgeois Subject." *Literature and History* 10:1, 1984.

White, Eric Charles. "Negentropy, Noise and Emancipatory Thought." *Chaos and Order: Complex Dynamics in Literature and Science*. Ed. N. Hayles. Chicago: University of Chicago Press, 1991.

White, Gilbert. *The Natural History of Selborne* . London: Penguin, 1977.

White, Richard. "Work and Nature: Are you an environmentalist or do you work for a living?" *Uncommon Ground: Re-Thinking the Human Place in Nature*. Ed. William Cronon. New York: Norton, 1996.

White, Lynn. "The Historical Roots of Our Ecological Crisis." *Science* 55, 1967.

Whitebook, Joel. "The Problem of Nature in Habermas." *Minding Nature: The Philosophers of Ecology*. Ed. David Macauley. New York: The Guilford Press, 1996.

Whiteside, Kerry *Divided Natures: French Contributions to Political Ecology*. Cambridge: MIT Press, 2002.

Whittle, Alasdair. "The First Farmers." *The Oxford Illustrated Prehistory of Europe*. Ed. Barry Cunliffe. New York: Oxford University Press, 1994.

Wilkins, Thurman. *John Muir: Apostle of Nature*. Norman: University of Oklahoma Press 1995.

Williams, Bernard. "Must a Concern for the Environment be Centered on Human Beings?" *Reflecting on Nature*. Eds. Lori Gruen and Dale Jamieson. New York: Oxford University Press, 1994 (2nd ed. Cambridge University Press, 1993 in Gruen and Jamieson 1994.

Williams, Dennis. *God's Wilds: John Muir's Vision of Nature*. College Station, Texas: Texas A&M Press, 2002.

Wilson, Bee. "The Last Bite." *The New Yorker*, May 19, 2008.

Wilson, Edward. *The Future of Life*. New York: Vintage, 2002.

Wilson, James. *The Moral Sense*. New York: Free Press, 1993.

Wilson, Peter. *The Domestication of the Human Species*. New Haven: Yale University Press, 1988.

Wise, Stephen. *Rattling the Cage: Toward Legal Rights for Animals*. Cambridge, Mass.: Perseus Publishing, 2000.

Wispe, Lauren. *The Psychology of Sympathy*. New York: Plenum Press, 1991.

Wokler, Ribert. "Perfectible Apes in Decadent Cultures: Rousseau's Anthropology Revisited." *Daedalus* 107, 1978.

Wolfe, Carey. *Animal Rites*. Chicago: University of Chicago Press, 2003.

——. "Learning from Temple Grandin, Or, Animal Studies, Disability Studies and Who Comes after the Subject." *New Formations* 64, 2008.

Wollstonecraft, Mary. *Travels in Sweden, Norway and Denmark*. London: Penguin, 1987.

Wood, Allen. "Kant on Duties Regarding Nonrational Nature." *Proceedings of the Aristotelean Society* 72, 1998.

Woodburn, James. "Egalitarian Societies." *Limited Wants, Unlimited Means*. Ed. John Gowdy. Washington, D.C.: Island Press, 1998.

BIBLIOGRAPHY | 307

Wooten, David. *Candide and Related Texts.* Indianapolis: Hackett, 2000.

Wordsworth, William and Dorothy Wordworth. *Home At Grasmere.* Ed. Colette Clark. London: Penguin Books, 1960.

Wordsworth, William. "Lines Written in Early Spring." *William Wordsworth.* New York: Oxford University Press, 1984.

———. *The Prelude*, Book III. London: Penguin, 1995.

Worster, Donald. *Nature's Economy.* Cambridge: Cambridge University Press, 1977.

Wright, Robert. *The Moral Animal.* New York: Vintage, 1994.

Zee, A. "On Fat Deposits around the Mammary Glands in the Females of Homo Sapiens." *New Literary History* 32, 2001.

Zerzan, John. "Why Primitivism?" *Telos* 124, Summer 2002.

Zimmerman, Michael. "Martin Heidegger: Antinaturalistic Critic of Technological Modernity." *Minding Nature: The Philosophers of Ecology.* Ed. David Macauley. New York: The Guilford Press, 1996.

Zimmerman, Michael et al. eds. *Environmental Philosophy: From Animal Rights to Radical Ecology.* Englewood Cliffs: Prentice-Hall, 1993.

INDEX

Abbey, Edward *224f*

Abrams, David *185, 213, 270, 273*

Adorno, Theodor *118, 231, 251f, 266*

Aeschylus *83*
  Oresteia *105f*
  Prometheus Bound *153*

alternative technology *218*

altruism *35, 135, 194-95, 200, 276-77*

animal rights *19, 35, 100, 123, 126, 196, 200, 269, 275*

anthropocentrism *7, 12, 22f, 17, 99, 194, 198, 207, 236, 256, 265, 270, 274-76*
  and modernity *69, 81, 131-32, 160*
  and Plato/Aristotle *82-3, 88*
  as teleology *120, 131*
  challenge to *196, 205*
  Muir and *188*

Antigone *83, 85-7, 164*

Appolonian *83*

Aquinas *118, 120, 129*

Arendt, Hannah *240*

Aristotle *82, 85, 88, 90, 92-4, 120, 139f, 164, 279*
  and natural law *121*
  and sociality *177f*

aryan civilization *79*

Athens *79-81*

atomism *93-4, 96-8, 184*

Attfield, Robin *82*

Augustine *81, 118, 138f*

Austen, Jane *260-61*

authenticity *3, 98 112, 213, 241*

Bacon, Francis *11, 18. 122-24, 127, 130-33, 155, 160, 166, 173, 184, 192, 204, 210, 237, 244*
  Novum Organum 126

Baier, Annette *242*

Bate, Jonathan *37, 135, 225f, 235*

Bauman, Zygmunt *133, 142f, 211*

Baxter, Brian *xvi, 262, 288f*

Bell, Derek *270*

Bentham, Jeremy *120, 197*

Berlin, Isaiah *135, 171*

Berman, Morris *3, 43, 45, 49, 53, 55, 128f, 104f,*

138f, 245
  on the neolithic *69, 75*
  and genetic memory *148*

Berry, R.J. *56f*

Berry, Thomas *vii, 192*

Berry, Wendell *57, 206, 210, 212, 215, 233*

Bird, David *54*

biocentric *160, 184, 205, 228, 207, 224f, 266*

biophilia *10, 57f, 283, 285*

bioregionalism *19*

Bookchin, Murray *12-17, 39, 50-52, 232, 260, 263, 270, 287f*
  and anthropocentrism *126, 265*

Botkin, Daniel *xvii-viii, 247, 253f, 262, 287f*

Boyden, Stephen *xvi, 4, 50*

Brennan, Andrew *289f*

Brody, Hugh *75*

Brown, Lester *xiii*

Bruntland Commission *13m, 218-19, 261, 273*

Buber, Martin *10-11, 40, 54, 56f, 275, 281*

buddhist *15, 19, 141f, 164, 178f, 203-04, 217, 218, 234, 249*
  and Muir *189*
  and nature *221f*

Buell, Laurence *21f*

Butler, Marilyn *xxif*

Calarco, Matthew *269*

Caldwell, Lynton *xviii, 22f*

Callicott, J. Baird *273, 275, 285*

Camus *284*

Capra, Fritjof *141f, 234, 246, 249, 253f*
  on buddhism *218*

Carr, Mike *213, 218, 232*

Carson, Rachel *xiv, xix, 12, 209, 232, 249*
  web of life *14*

Cartmill, Matt *139f*

Casas, Bartolome de las *167*

chain of being *117, 121, 160, 203*

chaos theory *20f, 41, 244*

Charles, HRH Prince *21f, 193*

Chaucer *161*

Childe, Gordon *69*

310 | INDEX

christianity *xix*, 82, 119, 133, 157
  fundamentalist 6
  and anthropocentrism 117-19
  and stewardship 206
  and Muir 191-92
Cicero 92, 126
citizen/citizenship *viii*, 5, 17-19, 80-5, 87, 168,
  212, 240, 250, 255-58, 267-71, 280-83
  Greeks on 80-1, 83, 88-9, 94, 165, 279
  nature as 209, 267, 282-83
Clare, John 52, 125-26, 160, 232
Clastres, Pierre 46-8, 105*f*
climate change 16, 18, 41-2, 44, 50, 68, 76,
  205-06, 232, 234, 253*f*
clinamen 97
Coetzee, J. M. 53, 57*f*, 200
Cohen, Michael 187
Coleridge, Samuel Taylor 64-5, 147, 165-66,
  193, 204, 236, 248, 284
  and pantheism 166, 214
Colletti, Lucien 155
Commoner, Barry 14, 22*f*, 101, 263
confucian 164, 170, 202-04
conservationism 10, 19, 206-10, 257, 273, 289*f*
consumerism 5, 205-06, 229, 218
Cooper, James Fenimore 168
Copernicus 127
Cronon, William 180*f*
Crosby, Alfred 103*f*
Daly, Herman 3, 177*f*, 214, 216, 225*f*
Damasio, Antonio 199
Danford, John 244
Darfur *xiv*, 75
Darwin, Charles 26, 31-3, 35, 44, 54, 134, 194,
  232
Darwinian 14, 34, 31, 34, 43, 48, 67, 75, 99, 121,
  135, 152, 222*f*, 256, 281
deep ecology 17, 186, 250, 299 271, 277
Democritus 117, 120, 208
Derrida, Jacques 200
Descartes, Rene 11, 18, 141*f*, 145, 157, 160,
  183-84, 192, 210
  and anthropocentrism 123
  and animals 124-28, 131, 140*f*
  and instrumental reason 150-53, 154,
    165*f*
  and religion 126

deVries, B 103*f*
Diamond, Stanley 3, 130
Dickens, Charles *xiv*, 128
Dilthey, Wilhelm 112
dionysian 80, 83
Dobson, Andrew 251*f*, 255, 264-65
Dryden, John 167
dualism *xix*, 3, 5, 8-9, 11, 20*f*, 75, 80, 107,
  123-25, 145, 151, 158, 164, 219, 230-33, 236,
  243, 252*f*, 256, 264, 271, 275, 290*f*
Ducks Unlimited 236
Dyson, Freeman 77
Eckersley, Robyn 281, 290*f*
ecocentrism 12-15, 184, 186, 212, 201-03,
  270-71, 290*f*
ecofeminism *viii*, 261, 241-45, 271
ecological citizenship 264-65
ecological management 208, 264, 266-67, 277,
  285
ecological stewardship 264
ecological sustainability 4, 12, 18, 136, 164,
  183-84, 205, 216, 240, 260, 264, 268
Edwards, Jonathan 187-88
Eisenberg, Evan *xxiif*, 226*f*
Eisley, Loren 204
Elias, Norbert 86, 234
  on civilization 74
Eliot, T.S. 151
Emerson, Ralph Waldo 112, 134
Empedocles 120, 198
Engels, Friedrich 78, 105*f*
enlightenment *viii*, 14, 25, 98, 114-15, 131-35,
  138, 146, 153, 183, 226*f*, 229-33, 250, 280
  Shelley and 159
environmental citizenship 262, 265
environmental crisis *xvi-xviii*, 7, 16, 19, 80,
  218, 234-35, 261, 273
Epicureans 82, 93-9, 107*f*, 158, 164, 202
  and Muir 191
  and ecocentrism 202
  and science 246
Epicurus 12, 15, 19, 80, 83, 129*f*, 93-8, 105*f*, 166,
  183-84, 235, 239, 249
  his death 106*f*
  and desire 163
  and friendship 164-65
Erasmus 120, 198

INDEX | 311

ethics of care 266, 278, 282
eudaimonia 242
Euripides 80, 88
  Medea 80-83, 88-9
Evernden, Neil xvii, xix, 7, 17, 232, 235, 273,
    279
  and phenomenology 203
  intrinsic value 273
factory farming 200
Feenberg, Andrew 264-65, 272, 279, 288f
Fernandez-Armesto, Felipe xix, 53, 102f, 170
  on humans and nature 87, 139f, 140f, 215,
    291f
  on modernity 137
Ferry, Luc 256, 266, 283
Finlayson, Clive 59f, 74
Flannery, Tim xiv
Fletcher, John 248
Foley, Robert 56f
Fox, Michael 273, 290f
Fox, Stephen, 190
Francis, St. 82, 117-18, 122, 160, 183, 235, 249
Freud, Sigmund 11, 54-5, 113-17, 136-37, 157,
    174, 261
  and oceanic feelings 26, 128, 193
  mastery 51, 122
  and human nature 21f, 35, 122, 139f, 154, 237
  as scientist 113
  Future of an Illusion 114-15
  Civilization and Its Discontents 116-17
Frost, Robert 65
Galileo 127-28, 130, 183
Geertz, Clifford 121
Genesis, Book of 81-2, 106f, 132, 162, 263, 265
Gilgamesh 72
Glavin, Terry 9, 167, 287f
global warming xiv, xvii, 5, 18, 29, 74, 213, 219,
    262
Godwin, William 2-3, 165, 166
Goethe, Wolfgang von 112, 129, 148, 229
golden age 71, 79, 147, 176, 150-53, 160-66, 173,
    180f, 183
Golding, William 51, 114
Goldblatt, David 288f
Goodall, Jane 199
Goodin, Robert 274
Goody, Jack 102f

Grandin, Temple 198
Greenpeace xiii, 100, 209, 232, 236
green romanticism 262, 264, 271
Guttmann, Amy 200
Habermas, Jurgen 255, 266, 283
Hacking, Ian 279
Hadot, Pierre 112
Hardin, Garrett 272
Harlan, Jack 76
Harrison, Robert 72, 74-05
Havel, Vaclav 260
Heidegger, Martin 196, 203-04, 211, 218
Heisenberg, Werner 141f, 244-45
Hesiod 162-3
Hinchman, Lewis and Sandra 227f
Hobbes 35, 46, 51, 114, 122-23, 157, 170, 176f,
    276, 280
  on state of nature 154-56
holism 10-11, 19, 26, 131, 138f, 146, 159, 161,
    198, 231, 236, 239, 256, 271, 275, 277-79
holocaust 200, 224f
homo economicus 77, 118, 136, 155
homo erectus 42, 50, 53, 59f
homo habilus 42
homo sapiens 15, 27, 29, 32, 34-6, 41-6, 53, 59f,
    60f, 63, 67-72, 123, 156-7, 194, 196, 199,
    230, 271, 275
  and ice age 74
  and mastery 99, 209, 231, 256, 263, 266
Hopkins, Gerald Manley 38
Horkheimer, Max xix, 118, 133, 231, 251f
Houellebecq, Michel 226f
human nature vii-viii, xix, 4, 8-11, 13, 18,
    32-9, 46-7, 53-4, 57f, 73
  debates about 121-25
  and reason 113, 124, 126, 136-37
  and mastery 154
  and cooperation 157-58
Schiller and 230
  and pleasure 237-38
  human rights 3, 269, 280-81
  humanism xix, 92, 168, 196, 236, 255, 264
Hume, David 14, 125, 140f, 169, 190, 244-45,
    279
hunter-gatherer 4, 8, 52, 11, 18, 28-30, 44-9,
    55, 66, 72, 75, 95
  and nature 9, 136

INDEX

and conflict 47-8, 49, 60f, 194
economy 45, 49, 69, 77, 99, 168, 238
contemporary 53, 76, 104f, 117, 123, 128
Hutcheson, Francis 169
Ignatieff, Michael 80-1, 160
individualism 73, 219, 279
   Rousseau and 158, 165
   Tocqueville and 166
   Charles Taylor and 240
instrumental reason 18, 118-19, 132-36,
   260-61, 141f, 149, 172, 237, 261
   Charles Taylor and 139f, 142f, 240
   Descartes and 126-31, 130
   critique of 133, 196, 211
intelligent design 31
Jefferson, Thomas viii, 212
Jeffersonian 166, 206, 212-14
Jolly, Alison 40, 55, 60f
   on altruism 194
Jung, Carl 192
Kaczynski, Ted 220
Kant, Immanuel 42, 118, 131, 152, 160-61, 210,
   230, 243, 256, 275
   and reason 132-33, 172, 242, 268
   and animals 131-32, 198
   and human nature 156, 255
   and the sublime 159, 177f
   and sympathy 169
kantian project 17-8, 268, 279, 281
Kingdon, Jonathan 50-1
Kinship 5, 45, 47-8, 80-2, 84-6, 87, 155-56,
   164, 192, 202, 223f, 234, 238, 271
Kohak, Erazim 202, 275, 278, 281, 285
   reason and animals 198
Kolbert, Elizabeth 226f, 262
Kretch, Shepard 179f
Kyoto Accord xix, 261, 273
Lafleur, William 221f
Lasch, Christopher 239
Lauck, Joanne 200
Lawson, John 21f
Leacock, Eleanor 47, 55
Lee, Richard 12, 45-6, 49, 53, 55, 76, 95
Leiss, William xix, xxif, 19, 127, 131, 134,
   264-65, 272, 279, 285
   on Francis Bacon 134
Leopold, Aldo vii, 10-12, 40, 77, 90, 190,

   232-35, 238, 248, 256-61, 275, 287f
   land ethic 209-12, 257, 283
Levi-Strauss, Claude 26-8, 35, 51, 101, 193
   on the neolithic 71
   and sympathy 169
Lewis-Williams, D. 102f, 103f
Light, Aandrew 262-4
limits to growth 213, 217-19
Livingston, John 9, 15-6, 22f, 195, 210, 225f,
   232, 289f
Locke, John 128, 133-34, 152, 155-56, 280
love canal 217
Lovejoy, Arthur and George Boas 27
Lovelock, James xviii, 14, 29, 247, 251f, 271, 275,
   284-85
Lucretius 12, 93-5, 97-100, 156, 163, 190,
   183-85, 235, 238
   on state of nature 153-54
Lyell, Charles 134
Machiavelli 63, 130
Marcuse, Herbert 3, 180f, 231, 266, 288f
Marsh, George Perkins 207
Marx, Karl 20f, 63, 204f, 216
   on epicureanism 106f
Masson, Jeffrey 198, 200
mastery agenda viii, xv, xviii-xix, 4, 6, 8-10,
   14, 16, 19, 51, 53, 91, 99-100, 117-19, 134,
   135-37, 145-47, 161, 173, 236, 239, 242, 250,
   263, 284-85
   Elias on 74
   reason and 89, 111, 113,
   and Bacon 126
   and Kant 132
   and Christianity 82
McEwan, Ian 100-01, 169
McIntyre, Alastair 199
McKibben, Bill xiii
McNeill, William 246
Medea 80, 83, 88-9
Megarry, Tim 58f, 60f, 61f, 157
memes 26, 192
Mench, Joy 200
Mencius 170
Merchant, Carolyn xix, 7, 12 133, 184 242, 252f
   on epicureanism 96
   on christianity 107f, 118, 138f
   on Descartes 124

partnership ethic 277-78, 282
mesolithic 70
Midgely, Mary 6, 9, 132
Miller, Sally 185
Mitchell, Alanna xxiif
Mitford, Mary 214
modern, reformulation of 6, 118
modernity viii, xiii-xv, 2-4, 5-8, 25-6, 30, 47,
    54-5, 69, 99, 117-20, 137, 155, 168, 171-74,
    212-14, 234-35, 236-40, 264-68
    Greek origins 80-2, 93
    and natural law tradition 120
    and reason 113-14, 130, 134-35, 149
    Wollstonecraft and 148
    critique of 145, 150, 156, 160, 164, 183-85,
       205
    Rousseau and 66
    Descartes and 131-32
Montaigne 119, 120
Montesquieu 164
moral sense 14, 57f, 133, 169-70, 240
Morgan, Lewis Henry 78
Morris, William 26, 215
Muir, John 10, 12, 26, 39, 112, 134, 185-95,
    209-10, 235
    and religion 186-90, 206
    and anthropocentrism 202
    and epicureanism 215
    and wilderness 134, 288f, 191, 210
    and preservationism 207
mutuality 8, 14, 16, 19, 47, 55, 81, 118, 189-90,
    267, 268, 273, 276
Naess, Arne 271
Nash, Roderick 248, 253f, 269, 274, 285
    island scenario 265
natural law 92, 120, 155
natural philosophy 129
natural theology 188
neanderthals 30, 42, 45, 59f, 123
Nearing, Helen and Scott 214
Neiman, Susan 177f
neolithic 8, 30, 70-78, 165, 231
    neolithic revolution 102f
Nesse, Randolph 222f
noble savage 51, 97, 104f, 150, 167-75, 179f, 183
Nussbaum, Martha 94, 107f, 161, 163, 180f,
    223f, 282, 290f

on stoicism as therapy 94-5, 131f
and sympathy 170-71, 242
reason and animals 199
and eudaimonia 238-40
capabilities approach 274-75
oceanic feeling 26-7, 193-94
    and Freud 117
    and Wollstonecraft 148
    and Muir 191
Oerlemans, Onno 160
Oelschlager, Max 12, 52, 188, 189
Ovid 162
Owenite 165-66
Pagden, Anthony 105f
paleolithic xviii, 6, 8, 19, 27, 30, 44, 46, 50, 55
    63, 230-31, 256
    gender in 77-8
pantheism 19, 89, 187
pantisocracy 167, 214
Park, Mungo 187
Pascal's Wager 16
pastoralism 8, 75, 99, 104f, 207
pathetic fallacy 258
PETA 119-20, 197
Penn, William 167
Pinchot, Gifford 207
Pinker, Stephen 45
Plato 83, 89-95, 120, 164, 231
    his Timaeus 89, 91
    his Phaedrus 89-90, 123
    and anthropocentrism 91
    and the golden age 152
Plumwood, Val xxif, 4, 7, 184, 243, 259,
    269-72, 275, 278, 282, 285, 289f
    on dualism 9
    on modernity 120
    on spacial remoteness 211, 215-16
    on citizenship 268
    and Kant 268
    and religion 288f
polis 80
Post, Laurens van der 192
post-modern 3, 6, 12-5, 34, 121, 147, 137, 200,
    211, 245
poverty xviii, 5, 34, 146, 239
preservationism 19, 207-08, 224f
primitive viii, 2-4, 10-12, 19, 20f, 28-31, 44,

314 | INDEX

49-51, 67-71, 85, 99, 123, 138f, 148, 152,
168, 224, 232, 256, 261
Shepard and 151
Freud and 26, 115-17, 137
Rousseau and 27, 79, 97, 111, 140f, 150, 158,
165, 167
Levi-Straus and 101
christianity and 133
Shelley and 174
Muir and 192-93
primitive communism 12, 45
primitivism 6, 26-7, 31, 102f, 114-15, 171, 205,
220, 265-66
progress xix, 2-6, 17-19, 31-3, 98, 114, 118-19,
124, 132, 162-63, 193, 205-06, 218, 231, 237,
244, 272
Wollstonecraft on 147-49, 172
Shelley and 159
Rousseau and 169
Pythagoras 119, 123, 198
quantum physics 15, 244-45
Quilley, Stephen 222f, 259
Rad, Gerhard von 106f
Raskin, Paul 206
rationalism 80, 126, 128, 286
Readings, Bill 171
reciprocal altruism 34-5, 276
reciprocity 7, 19, 47-8, 53, 57f, 232, 249, 273,
285
Rees, Bill xiii
Ridley, Matt 48, 58f
Roach, Catherine 251f
Roberts, Hugh 20f, 56f
Rolston, Holmes xvi, 33, 275
romanticism 146, 150, 162, 245
green 262, 264, 267, 271
romantics 19, 29, 39, 66, 111-12, 116, 151, 158,
230, 246
and Lucretius 154
and modernity 170
Rorty, Richard 34, 120, 180f
Rousseau, Jean-Jacques 2, 9, 14 48-50, 55, 69,
73, 113, 119, 158-60, 198, 210, 230, 248, 276
on sentiment of existence 25-9, 38, 56f, 66,
90, 111, 164, 238
on sympathy 35, 169, 190
on state of nature 46, 66, 68, 104f, 150,

155-59, 177f, 214
on the golden age 78-9, 147, 164-66
on the neolithic 78-9
on the primitive 2, 48, 97, 137, 156, 167-68,
179f
Kant and 131-32
Wollstonecraft and 148
and autonomy 157, 279
on desires 163, 178f
and happiness 55, 113, 140f, 164, 178f, 215
on progress 218
on simplicity 239
Rowe, Stan 9, 54, 193, 203-04, 212, 232, 270, 278
Ruskin, John 26, 39, 214
Russell, Bertrand 215
Sahlins, Marshall 47, 53, 55, 95
Sale, Kirkpatrick 47, 213, 215
Sartre, Jean Paul 35
Scarry, Elaine 112
Schermer, Michael 269-70
Schiller, Friedrich 229-33, 235, 273
Schumacher, E.F. 217-18, 249
Schweitzer, Albert 274
scientific method xxif, 127-28, 130, 245
Seneca 93-4, 106f, 158, 163, 166
on the golden age 153-54
sentience 33, 132, 191, 196-99, 274-75, 278
sentiment of existence 39, 67, 100, 102f, 111,
157, 164, 285
Serres, Michel xvii, xxiif, 68, 93, 184, 284-85
on epicureanism 94, 96-7
on Lucretius 185
on mastery 135
natural contract 158
shadow modernity viii, 2, 7-8, 66, 80, 160,
183-84, 205, 235-36, 248
Shakespeare 90
Shaner, David 178f
Shelley, Mary viii, xv, xix, 153, 156, 159, 171-74
Frankenstein xv, 152, 156, 173-4, 244
Shelley, Percy viii, xv-xvi, 1-2, 12, 14, 111, 113,
132, 138f, 165, 185, 211, 231, 248, 251f, 285
and animals 120, 126, 174, 198-200
and Prometheus 153
Mont Blanc 158-60
Ode to the West Wind 174
and vegetarianism 22f, 201

Sheldrake, Rupert 12
Shepard, Paul 11, 36, 58f, 102f, 104f, 193
 the neolithic 71
 pastoralism 75
 gender and farming 77
 the primitive 151
Sierra Club 112, 185, 192, 207, 236
Simmons, Alan 102f
Simons, Julian 219
simplicity 6, 66, 124, 147, 149, 154, 163-64,
 172-73, 213-14, 219, 239, 249
Singer, Peter 139f, 197, 223f, 224f
Smith, Adam viii, xix, 14, 259
 and sympathy 169, 278
 and division of labour 184
Smith, Kimberly 224f, 269
Snyder, Gary xvi, 193, 203, 209-10
Socrates 80, 89-90, 92, 94
socratic 164
Soper, Kate 177f, 223f, 240, 252f
Sophocles 85
Southey, Robert 166, 214
Spinoza 132, 202
speciesism 196, 200, 282
state of nature 46, 121, 151-58, 167
 Rousseau and 46, 67, 151
steady-state 5, 19, 205-06, 213, 216-18
Stegner, Wallace 11, 52, 194, 212
Steiner, George 203
stewardship 163, 205-07, 264, 266, 277
 christian basis 81, 99-100
stoicism viii, xix, 94, 107f
 self-command 164, 261
stoics 81-2, 94, 107f, 158, 164, 239
Stone, Christopher 22f, 281-82, 285, 290f
subsidiarity 214
sustainable development 13, 16, 208, 218, 261
Suzuki Foundation xiii
Swift, Jonathan 167
sympathy (empathy) 7, 35, 47, 156-57, 169-70,
 173, 179f, 219, 201, 245, 249, 276-78
 and ecofeminism 242-43
Tacitus 167
Tainter, J.A. 6, 60f
taoist 19, 141f, 193, 203-04, 234, 253f
Taylor, Charles 91, 112, 252f
 on instrumental reason 133, 139f, 142f

 and sympathy 170
 on steady state 218, 240
Taylor, Paul 15, 276
Theodicy 191
Thiele, Leslie 17
Thomas, Keith 52, 140f, 160, 223f
Thompson, E.P. 149, 166
Thoreau, Henry 3, 12, 26, 113, 184, 191-92, 203,
 212, 214, 256, 232, 239
Tobias, Philip 58f, 74
Tocqueville, Alexis de 166, 213
Todorov, Tzvetan 157
 on mastery 132
Torgovnick, Marianna 105f
Tuan, Yi-Fu 11, 34
Turner, Frederick 188, 221f
Turtle Island 92
Vanderheiden, Steve xxif
Varro 73
vegetarianism 22f, 125-26, 156, 174, 200, 273,
 289f
Volney 168
Voltaire 150, 160, 168
Waal, Frans de 199
Washoe 283
web of life 14, 189, 247
Weber, Max 77, 215
 and disenchantment 133
Weiming, Tu 135, 222f
Weisman, Alan 15
White, Gilbert 210, 214, 232
White, Lynn 7, 235
White, Richard 226f
wilderness xxiif, 188, 209-10, 216, 245, 248,
 253f, 259
Williams, Bernard 255, 275
Wilson, Edward xvii, xxiif, 6-7, 10, 16, 36, 50,
 54, 77, 134, 224f, 232, 287f
 biophilia 201, 260, 282
Wilson, James Q. 57f, 60f, 268
 and sympathy 169
 on altruism 195
Wilson, Peter 41
wise use 13, 207
Wispe, Lauren 179f
Wolfe, Carey 200, 222f, 275, 283
Wollstonecraft, Mary 12, 38, 101, 120, 159,

316 | INDEX

171–73, 202, 229
A Short Residence in Sweden 146–49
and Rousseau 172
and religion 176f
Wong, Milton 260
Wordsworth, William *xiv*, 112, 166–67, 190,
202, 212, 214, 238–39, 241
Ruined Cottage 101
Wright, Robert 34, 104f
Yosemite 134, 185, 189–91, 207, 209–10
Zernan, John 176f
Zimmerman, Michael 222f